# THE COSMIC MICROWAVE BACKGROUND

## From Quantum Fluctuations to the Present Universe

This volume presents the lectures of the nineteenth Canary Islands Winter School, dedicated to the Cosmic Microwave Background (CMB). This relic radiation from the very early Universe provides a fundamental tool for precision cosmology. Prestigious researchers in the field present a comprehensive overview of current knowledge of the CMB, reviewing the theoretical foundations, the main observational results, and the most advanced statistical techniques used in this discipline.

The lectures give coverage from the basic principles to the most recent research results, reviewing state-of-the-art observational and statistical analysis techniques. The impact of new experiments and the constraints imposed on cosmological parameters are emphasized and put into the broader context of research in cosmology. This is an important resource both for graduate students and for experienced researchers, revealing the spectacular progress that has been made in the study of the CMB within the last decade.

*Canary Islands Winter School of Astrophysics*

Volume XIX

*Editor in Chief*
F. Sánchez, *Instituto de Astrofísica de Canarias*

*Previous books in this series*

    I. Solar Physics
   II. Physical and Observational Cosmology
  III. Star Formation in Stellar Systems
  IV. Infrared Astronomy
   V. The Formation of Galaxies
  VI. The Structure of the Sun
 VII. Instrumentation for Large Telescopes: a Course for Astronomers
VIII. Stellar Astrophysics for the Local Group: a First Step to the Universe
  IX. Astrophysics with Large Databases in the Internet Age
   X. Globular Clusters
  XI. Galaxies at High Redshift
 XII. Astrophysical Spectropolarimetry
XIII. Cosmochemistry: the Melting Pot of Elements
XIV. Dark Matter and Dark Energy in the Universe
 XV. Payload and Mission Definition in Space Sciences
XVI. Extrasolar Planets
XVII. 3D Spectroscopy in Astronomy
XVIII. The Emission-Line Universe

Participants of the XIX Canaray Islands Winter School of Astrophysics, together with the lecturers of the first week, in front of the Congress Center in Puerto de la Cruz, Tenerife.

Same as above, but with the lecturers of the second week.

# THE COSMIC MICROWAVE BACKGROUND: FROM QUANTUM FLUCTUATIONS TO THE PRESENT UNIVERSE

*XIX Canary Islands Winter School of Astrophysics*

Edited by

J. A. RUBIÑO-MARTÍN,
R. REBOLO AND E. MEDIAVILLA

*Instituto de Astrofísica de Canarias, Tenerife*

# CAMBRIDGE
## UNIVERSITY PRESS

University Printing House, Cambridge CB2 8BS, United Kingdom

One Liberty Plaza, 20th Floor, New York, NY 10006, USA

477 Williamstown Road, Port Melbourne, VIC 3207, Australia

314-321, 3rd Floor, Plot 3, Splendor Forum, Jasola District Centre, New Delhi - 110025, India

103 Penang Road, #05-06/07, Visioncrest Commercial, Singapore 238467

Cambridge University Press is part of the University of Cambridge.

It furthers the University's mission by disseminating knowledge in the pursuit of education, learning and research at the highest international levels of excellence.

www.cambridge.org
Information on this title: www.cambridge.org/9780521764537

© Cambridge University Press 2010

First published 2010
First paperback edition 2013

*A catalogue record for this publication is available from the British Library*

ISBN 978-0-521-76453-7 Hardback
ISBN 978-1-107-69561-0 Paperback

Cambridge University Press has no responsibility for the persistence or accuracy of URLs for external or third-party internet websites referred to in this publication, and does not guarantee that any content on such websites is, or will remain, accurate or appropriate.

# Contents

List of contributors                                                    *page* viii

List of participants                                                    ix

Preface                                                                 xi

Acknowledgments                                                         xiii

1   CMB Observations and cosmological constraints                       1
    *R. Bruce Partridge*

2   The inflationary universe                                           40
    *Sabino Matarrese*

3   CMB theory from nucleosynthesis to recombination                    70
    *Wayne Hu*

4   CMB fluctuations in the post-recombination Universe                 108
    *Matthias Bartelmann*

5   Statistical techniques for data analysis in cosmology               128
    *Licia Verde*

6   Gaussianity                                                         161
    *Enrique Martínez-González*

7   Galactic and extragalactic foregrounds                              192
    *Rodney D. Davies*

8   Probes of fundamental cosmology                                     227
    *Malcolm Longair*

# Contributors

MATTHIAS BARTELMANN, Zentrum für Astronomie der Universität Heidelberg, Institut für Theoretische Astrophysik, Heidelberg, Germany

ROD D. DAVIES, Jodrell Bank Observatory, University of Manchester, UK

WAYNE HU, University of Chicago, Chicago, IL, USA

MALCOLM LONGAIR, Cavendish Laboratory, University of Cambridge, UK

ENRIQUE MARTÍNEZ-GONZÁLEZ, Instituto de Física de Cantabria, Santander, Spain

SABINO MATARRESE, Universita degli Studi di Padova and INFN Sezione di Padova, Padova, Italy

BRUCE PARTRIDGE, Haverford College, Haverford, PA, USA

LICIA VERDE, Institute of Space Sciences (ICE) IEEC/CSIC, Barcelona, Spain

# Participants

Participants (students and lecturers) of the XIX Canary Islands Winter School, held in Puerto de la Cruz (Tenerife, Canary Islands, Spain), from 19 November to 30 November 2007

| | |
|---|---|
| Aich, Moumita | IUCAA, India |
| Bartelmann, Matthias | Institut für Theoretische Astrophysik, Germany |
| Beltrán Jiménez, José | Universidad Complutense de Madrid, Spain |
| Bessada, Dennis | INPE – National Institute for Space Research, Brazil |
| Bottino, Maria-Paola | Max Planck Institute for Astrophysics, Germany |
| Buitrago Alonso, Fernando | University of Nottingham, UK |
| Catalano, Andrea | Observatoire de Paris, France |
| Cruz Rodríguez, Marcos | Instituto de Física de Cantabria (CSIC-UC), Spain |
| Cruz-Dombriz, Alvaro | Universidad Complutense de Madrid, SPAIN |
| Cuesta Vazquez, Antonio | Instituto de Astrofisica de Andalucia, Spain |
| Curto Martín, Andrés | Instituto de Física de Cantabria (CSIC-UC), Spain |
| Dantas, Maria Aldinêz | Observatório Nacional – MCT, Brasil |
| Davies, Rod D. | University of Manchester, UK |
| Fauvet, Lauranne | Université Joseph Fourier, France |
| Flores Cacho, Inés | Instituto de Astrofísica de Canarias, Spain |
| Font Ribera, Andreu | Instituto de Ciencias del Espacio (ICE-CSIC), Spain |
| Fraisse, Aurelien | Princeton University, USA |
| Franzen, Thomas | University of Cambridge, UK |
| Galaverni, Matteo | INAF-IASF Bologna / Ferrara University, Italy |
| Génova Santos, Ricardo | University of Cambridge, UK |
| Ghosh, Tuhin | IUCAA, India |
| Groeneboom, Nicolaas | University of Oslo, Norway |
| Gruppuso, Alessandro | INAF- IASF BO, Italy |
| Hildebrandt, Sergi | Instituto de Astrofisica de Canarias, Spain |
| Hu, Wayne | University of Chicago, USA |
| Jagannathan, Preshanth | University Würzburg / IUCAA, Germany |
| Jiménez, Raúl | Instituto de Ciencias del Espacio (IEEC/CSIC), Spain |
| Jiménez Teja, Yolanda | Instituto de Astrofísica de Andalucía (CSIC), Spain |
| Kristiansen, Jostein | University of Oslo, Norway |
| Lanz Oca, Luis Fernando | Instituto de Física de Cantabria (CSIC-UC), Spain |
| Le Jeune, Maude | AstroParticule et Cosmologie Paris VII, France |
| Longair, Malcolm | University of Cambridge, UK |
| López-Caniego, Marcos | University of Cambridge, UK |

| | |
|---|---|
| Mangilli, Anna | Universitá degli Studi di Padova, Italy |
| Martínez Glez, Enrique | Instituto de Física de Cantabria, Spain |
| Matarrese, Sabino | Universitá di Padova, Italy |
| Miranda Ocejo, B. Selene | Universidad Nacional Autónoma de México, México |
| Molino Benito, Alberto | Instituto de Astrofisica de Andalucia, Spain |
| Muya Kasanda, Simon | University of KwaZulu-Natal, South Africa |
| Paci, Francesco | University of Bologna / INAF–IASF Bologna, Italy |
| Padilla-Torres, Carmen P. | Instituto de Astrofísica de Canarias, Spain |
| Partridge, Bruce | Haverford College, USA |
| Quealy, Erin | University of California Berkeley, USA |
| Rebolo, Rafael | Instituto de Astrofísica de Canarias, LOC |
| Rubiño-Martín, José Alberto | Instituto de Astrofísica de Canarias, LOC |
| Ruiz-Granados, Beatriz | Universidad de Granada, Spain |
| Sánchez Colin, Angel | Instituto de Física de Cantabria, Spain |
| Trigueros Páez, Emilio | University of La Laguna, Spain |
| Tucci, Marco | Instituto de Astrofisica de Canarias, Spain |
| Verde, Licia | Instituto de Ciencias del Espacio (IEEC/CSIC), Spain |
| Wong, Wan Yan | University of British Columbia, Canada |

# Preface

The cosmic microwave background (CMB) is one of the basic pillars of modern cosmology. Discovered in 1965 by A. A. Penzias and R. W. Wilson, it was soon recognized as evidence of cosmic evolution. At some time in the remote past, our Universe went through a very hot phase where matter and radiation were in very intense interaction. The CMB is a relict of the Big Bang that pervades the Universe as the major contributor to the energy density of radiation. It offers an image of how the Universe was at the age of 380 000 y, long before the formation of the very first stars and galaxies. The discovery of the CMB led to the award of the Nobel Prize for Physics in 1978. The accurate measurement of the CMB spectral energy distribution, a (nearly) perfect blackbody curve, and the detection in 1992 of anisotropies of cosmological origin by NASA's COBE (Cosmic Background Explorer) resulted in the award of the Nobel Prize to J. C. Mather and G. F. Smoot in 2006.

Soon after the discovery of the CMB it was understood that small matter/energy density fluctuations in the very early Universe, the seeds of the large structures we see today (galaxies, clusters, voids and superclusters), should have imprinted small temperature differences in this radiation. For decades the experimental search for the expected tiny temperature anisotropies only led to upper limits (e.g. the Tenerife Experiment). In 1992, COBE's discovery of CMB anisotropies on large scales provided strong support for inflationary Big Bang models and to the gravitational instability scenario for the development of structures in the Universe. The precise measurement of the intensity of the radiation along different lines of sight showed small amplitude fluctuations of the order of one part in 100 000 that are extremely important in understanding the global properties of the Universe. More recently, measurement of the CMB angular power spectrum has become an essential tool for cosmology. With the latest results from ground-based and balloon-borne experiments (e.g. BOOMERanG, MAXIMA, VSA, CBI, DASI, ARCHEOPS, ACBAR) and especially from the Wilkinson Microwave Anisotropy Probe (WMAP) satellite, it is now possible to constrain the cosmological parameters describing the Standard Cosmological Model with an accuracy better than 5 percent.

This field is still evolving very rapidly and in the coming years we shall certainly see significant progress. For instance, the launch of the European Space Agency's Planck satellite will be a major step forward. Its sensitivity will improve accuracy in the determination of cosmological parameters, getting close to 1%.

Another experimental path that Planck and the next generation of experiments will pursue is the study of CMB polarization. In the Standard Model the CMB is linearly polarized, Thomson scattering during recombination being the physical mechanism responsible for this polarization. One of the key goals in the study of CMB polarization is the detection of rotational modes (also known as $B$-modes) in the maps. Vector and tensor perturbations, such as those due to gravitational waves in the primordial Universe, are the only known mechanisms that could generate $B$-modes in the CMB polarization on large angular scales. These modes offer a unique way of carrying out a detailed study of the inflationary epoch. However, their detection represents an enormous effort in observational cosmology since it will be necessary to measure a large portion of the sky with extremely sensitive radiometers and bolometers with small and well-defined systematic effects. Likewise, it will also imply an important effort in our understanding of Galactic emission. Polarized Galactic foregrounds are not yet sufficiently well known to be adequately subtracted from the CMB polar signal and they must be measured and mapped to a sufficient level in order not to cause confusion in cosmological measurements.

In *The Cosmic Microwave Background: From Quantum Fluctuations to the Present Universe*, eight world experts review the observational and theoretical status of CMB research, offering an updated and rigorous introduction to the field. This includes the theoretical basis of the generation of quantum perturbations in the inflationary Universe and the production of CMB anisotropies, and the experimental effort to measure its properties accurately, including state-of-the-art experimental and data analysis techniques, as well as an introduction to our current knowledge of Galactic emission. This is complemented by a review of other major observational results in cosmology. Doctoral (Ph.D.) students and researchers in cosmology will find this compendium of lectures a valuable tool for understanding the basic aspects of the CMB.

The Editors
La Laguna, Tenerife

# Acknowledgments

The editors would like to express their sincere gratitude to the lecturers for their participation in the XIX Canary Islands Winter School. Those two weeks (November 19–30, 2007) were a memorable experience for us and for all the students. This was not only our impression but the written opinion that we collected from them at the end of the school. Special thanks also to Raúl Jiménez for the preparation and realization of two tutorials during the School, which were devoted to the use of "standard software tools" in CMB research. The list of activities which were carried out during those tutorials is summarized at the end of Chapter 5 in the book.

We also wish to thank to all IAC people who made this school possible. First of all, to Nieves Villoslada, because of her diligence and help during all stages of the preparation and the celebration of the event. Her colleague Lourdes González was also very helpful on several occasions. We thank to Jorge Andrés Pérez Prieto for his help in the preparation of the Winter School website. The school poster was prepared by Ramón Castro and Gabriel Pérez. Carmen del Puerto, Natalia R. Zelman, Iván Jiménez, Miguel Briganti (SMM/IAC) and Terry Mahoney did an excellent work in making possible a very successful Press (or should we say "pressing"?) Room during the School. All the interviews, press releases and videos are available at the IAC website (see Educational Outreach area at http://www.iac.es, and also http://www.iac.es/winters/2007/winschool2007/Index07.htm).

Finally, we also acknowledge the financial support from the Spanish Ministerio de Educación y Ciencia (MEC), the Cabildo de Tenerife and the Gobierno de Canarias.

In the scientific part, some of the images of this book made use of the HEALPIX software package (Gorski et al., 2005; *Astrophys. J.* **622**, 759; website http://healpix.jpl.nasa.gov). This was a very useful tool also during the three tutorials that we had in the Winter School. We also acknowledge the use of the Legacy Archive for Microwave Background Data Analysis (LAMBDA) during these tutorials. Support for LAMBDA is provided by the NASA Office of Space Science.

# 1. CMB Observations and cosmological constraints

## R. BRUCE PARTRIDGE

## Abstract

Measurements of the spectrum of the cosmic microwave background (CMB) and of the power spectrum of fluctuations in its intensity across the sky have allowed us to refine crucial cosmological parameters, and have opened up the age of "precision cosmology." Recent spectral and anisotropy measurements of the CMB are discussed; some of the observational difficulty faced in making such observations is noted, from an experimentalist perspective. Then some of the conclusions which can be drawn from measurements of the spectrum and of the power spectrum of fluctuations, as well as from increasingly sensitive measurements of polarized fluctuations in the CMB, are presented.

## 1.1 Introductory remarks

My aim in the talks given at the XIX Canary Islands Winter School of Astrophysics was to present recent observational work on the cosmic microwave background (CMB), to sketch some of the observational difficulties in making CMB measurements and to list the cosmological constraints that those measurements established. I included a good deal of pedagogical material which I have reduced, given limits of space, in this written version. The written version relies more on references to the literature and emphasizes the observational constraints that several decades of careful measurements of the CMB have established. I continue to give the material an observational or experimental flavor, since most of the other participants in the Winter School were theorists. I also retain the informal style of my talks in Tenerife.

## 1.2 The spectrum of the CMB and what it tells us

The story of the discovery of the cosmic microwave background radiation (CMB) is well known, and frequently recounted as an example of serendipity in science. In 1964, Arno Penzias and Bob Wilson were faced with the problem of "excess temperature" in a communications antenna operating at a wavelength of 7 cm. The signal was difficult to understand because it was substantial in amplitude (equivalent to 3.5 K) and "isotropic, unpolarized and free from seasonal variations," to quote their 1965 paper (Penzias and Wilson). Note that Penzias and Wilson had already remarked on the isotropy of what we now know as the CMB. The actual interpretation of the signal as cosmic in origin was not made by Penzias and Wilson, but in a companion paper by Dicke *et al.* (1965), who argued that it is radiation left in the Universe from an earlier high-density, high-temperature, ionized phase.

That interpretation was not immediately accepted by all astronomers. We needed to show the ∼3 K signal was cosmic, not local, in origin. There were two important tests of the cosmic hypothesis: the spectrum should be blackbody, and the radiation should be largely isotropic. The isotropy follows from the approximate isotropy and homogeneity of the Universe as a whole.

1

Now let us explore why a blackbody spectrum would be expected from a hot early phase. Independent of any particular value for the current baryon density, there must have been a time early in the history of the Universe when the timescale for thermal processes was shorter than the expansion timescale, $H^{-1}$. For reasonable values of the current baryon density, it can be shown that the corresponding redshift was $\sim 2 \times 10^6$ (Burigana et al., 1991; see also chapter 5 of Partridge, 1995). Unless pathological initial conditions are assumed, the Universe would have had time to reach thermal equilibrium by a time corresponding to that redshift, or roughly one month after the Big Bang. Both scattering processes, like the Compton and inverse Compton effects, and photon-generating processes, such as thermal bremsstrahlung and the radiative Compton effect, were involved. Thermal equilibrium in turn generates a blackbody spectrum in the radiation field.

Because the expansion of the Universe is adiabatic, a blackbody spectrum, once established, is maintained, and expansion affects only the temperature as $T(z) = T_0(1 + z)$. The argument that adiabatic expansion preserves a blackbody spectrum is made in Weinberg's book on *Gravitation and Cosmology* (1972), and goes back to much earlier work by Robertson.

By 1967, observations had shown that the radiation was isotropic to a few tenths of a percent, and that its spectrum was consistent with a blackbody curve with a temperature of $\sim 2.75$ K over a range of wavelengths exceeding a factor of ten (see e.g. Stokes et al., 1967; Wilkinson, 1967 and a summary of measurements using interstellar cyanogen by Thaddeus, 1972). The interpretation of Dicke and his colleagues had passed both tests. In addition, of course, attempts to improve the spectrum led to a more precise value of the present temperature, $T_0$.

As the cosmic interpretation of the microwave excess took hold, increasing interest arose in small departures from perfect isotropy and small departures from a perfect blackbody spectrum. In this section, I explore the latter.

### 1.2.1 Varieties of equilibria, and the traces they leave in the CMB spectrum

I will condense here material discussed in somewhat more detail in Chapter 5 of Partridge (1995). First, *kinetic equilibrium* can be established by any scattering process with a timescale substantially less than $H^{-1}$. Under the conditions of kinetic equilibrium, as opposed to true thermal equilibrium, the occupation number of photons can be written as $\eta = \left[\exp\left(\frac{h\nu}{kT} + \mu\right) - 1\right]^{-1}$ where $\mu$ is the "chemical potential." The resulting spectrum is the Bose–Einstein spectrum. Clearly, $\mu$ plays only a small role at very high frequencies, where the above expression for $\eta$ relaxes to the Planckian form; the discrepancy becomes larger and larger as the frequency drops.

True *thermal equilibrium* requires the creation and destruction of photons as well as energy redistribution by scattering (see, for instance, Kompaneets, 1957). In the early Universe, both radiative Compton and bremsstrahlung processes permit the generation of photons needed to ensure true thermal equilibrium. While the former is likely to be dominant with current values of the baryon density (Burigana et al., 1991), the latter process generates a characteristic spectral feature worth exploring. The bremsstrahlung cross section is proportional to $\lambda^2$, and as a consequence bremsstrahlung is most effective at creating thermal equilibrium, and a consequent blackbody spectrum, at large wavelengths. Now consider the possibility that bremsstrahlung has not completed the process of thermal equilibration. The result will resemble Fig. 1.1, where the dashed line is a Bose–Einstein spectrum characterized by a positive chemical potential, $\mu$, and the solid line shows the effect of partial equilibration by bremsstrahlung at low frequencies.

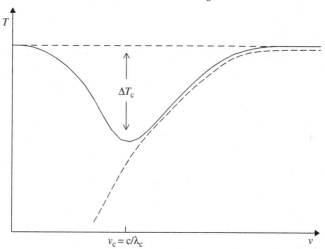

FIGURE 1.1. The "$\mu$-distortion" of the CMB spectrum described in the text; the dashed line shows a pure Bose–Einstein spectrum.

The claim that thermal equilibrium is established at the high redshift of $\sim 2 \times 10^6$ is equivalent to the claim that $\mu$ is driven essentially to zero by that redshift. However, if energy is added to the CMB radiation field *after* an epoch corresponding to a redshift of $\sim 2 \times 10^6$, there may still be time to reintroduce kinetic equilibrium, but not full thermal equilibrium. In this case, the spectrum shown by the solid line in Fig. 1.1 will result. We refer to the departure from a Planck spectrum at fixed $T$ as a "$\mu$-distortion." The center frequency (or wavelength, $\lambda_c$) and the amplitude of the evident dip depend on $\mu$ and on the baryon density (since only baryonic matter is involved in the thermalizing processes). For instance, $\lambda_c \sim 2.2(\Omega_b h^2)^{-2/3}$ cm. And $\Delta T_c$, the amplitude of the dip, is given by $\Delta T_c/T_0 = 2.3\,\mu(\Omega_b h^2)^{-2/3}$ (from Illarionov and Sunyaev, 1975; and Burigana *et al.*, 1991). Since $\lambda_c$ depends only on $\Omega_b$ and $h$, and not $\mu$, we can evaluate it: $\lambda_c \sim 30$ cm for currently accepted values of the cosmological parameters.

Now let us ask about scattering processes, those that conserve photon number, but readjust photon energy. By far the dominant process is inverse Compton scattering, in which the energy of a CMB photon is "boosted" by collision with hot electrons in the ionized material. This is the effect explored by Sunyaev and Zel'dovich (1969 and 1980a,b). This scattering process in a uniform plasma affects the overall spectrum of the CMB. The resulting spectral distortion can be written as

$$\frac{\delta I(\nu)}{I(\nu)} = -y\frac{xe^x}{e^x - 1}\left[4 - x\coth\left(\frac{x}{2}\right)\right],$$

derived, for instance, by Peacock (1999) from the Kompaneets equation. Here $x = \left(\frac{h\nu}{kT_0}\right)$ and the $y$ parameter involves the integral along the line of sight of the electron pressure,

$$y = \int \frac{kT_e n_e \sigma_T}{m_e c^2}\,\mathrm{d}\ell. \tag{1.1}$$

In this expression $T_e$, $n_e$ and $m_e$ are the temperature, number density and mass of the electrons, respectively, and $\sigma_T$ is the Thompson scattering cross section.

### 1.2.2 Experimental issues

To look for either form of distortion, sensitive measurements of the CMB spectrum over a substantial range of wavelength are required. Here, I discuss briefly some of the observational difficulties in making such measurements. First, for an *absolute* measurement of $T_0$ at a given wavelength, one needs some form of absolute calibration. While in principle a blackbody at any known temperature could be used, it is strongly advantageous to have a calibration blackbody at a temperature close to $2.7\,\text{K}$, so that non-linearities in the response of the detector do not bias the measurement. For that reason, all CMB spectral measurements have employed cryogenic calibrators, referred to (in terminology introduced 60 years ago by Dicke) as "cold loads." An ideal cold load would enclose a measuring antenna with a perfect blackbody at a precisely measured temperature. Ground-based measurements have used large dewars containing liquid helium and a (nominally) perfect blackbody immersed in it (e.g. Smoot *et al.*, 1985). The FIRAS instrument aboard the Cosmic Background Explorer (COBE) satellite, to be discussed below, employed a somewhat similar system, but allowed for small temperature adjustments in the cold load to equal and thus null out the CMB signal.

If an accurate calibration is achieved, the remaining problems are sources of foreground radiation that add to the CMB signal entering an antenna viewing the sky. These take three forms: radiation from the ground and nearby structures that diffracts into the antenna mouth; radiation from the Earth's atmosphere; and radiation from the Galaxy. The first of these can be mitigated by the use of ground screens, reflective surfaces that prevent direct radiation from the hot ground into the antenna mouth. Ground screens are now routinely used for *all* CMB observations, whether they are spectral or isotropy observations. Emission from the atmosphere can be measured and subtracted by using the zenith angle dependence of atmospheric emission. In the case of a thin, plane-parallel atmosphere, the atmospheric signal scales simply as $T_{\text{atm}} \propto \sec z$, where $z$ is the zenith angle. Atmospheric emission can also be reduced by observing at frequencies between the strong water-vapor and oxygen lines that dominate microwave emission of the atmosphere. The presence of the strong $H_2O$ line at $22\,\text{GHz}$ and $O_2$ lines at 60 and $120\,\text{GHz}$ explain why so many ground-based observations are made at frequencies well below $22\,\text{GHz}$, or in "valleys" between the lines, that is near 30 or $90\,\text{GHz}$. Atmospheric emission is discussed in more detail in the review by Weiss (1980) and in Danese and Partridge (1989). Some rough numbers show the magnitude of the problem presented by the atmosphere: at sea level $T_{\text{atm}} \gtrsim 3\,\text{K}$ (or $\gtrsim T_0$) for $\lambda \lesssim 3\,\text{cm}$. Even at the best, high altitude sites, $T_{\text{atm}} \sim 2\,\text{K}$ for $\lambda = 3\,\text{mm}$.

Note that atmospheric emission pushes one to observe at relatively low frequency. Unfortunately, the Galactic background favors observations at high frequency. In the microwave region, the dominant source of Galactic emission is the synchrotron process, with a temperature spectrum going roughly as $T_G \propto \nu^{-2.8}$, a proportionality that holds for frequencies below or about $90\,\text{GHz}$. At higher frequencies, reemission from Galactic dust becomes important, as does free–free emission with a flatter spectrum, $T_G \propto \nu^{-2.1}$. We will return to Galactic emission in Section 1.3.3.2. Of course, Galactic foregrounds can be minimized by making spectral measurements near the Galactic poles.

### 1.2.3 COBE

By the mid 1970s, it was clear that improved spectral and anisotropy measurements, at least on a large scale, required a space platform. The COBE satellite, launched in 1989, was designed to provide both kinds of measurements, and to probe far infrared radiation from the Galaxy as well. The spectrum was measured by the FIRAS instrument, the "Far Infrared Absolute Spectrometer." Unlike earlier ground-based measurements, it

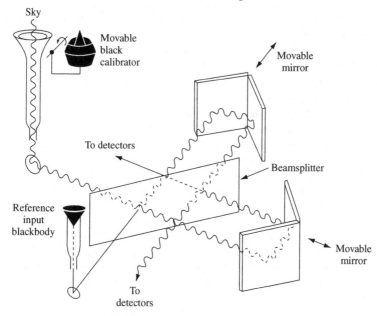

FIGURE 1.2. Block diagram of COBE's FIRAS instrument. (From COBE slide set. NASA image.)

employed a Michelson interferometer to measure temperature as a function of wavelength over a wavelength range of roughly 0.05–0.5 cm. The instrument was carefully designed to make a null measurement, by comparing the observed temperature of the CMB with the temperature-controlled cold load. As a further check, the external antenna could be covered by a second temperature-controlled cold load, as shown in the schematic in Fig. 1.2. It is worth emphasizing that uncertainties in the final result from FIRAS depend more on understanding the two cold loads than they do on actual measurements of the sky.

The results of the first few minutes of COBE's observation were presented in January of 1990, to a standing ovation (Mather *et al.*, 1990). Subsequent analysis, again particularly of the cold load, has resulted in a slight improvement in the sensitivity of the experiment, and I give the final result (Fixsen and Mather, 2002):

$$T_0 = 2.725 \pm 0.001 \text{ K}. \tag{1.2}$$

At about the time the COBE results were first announced, a pioneering rocket experiment, also based on an interferometer, was measuring the CMB spectrum over a similar range of wavelengths. This important work by Gush *et al.* (1990) provided crucial confirmation of the COBE results.

The precision of the 2002 COBE result is astonishingly high – we know the temperature of the CMB, a crucial cosmological parameter, to better than 0.05%. However, the wavelength range covered by COBE was limited, particularly at the long-wavelength end. As a consequence, there have been and continue to be attempts to improve our knowledge of the CMB spectrum at wavelengths of 1 cm and longer. In particular, to reveal a possible $\mu$-distortion, observations at $\lambda > 10$ cm are needed.

### 1.2.4 Longer wavelength measurements

At wavelengths longer than about 3 cm, the atmospheric emission is sufficiently small to allow these experiments to be done from the ground, though balloons are also being

used, for instance by the ARCADE group (Kogut *et al.*, 2006). To date, the most precise
observations are those made by the TRIS team based in Milan. While these results have
not yet been published, Giorgio Sironi has kindly given me permission to quote some
numbers; for details see Gervasi *et al.* (2008).

- At 50 cm, the group found $T_0 = 2.685 \pm 0.038$ ($\pm 0.066$) K
- At 36 cm, $T_0 = 2.772 \pm 0.012$ ($\pm 0.40$) K
- At 12 cm, $T_0 = 2.516 \pm 0.107$ ($\pm 0.28$) K

Here, the numbers in parenthesis include systematic error, largely due to correction for
Galactic emission. To these we may add the following observations at slightly shorter
wavelengths:

- At 3 cm, $T_0 = 2.730 \pm 0.014$ K (Staggs *et al.*, 1996)
- At $\sim$4 cm, $T_0 = 2.84 \pm 0.014$ K (Kogut *et al.*, 2006)

### 1.2.5 Constraints on spectral distortions

We have known for more than a decade (see summary in chapter 4 of Partridge, 1995)
that any distortions in the CMB spectrum are small. Let us now examine the best current
limits on distortions and what they imply.

**$y$-distortion.** The COBE team (Fixsen *et al.*, 1996) fixed a tight upper limit on the $y$
parameter: $y < 1.5 \times 10^{-5}$. This in turn limits the amount of inverse Compton scattering
by a generally distributed plasma at $z \lesssim 10^5$ (local concentrations of plasma, such as the
intergalactic medium (IGM) in clusters of galaxies, can produce much larger values of $y$;
these are discussed in Section 1.4 below).

The refinement of the COBE data in later papers by the COBE team (e.g. Fixsen and
Mather, 2002), and the addition of long-wavelength measurements allowed the TRIS team
to establish improved limits (Gervasi *et al.* 2008). They claim $-5 \times 10^{-6} < y < 3.5 \times 10^{-6}$.
Since there is a direct connection between $y$ and the excess energy added to the radiation
field $\Delta u$, we can in turn limit $\Delta u$:

$$\Delta u / u = 4y, \text{ so that} \qquad \Delta u \lesssim (2 \times 10^{-5})u, \tag{1.3}$$

where $u$ is the energy density of the CMB radiation field.

One potential source of a $y$-distortion is a thin, more-or-less uniform, intergalactic
plasma. There was a vogue some years ago for explaining a portion of the X-ray background as thermal emission from such a hot IGM, with $T \sim 10^6$ K. The observed limits
on $y$ set extremely tight limits on the density of such a plasma (this is left as an exercise;
see Wright, 1994, for some guidance).

**$\mu$-distortion.** Here the constraint from TRIS and COBE measurements is $\mu < 7 \times 10^{-5}$.
For small values of $\mu$, $\Delta u / u \sim 0.7 \mu$ (Illarionov and Sunyaev 1975) so we find $\Delta u \lesssim$
$5 \times 10^{-5} u$. This limits energy release (say, by decay of exotic particles or the evaporation
of low-mass black holes by the Hawking mechanism) at epochs less than $\sim$30 y.

## 1.3 Measurements of CMB anisotropies on large angular
##      scales ($\ell \lesssim 300$)

Those who attended the Winter School in Tenerife will recognize that I have changed
the order of this section. This was done largely to remove some overlap between my
presentation, Wayne Hu's, and to a lesser extent Sabino Matarrese's.

### 1.3.1 A cartoon view of a power spectrum of anisotropies

I will begin by presenting a cartoon version of the power spectrum of CMB anisotropies; for details see the much more careful presentations by Hu and Matarrese. This quick outline is designed to highlight the key physical processes responsible for features in different ranges of spatial frequency, $\ell$, in the power spectrum of the CMB (Fig. 1.3). This approach allowed me to divide up the presentation of CMB anisotropies into two sections, corresponding to two talks at the Winter School. Recall that $\ell \sim 180°/\theta$.

$\ell \lesssim$ **50–100.**  The angular scale of a causal horizon on the surface of last scattering (at $z \sim 1000$) corresponds approximately to 2 degrees. At angular scales larger than this, or $\ell \lesssim 100$, we thus see the power spectrum imprinted during the inflationary epoch, unaffected by later, causal, physical processes. This point is explained in detail in Matarrese's article here. He also explains why the shape of the power spectrum is expected to have a power-law index close to $n_S = 1$, which in turn translates into a nearly flat power spectrum at values of $\ell \lesssim 100$.

At these low values of $\ell$, we also need to consider cosmic variance. This is perhaps a slight misnomer: the variance is not in the cosmos but in the models. Consider a specific cosmological model, with chosen values for the various cosmological parameters. The model parameters can be used, for instance in programs like CMBFAST, to calculate the CMB power spectrum. At low values of $\ell$, however, there is some intrinsic uncertainty in the model predictions, depending on which particular realization of the low-order multipoles is selected. Thus there is some variance in the *predicted* amplitudes of low-$\ell$ signals, even with a very specific model. Since our Universe represents only a single instance, the measured low-order multipoles could fall anywhere in the range projected by various realizations of the model. This range is cosmic variance. It is worth going on to make the obvious point that, in the range of $\ell$ dominated by cosmic variance, further improvements in the accuracy of observational results will not tighten constraints on the cosmological model or on cosmological parameters. We shall see that the results from the COBE satellite, launched nearly twenty years ago, have determined the low-order multipoles to adequate accuracy. Subsequent improvements in precision have resulted from better control of systematics and foregrounds.

$\ell \gtrsim$ **100.**  On scales below 2 degrees, the rich phenomena described so ably by Wayne Hu come into play. Density perturbations on the surface of last scattering produce a series of acoustic peaks in the CMB power spectrum, with the fundamental at $\ell \sim 200$. The

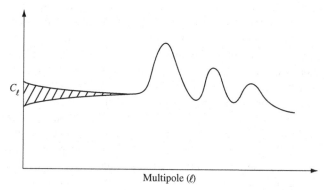

FIGURE 1.3. A cartoon version of the CMB power spectrum showing the various physical mechanisms discussed above.

physics explaining these peaks is quite well understood (but see some cautionary remarks by Wayne Hu), or at least well enough understood to give us reasonable confidence in the angular scale and relative amplitude of these acoustic peaks. These quantities in turn depend in predictable ways on crucial cosmological parameters such as the curvature, $k$, the baryon density, $\Omega_b$, the Hubble parameter and so on. In this section and the next, we will explore some of the constraints on such parameters set by the observations of the acoustic peaks.

$\ell \gtrsim 500$. As $\ell$ increases, the amplitude of the acoustic peaks is damped or cut off by two processes, Silk damping and the averaging of fluctuations of different sign resulting from the finite thickness of the last scattering surface. Both are described in Hu's article. Here, I use the latter effect to fix an approximate angular scale for the onset of this damping: the rough scale is given by $\theta \sim \frac{\Delta z}{z}(2°)$, or something like 10 arc minutes, with $\Delta z \sim 150$ for the thickness of the last scattering surface. Thus these effects appear predominately at large values of $\ell$.

In this section, I will treat measurements of the power spectrum on scales $\ell \lesssim 300$. That range includes the predicted plateau at low $\ell$ and the first acoustic peak, with emphasis on the angular scale of the latter. Although its amplitude owes primarily to local effects, rather than cosmological ones, I will also describe the dipole moment of the CMB. In Section 1.4, I will treat measurements of the entire series of acoustic peaks, evidence for Silk damping and the overall slope of the initial power spectrum in order to look for small departures from the $n_S = 1$ expectation.

### 1.3.2 Some observational issues

I recognize that the division between $\ell \lesssim 300$ and $\ell \gtrsim 300$ is slightly artificial. It is inspired in part by the difference in observational techniques generally employed in these two regions of angular scale. Roughly speaking, as we shall see, the low $\ell$ measurements must be made from space or at least above a substantial fraction of the Earth's atmosphere, whereas ground-based measurements are possible at the smaller angular scales, $\ell \gtrsim 300$.

#### 1.3.2.1 Comparative measurements

Unlike the case of spectral measurements discussed in Section 1.2 isotropy measurements can be made *comparatively*; absolute calibration is less pressing. The basic technique is to compare measurements made at two different positions in the sky. This can be done by fixing the apparatus and allowing the sky to pass overhead; by scanning the sky rapidly; or by beam switching, that is, comparing the difference in intensity observed by either two antennae pointed in different directions, or two beams from a single antenna employing a moving secondary or two feed horns. In practice, some or all of the techniques are combined to further reduce systematic errors. But the basic point is that anisotropy measurements can be made on the comparative basis, which is the main explanation for their far greater sensitivity than absolute measurements. MicroKelvin (μK) accuracy, rather than milliKelvin (mK) accuracy is required and can be attained. The Differential Microwave Radiometer (DMR) instruments aboard COBE, shown below in Fig. 1.4a, provide a good example. Matched pairs of microwave antennae, with their optical axes separated by 60°, fed signals into a comparative radiometer. Thus the difference in observed intensity from the sky was measured at the switching frequency, which was kept high enough to avoid gain drifts in the receiver. In addition, as the satellite rotated about its body axis, the two directions scanned circles in the

(a)

(b)

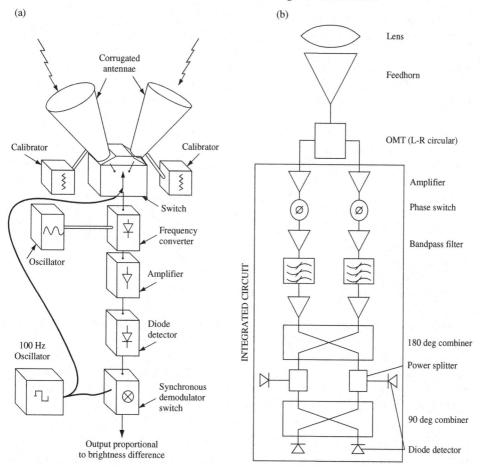

FIGURE 1.4. (**a**) Block diagram of the DMR instrument on COBE, showing symmetrical horns and calibrators. (From COBE slide set. NASA image.). (**b**) A pseudo-correlation receiver for polarization measurements (diagram from Todd Gaier). Signals from the horn are split in an orthomode transducer (OMT), amplified and recombined in such a way as to produce two linearly polarized outputs. Both polarizations pass through both amplifiers; thus variations in the gain average out.

sky. The annual passage of the satellite around the sun shifted the scanning directions, so that a complex, interlocking series of measurements of temperature differences in the CMB could be made. Unlike DMR, the Planck receivers will compare measurements of the sky with a controlled temperature cold load. The Planck satellite thus relies on the rapid rotation of the spacecraft to control gain drifts.

### 1.3.2.2 Calibration

To ensure that measured differences can be expressed in temperature terms, to compare to the 2.725 Kelvin background temperature, good calibration is required. It need not be at the same precision as the calibration required for absolute temperature measurements; they require precision of $\lesssim 1\,\mathrm{mK}$ for calibration. In FIRAS this accuracy, or $\sim 4 \times 10^{-4}$ of the signal, was achieved. The same relative precision in calibration would allow us to reach $1\text{--}2\,\mu\mathrm{K}$ for a comparative measurement.

It is interesting to note that the most recent large-scale anisotropy experiments employ the COBE-measured dipole as the primary external calibration. The sensitivity of modern experiments, such as the Wilkinson Microwave Anisotropy Probe (WMAP) satellite to be discussed later, is so high that they can quickly detect the CMB dipole first accurately measured by COBE, and use it to calibrate the instruments. This turns out to be more accurate than the use of "point" sources, such as planets, as calibrators.

### 1.3.2.3 Other instrumental effects

To make accurate comparative measurements, we need to ensure that there are no instrumental offsets, and more importantly that instrumental offsets do not change with time. For instance, for observations made from beneath the atmosphere, the two directions being compared must lie at the same elevation, to avoid differential atmospheric emission. Equally, care must be taken to ensure that any signals diffracted into the antennae be independent of antenna alignment and preferably not time variable. Great care must also be taken to ensure that the gain and any loss in the receiver is symmetrical for the two directions being compared. In this regard, the use of a pseudo-correlation receiver, one that explicitly mixes the two signals, is much preferred. The design of such a receiving element is included as Figure 1.4b, and is explained further in the caption.

While eyes may glaze over with the discussion of some of these instrumental effects, controlling them is crucial to ensure the accuracy of CMB measurements. A measure of their importance can be seen in the series of long explanatory articles published as part of the analysis of the first three years of WMAP data (see below). Concern about instrumental effects also has an important impact on the *design* of CMB experiments, favoring those with a high degree of symmetry (as is true for WMAP).

### 1.3.3 Foreground emission

Even a perfect instrument faces the problem of foreground emission. The two major sources are the Earth's atmosphere and the Galaxy. I will defer discussion of a third source of foreground contamination, emission by radio galaxies and dusty galaxies, to Section 1.4.

### 1.3.3.1 Atmospheric fluctuations

If the atmosphere were an absolutely homogeneous, plane-parallel slab, as implicitly assumed in Section 1.2, atmospheric emission would simply cancel out if we compared two sky regions at the same elevation. But the atmosphere is turbulent, and turbulence introduces fluctuations, particularly on large angular scales. Indeed, it is the scale-dependence of atmospheric fluctuations that makes it extremely difficult to carry out CMB observations at $\ell \lesssim 300$ from the surface of the Earth.

First, let us look at the spatial dependence of atmospheric fluctuations. To first order, they can be modeled as a Kolmogorov process. For the conditions obtaining in the Earth's atmosphere, that predicts a power-law dependence of root-mean-square (rms) noise on linear scale as $d^{\frac{5}{6}}$. If we then make the assumption that these fluctuations are carried past an antenna beam at a uniform velocity, the rms fluctuations will scale with *time* to the same power. Some observational work by Smoot *et al.* (1987) finds time variation with an index of approximately 0.7, in rough agreement with the simple predictions. If we again assume a constant wind speed, we can back-calculate from this result to show that the amplitude of fluctuations scales with their length as $d^{0.7}$. Thus a high price is paid for observations on large angular scales. It is true that atmospheric fluctuations can be mitigated by working at high altitude (particularly important to reduce the effect of water-vapor fluctuations, since their scale height is small), and at cold and dry sites (again to reduce the overall amplitude of water-vapor fluctuations). It is no surprise that many recent experiments to determine the large-scale anisotropy of the CMB have operated at balloon altitudes. This is particularly true for those experiments using bolometers as detectors. Bolometers work best at relatively high frequency, and we have already seen that atmospheric emission is substantially stronger at high frequencies than low.

### 1.3.3.2 Galactic emission

Atmospheric fluctuations favor observations at low frequency; unfortunately, Galactic emission favors observations at high frequencies. Emission from our Galaxy has proven to be the primary source of foreground contamination in the WMAP observations, and is certain to dominate the Planck observations as well. Indeed, one reason for including as many frequency bands as we did in the Planck experiment was to control Galactic emission. The crucial importance of the Galaxy is well reflected in Figure 1.5, the WMAP image of the sky made at 23 GHz, its lowest observing frequency. That Galactic emission is strong and spatially variable is apparent.

There are four basic mechanisms for producing microwave emission from the Galaxy. At frequencies below roughly 100 GHz, the dominant one is synchrotron emission. The temperature spectral index for synchrotron emission is given approximately by $T(\nu) \propto \nu^{-2.8}$. Unfortunately, both the surface brightness and the spectral index vary from location to location in the Galaxy. Synchrotron emission can also be polarized. All these make the subtraction of Galactic emission difficult – as Rod Davies shows in much more detail later in this volume. A frequently used solution – by the WMAP team, for instance – is simply to mask out the Galactic plane entirely.

Unlike synchrotron emission, free–free (bremsstrahlung) emission has a fixed spectral index: $T(\nu) \propto \nu^{-2.1}$. Unfortunately, free–free emission is generally strongly dominated at $\nu \lesssim 50$ GHz by synchrotron emission, and at $\nu \gtrsim 150$ GHz by re-emission from warm dust. That makes free–free emission difficult to measure unambiguously and then subtract. Adding to the difficulty of controlling for free–free emission is our spotty knowledge of its distribution in the Galaxy. To some degree, it correlates with H$\alpha$ emission from warm

FIGURE 1.5. The WMAP 23 GHz sky image, in the usual projection with the Galactic plane running horizontally through the center. The dominance of Galactic emission is evident. (WMAP Data Product Images; NASA and the WMAP Science Team.)

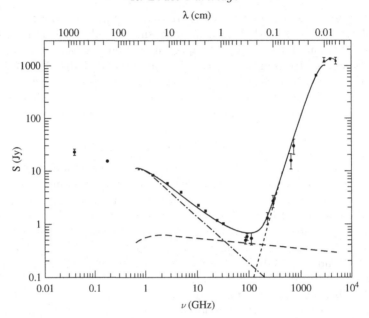

FIGURE 1.6. Microwave and far-IR spectrum of a galaxy, in this case the starforming system M82 (Condon, 1992). At low $\nu$, synchrotron emission dominates; at high $\nu$, re-emission from warm dust. Reprinted, with permission, from the *Annual Review of Astronomy and Astrophysics*, volume 30 (c) 1992 by Annual Reviews (www.annualreviews.org).

interstellar gas (Bennett *et al.*, 2003); Kogut *et al.* (1996) also show some correlation between free–free and dust emission.

Warm dust at temperatures around 20 K re-radiates a modified blackbody spectrum,[1] with a long-wavelength tail that produces a surface brightness equivalent to synchrotron emission at $\nu \sim 150$ GHz. Fig. 1.6 shows the overall emission spectrum of a galaxy; note that dust re-emission clearly dominates at $\nu > 200$ GHz. M82 is a star-forming galaxy, so dust re-emission is more prominent than average. Nevertheless, the figure demonstrates why most large-scale CMB anisotropy measurements are carried out in or near the "valley" in Galactic emission at 50–150 GHz.

An aside: there is an interesting observational question here, which CMB experts have debated at length. Is it best to work at the very minimum of Galactic emission (but pay the price that at least two components are contributing more or less equally to the foreground contamination)? Or is it better to work at a slightly lower $\nu$, where dust re-emission is negligible; or higher $\nu$, where synchrotron emission falls off steeply (but then pay the price of larger amplitude contamination)?

Complicating this question (and the correction for Galactic foreground emission generally) is microwave emission by spinning dust, a mechanism explored by Draine and Lazarian (1998, 1999). The spectral signature does not follow a power law, but peaks at $\nu \sim 10$–30 GHz. Spinning dust will also be discussed more fully in Rod Davies' contribution (Chapter 7); see also Boughn and Pober (2007).

Next, let us consider briefly the angular scale dependence of Galactic emission. This naturally depends on the mechanism – but broadly speaking, we expect fluctuations in Galactic emission to scale with $\theta$ as roughly $\theta^{3/2}$ (see Davies, Chapter 7, for further

---

[1] $T(\nu) \propto \nu^3$ or $\nu^4$, depending on the frequency dependence of the emissivity of the dust; see Davies, Chapter 7, for details.

discussion). As a consequence, emission from the Galaxy plays a smaller role in measurements of the CMB power spectrum at large $\ell$.

### 1.3.4 What we can learn from the power spectrum at $\ell \lesssim 300$

Despite these observational difficulties, many groups have worked hard over decades to measure the low-$\ell$ power spectrum of the CMB because it provides so much information about fundamental cosmology.

#### 1.3.4.1 The dipole moment

The CMB dipole, first firmly detected in 1976 (for the early history, see Peebles *et al.*, 2009), is not strictly cosmological in origin. There is no doubt an intrinsic $\ell = 1$ component (presumably of amplitude approximately equal to both the quadrupole moment and to cosmic variance at $\ell = 1$), but this is swamped by the dipole induced by the Doppler effect due to our motion:

$$\Delta T/T_0 = \frac{v}{c}\cos\theta + \frac{1}{2}\left(\frac{v}{c}\right)^2 \cos 2\theta + O\left(\frac{v}{c}\right)^3, \tag{1.4}$$

where $\theta$ is the direction of our motion.

First, here's a brief, historical perspective. When the dipole was first measured with reasonable accuracy, its amplitude was no surprise, but its direction was. It aligns very poorly with the inferred solar velocity in the Galaxy. Indeed the direction is close to opposite the expected vector of solar motion. The only explanation is that the center of mass of the Galaxy is in motion at $v \sim 600\,\mathrm{km\ s^{-1}}$ relative to the CMB. Such a high velocity required explanation. It is now accepted that our Galaxy has a peculiar velocity of this magnitude, gravitationally induced by relatively local mass concentrations, as outlined in chapter 8 of my book, *3 K: The Cosmic Microwave Background Radiation* (1995).

#### 1.3.4.2 The quadrupole moment

Limits on the quadrupole amplitude constrain anisotropic cosmological models. Anisotropic (yet homogeneous) models are perfectly consistent with General Relativity. A simple case is a model with slightly faster expansion along one axis than the other two. This produces an oblate distribution of observed temperature, and hence a quadrupole moment.

Homogeneous but anisotropic models were classified over a century ago by Bianchi (1898); different Bianchi classes leave different traces in the observed distribution of the CMB temperature. The very small amplitude of the quadrupole term (as given in Section 1.3.5.2 below) has largely damped interest in such models (for more details, see section 8.3 of my book, *3 K*).

There is a small quadrupole component with a precisely known alignment that arises from the Doppler effect:

$$\Delta T_2/T_0 = \frac{1}{2}\left(\frac{v}{c}\right)^2 \cos 2\theta, \tag{1.5}$$

but it can easily be shown that the amplitude is substantially smaller than the cosmic variance at $\ell = 2$.

#### 1.3.4.3 $\ell \sim 10$–$50$

In this range of $\ell$, we measure the primordial power spectrum, unaffected by later physical processes. Two quantities are of interest, the shape in the power spectrum and its overall amplitude in this range of $\ell$. The former helps constrain the index of scalar perturbations, $n_S$. As Matarrese explains, $n_S = 1$ is expected for a pure Harrison–Zel'dovich spectrum of scalar perturbations and $n_S \sim 0.95$ for many inflationary models. We return to $n_S$ in much more detail in Section 1.4.4.8.

Now let us look at the overall amplitude. We will do so by taking a brief detour to review in schematic form the growth of structure in the Universe.

In 1946, Lifschitz showed that (for matter-dominated expansion), the linear growth of amplitude in density fluctuations is

$$\Delta\rho/\rho \propto (z+1)^{-1}, \text{ or } \frac{\Delta\rho}{\rho}(t) = \left(\frac{\Delta\rho}{\rho}\right)_0 (z(t)+1)^{-1}. \tag{1.6}$$

In the currently favored cosmological model, matter domination gives way to exponential expansion at $z \sim 1$. Thus we expect $(\Delta\rho/\rho)_0 \approx 500(\Delta\rho/\rho)_{LS}$ when $(\Delta\rho/\rho)_{LS}$ is the amplitude of the perturbations on the surface of last scattering at $z \sim 1000$. On scales of 5–10 Mpc and below $(\Delta\rho/\rho)_0$ is now $>1$; thus we require density fluctuations on these scales of $\sim 0.2$ percent on the surface of last scattering.

If density fluctuations of this amplitude were present in the *baryon* content of the Universe, the resultant temperature fluctuations in the CMB would be of order $\Delta T/T \sim 10^{-3}$ or a bit less.[2] Well before the first detection of CMB anisotropies in the $\ell = 10$–50 range (by COBE in 1992; see Smoot *et al.*, 1992), it was recognized that $\Delta T/T$ was in fact $\lesssim 10^{-4}$ on a range of angular scales. The resolution of this apparent contradiction is simple: we invoke dark matter.

The basic physical picture is the following. Density fluctuations in dark matter are free to grow in amplitude from the time they enter the causal horizon until the epoch corresponding to last scattering. Hence they can reach the required amplitude ($2 \times 10^{-3}$, by the rough calculation cited above). The growth of fluctuations in the baryons, on the other hand, is impeded by interaction with the CMB radiation field itself. Thus the amplitude of density fluctuations in the baryons can be much smaller than the amplitude of fluctuations in dark matter.

There are two important consequences. First, temperature fluctuations linked to baryon perturbations can be substantially smaller than $10^{-3}$. Indeed, the amplitude of the temperature fluctuations is determined primarily by the interaction of the Sachs–Wolfe and Doppler effects, as described in detail by Wayne Hu in Chapter 3. Second, after last scattering, the interaction between radiation and baryonic matter essentially ceases, and the baryons are free to fall into the potential wells of larger-amplitude dark matter fluctuations. In loose language, the amplitude of baryon perturbations "catches up" with dark matter fluctuations.

We thus see that the amplitude of the power spectrum at low $\ell$ depends crucially on the dark matter content of the Universe. This is generally expressed as $\Omega_{dm}$ where $\Omega_{dm}$ is the ratio of the density of dark matter to the critical density, $\rho_c = 3H_0^2/8\pi G$.

### 1.3.4.4 The angular scale of the first acoustic peak

Next we consider the physical or angular *scale* of perturbations on the surface of last scattering, rather than their amplitude. Here, to save space, I will depend on the elegant and detailed presentation by Wayne Hu in Chapter 3 to draw the following basic conclusions. The physical scale of oscillations in density is well constrained. So too is the scale of the resulting temperature fluctuations imprinted on the surface of last

---

[2] To support this claim, consider the very simplest case, pure adiabatic fluctuations. Then ignoring all the details of both the Doppler and the Sachs–Wolfe effects, we have $\Delta T/T = 1/3\Delta\rho/\rho$, so we expect temperature fluctuations of amplitude roughly $\Delta T/T \sim 0.7 \times 10^{-3}$.

scattering. A specific length scale, call it $d$, is picked out on the surface of last scattering and so are harmonics of that scale, $d/2$, $d/3$, ... Equally, we know the distance $D$ to the surface of last scattering to high precision. It is therefore straightforward to calculate the corresponding *angular scale* (related to $d/D$) for any particular cosmological model. In the simplest case, a flat model with curvature $k = 0$ (or total mass energy density equal to the critical density, so that $\Omega_{TOT} = 1$), the angular scale of the first acoustic peak is at $\theta \sim 0.8°$ or $\ell \sim 220$. In positively curved space, where the sum of angles in a triangle exceeds $180°$, it is easy to show that $\theta$ is larger (or $\ell$ smaller), and vice versa for negatively curved space. For values of $\Omega_{TOT}$ not radically different from unity,

$$\theta = 0.8° \sqrt{\Omega_{TOT}}. \qquad (1.7)$$

Thus the position of the first acoustic peak in the power spectrum of the CMB gives a clear and straightforward measure of the curvature of the Universe, and hence of $\Omega_{TOT}$. As we will see below, the position of the first peak is able to fix $\Omega_{TOT}$ to an accuracy of a few percent or better.

### 1.3.4.5 The relative amplitude of the first acoustic peak

There is additional information in the amplitude of the first acoustic peak when compared to the $\ell = 10$–50 plateau. The amplitude of temperature fluctuations in the flat region $\ell = 10$–50 in the power spectrum, that is at angular scales $\gtrsim 3°$, is fixed by the amplitude of density perturbations induced during inflation, and is unaffected by later physical processes. The amplitude of the acoustic peak, on the other hand, is determined by physical processes at and before last scattering. Thus the relative amplitude of peak to plateau depends on cosmological parameters such as the equation of state of contents of the Universe. The relative amplitude, for instance, depends on the ratio $\Omega_{dm}/\Omega_b$.

More dramatically, if we move away from the now-accepted model of adiabatic density fluctuations as the seeds of present cosmic structure to a model in which cosmic defects account for structure, a very different power spectrum is predicted. In cosmic defect models, the acoustic peaks cancel out, and the power spectrum remains essentially flat to much higher values of $\ell$ (see Pen *et al.*, 1997, for a pre-WMAP treatment). The fact that the acoustic peaks are robustly detected (see below) killed off cosmic defect models (though of course one is free to argue that cosmic defects are responsible for some small fraction of the presently observed cosmic structure).

### 1.3.5 Observational results and constraints on cosmic parameters

Now what do the currently available results tell us? In this subsection, I will emphasize those results that depend crucially on observations at values of $\ell$ well below 300. I recognize that the distinction between high $\ell$ and low $\ell$ is in many ways artificial, and I will return to cosmic constraints set by the *overall* power spectrum in Section 1.4 below.

### 1.3.5.1 Early (pre-WMAP) results

The long-sought temperature fluctuations in the CMB were first detected by the DMR instrument on NASA'a COBE satellite (Smoot *et al.*, 1992). There had been hints of the detection of temperature anisotropies from ground-based and balloon-borne experiments earlier, but the groups responsible for these experiments conservatively reported only upper limits. The COBE detection released a flood of follow-up observations on a range of angular scales. Here it should be noted that the angular resolution of the DMR instruments, approximately $7°$, was not adequate to detect the acoustic peaks.

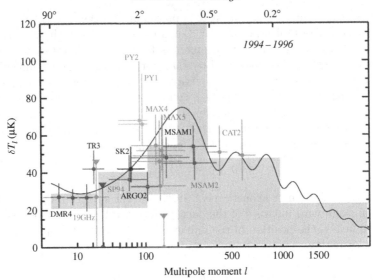

FIGURE 1.7. Measurements and constraints on the CMB anisotropy power spectrum available in 1996 (after *COBE*, but before refined ground-based measurements and WMAP). Figure from Peebles *et al.* (2009), with permission.

A number of ground-based and balloon-borne experiments soon confirmed the COBE results, extended the power spectrum to higher values of $\ell$, and produced tentative evidence for the existence of the first acoustic peak. These early experiments are beautifully summarized in the compendium prepared by Lyman Page for the volume titled *Finding the Big Bang* (Peebles *et al.*, 2009). In Fig. 1.7, I reproduce a power spectrum showing the measurements available in 1996. This includes those made within a few years following the COBE announcement.

A few observations at large $\ell$ also suggested the cutoff caused by Silk damping and thus the appreciable thickness of the surface of last scattering.

Although the details of the spectrum of acoustic peaks beyond the first spectrum was not yet clear, in many ways the situation in the late 1990s resembled the cartoon sketch shown in Fig. 1.3; all the major physical features explored in Section 1.3.1 were present.

### 1.3.5.2 WMAP Results

The NASA WMAP satellite was designed to determine the power spectrum of the CMB from $\ell = 2$ to nearly 1000, thus incorporating the low $\ell$ plateau and the first few acoustic peaks. Its symmetrical design, the scan strategy employed, and the placement of the satellite at the second Lagrange point of the Sun–Earth system were all designed to minimize systematic error. The WMAP power spectrum, first released in the spring of 2003 (Page *et al.*, 2003) revolutionized our view of CMB anisotropies. The precision of the WMAP results at $\ell$ up to about 1000 was higher than precursor observations, including both ground-based work and the COBE results. More important, the wide range of $\ell$ spanned by a single instrument gave us much greater confidence in the overall shape of the spectrum: to use the old analogy, we were seeing the elephant whole, not sampling its trunk, its feet, its tusks....

Most of the constraints I cite below are drawn from a series of papers (Hinshaw *et al.*, Page *et al.* and Spergel *et al.*), all published in 2007 in the *Astrophysical Journal*

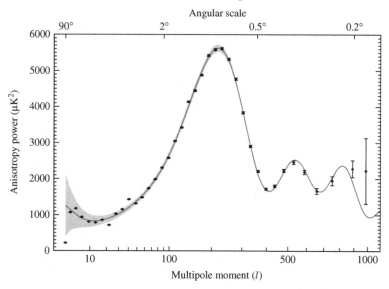

FIGURE 1.8. The power spectrum from three years of WMAP observations (from Hinshaw *et al.* 2007). NASA and the WMAP science team. Reproduced by permission of the AAS.

*Supplement.* These are based on the first three years of WMAP observations; by the time this volume appears, the four- or five-year results may be available (Figure 1.8).

Initially, the first acoustic peak is found by WMAP to lie at $\ell = 220.8 \pm 0.7$. Thus the WMAP results are fully consistent with the flat cosmological model. An initial value $\Omega_{\text{TOT}} = 1.02 \pm 0.02$ was later refined to $0.989 \pm 0.012$. Both values, of course, are consistent with 1.00, the value expected for $k = 0$ (i.e. a flat universe). The WMAP result constrains $\Omega_{\text{TOT}}$, that is the sum of the matter density and the mass-energy contributed by any cosmological constant term: $\Omega_{\text{m}} + \Omega_{\Lambda}$. Comparisons of the luminosity distance and redshifts of supernovae, on the other hand, constrain a different combination of $\Omega_{\text{m}}$ and $\Omega_{\Lambda}$, much closer to the difference of the two. Roughly, the supernova constraint is on $0.8\Omega_{\text{m}} - 0.6\Omega_{\Lambda}$.

If we thus plot the CMB limits from WMAP and the supernova limits of Riess *et al.* (2004) and Astier *et al.* (2006), for instance, the constraint regions are essentially orthogonal, as shown in Fig. 1.9. When we include the supernova results along with the WMAP findings, we can evaluate both the equivalent density of a cosmological constant term and the overall matter density, including both dark matter and baryonic matter. The constraints shown in the figure are independent of other measures of $\Lambda$ and dark matter derived from details of the CMB power spectrum. Using such an analysis, and more recent CMB and supernova results, Spergel *et al.* (2007) find $\Omega_{\Lambda} = 0.71 \pm 0.03$ and $\Omega_{\text{m}} = 0.26 \pm 0.03$.

The WMAP three-year results also refined the value of the dipole moment to $\Delta T_1 = 3.358 \pm 0.017\,\text{mK}$, corresponding to a velocity of $369\,\text{km s}^{-1}$ towards $\ell = 264.14°$ and $b = 48.26°$ in Galactic coordinates; the uncertainty attached to each number is $\sim 0.15°$. Converting to right ascension and declination, we have $\alpha = 11^{\text{h}}12^{\text{m}} \pm 1^{\text{m}}$ and $\delta = -7.1° \pm 0.2°$. Note that we know the overall velocity of the Earth with respect to the CMB to much higher precision than we know the solar motion in our own Galaxy. Nevertheless, with reasonable values for the latter, the velocity of the Galaxy is of order $500\text{–}600\,\text{km s}^{-1}$.

One of the most surprising results of the initial WMAP data release was the low amplitude of the quadrupole moment. That is confirmed in the three-year data. Indeed,

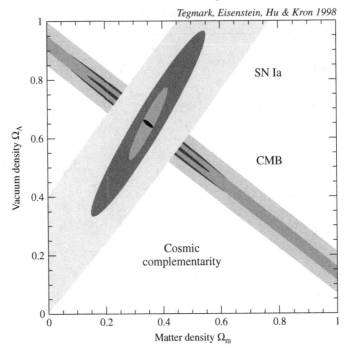

FIGURE 1.9. An early diagram from Tegmark *et al.* (1998) and Eisenstein *et al.* (1999) showing the essentially orthogonal constraints set by supernovae and CMB measurements. This figure does *not* include the WMAP results which sharpen the CMB constraints substantially. The figure was taken from M. Tegmark web page (http://space.mit.edu/home/tegmark).

as may be seen from Fig. 1.8, the observed quadrupole amplitude is barely consistent with cosmic variance. Nor can the second-order Doppler term explain the low value; since we know $v$ to high precision we can calculate the amplitude of the $(v/c)^2$ term and show that it is small compared both to the observed quadrupole amplitude and to the apparent discrepancy between the observed amplitude and the central ridge of the cosmic variance band (left as an exercise).

The three-year results from WMAP give $\Delta T_2^2 \sim 240\,\mu K^2$ or $\Delta T_2/T_0 \sim 5.7 \times 10^{-6}$.

The octopole ($\ell = 3$) amplitude is closer to the centroid of the cosmic variance band at $\Delta T_3^2 = 1050\,\mu K^2$ or $\Delta T_3/T_0 \sim 11.8 \times 10^{-6}$. While the amplitude is closer to expectations, there is also some indication that the octopole and quadrupole moments are aligned. Why should this be? This is a subject that has been discussed extensively (see Eriksen *et al.*, 2004, for instance, and the discussion in Chapter 6 by Enrique Martínez-González).

Both the alignment and the low amplitude of the quadrupole moment could be mere accidents. They could also be the result of an instrumental or systematic effect remaining in the data even after the careful subtraction of Galactic emission (as Fig. 1.5 shows, Galactic emission introduces a huge quadrupole signal). Or, as some of the authors cited above argue, the apparent alignment and small amplitude of the quadrupole moment could be a sign of new and interesting cosmic physics. Again, see Chapter 6 in this volume.

Let me mention a few other results from the WMAP observations at values of $\ell <$ 300. Most obviously, the overall amplitude is inconsistent with pure baryonic matter; as explained above, dark matter is required to explain the overall (low) amplitude of the power spectrum. Next, the presence and amplitude of the acoustic peaks sharply constrain cosmic defect models.

FIGURE 1.10. CMB fluctuations as revealed by the WMAP three-year data set (the image has been slightly reconstructed in the Galactic plane, which runs horizontally through the center of the image). (WMAP Data Product Images; NASA and the WMAP Science Team.)

I should mention also the controversial and much-discussed issue of the "cold spot" identified in the WMAP full-sky image shown in Fig. 1.10, and other possible non-Gaussian features. These too are discussed by Martínez-González; see also Marcos Cruz *et al.* (2007).

Finally, the (relatively) flat power spectrum at $\ell = 5$–30 is consistent with a Harrison–Zel'dovich spectrum as well as simple models of inflation which predict a value of $n_S$ slightly below 1.0. These findings hardly constitute a definitive test of inflation, but are comforting. The actual value $n_S$ is better defined by a consideration of the entire power spectrum. This is an example of ways in which my artificial division of the power spectrum into high $\ell$ and low $\ell$ portions is slightly misleading. At the end of Section 1.4, I will consider constraints set by the entire power spectrum including those on $n_S$.

## 1.4 Smaller-scale, primary and secondary, anisotropies in the CMB

Here I concentrate on small angular scale fluctuations, those at $\ell > 300$. While the division between high $\ell$ and low $\ell$ is artificial, as I have noted, there are observational or experimental reasons for making the distinction. For one thing, high-$\ell$ observations can be made from the ground and indeed, in the case of $\ell \gtrsim 300$, must be made from the ground because the high angular resolution implies large antennae. Another reason for drawing the distinction is that most sources of secondary anisotropies, those induced at redshifts well below the redshift of last scattering, appear at high $\ell$. Finally, the experimental techniques, sources of systematic error, and foregrounds are different.

In this section, I will first treat mechanisms that introduce small-scale fluctuations in the CMB, then turn to foregrounds and systematics that can mask or confuse CMB fluctuations. I will then briefly discuss some of the observational techniques used, and provide a partial listing of some of the active programs to measure CMB anisotropies on scales $\ell \sim 300$ and above. I will look at some of the implications for cosmology set by observations of small angular fluctuations. Finally, I will look at the entire assembly of data on the CMB power spectrum, including particularly the WMAP results.

### 1.4.1 Mechanisms for generating anisotropies at $\ell > 300$

It is conventional to divide CMB fluctuations into two classes. *Primary* fluctuations are those induced at or near the surface of last scattering at $z \sim 1100$. *Secondary* anisotropies, on the other hand, are produced at much lower values of redshift; these are the subject of Bartelmann's Chapter 4 in this volume.

### 1.4.1.1 Primary fluctuations at $\ell > 300$

The main feature in the power spectrum at $\ell > 300$ is the continuation of the acoustic oscillation spectrum, the "overtones" of the first acoustic peak at $\ell = 220$. Since the power spectrum measures the amplitude of temperature fluctuations, independent of their sign, there are acoustic peaks present for both compressions and rarefactions. The odd-numbered acoustic peaks, corresponding to compressional modes, and the even-numbered peaks corresponding to rarefaction modes, are differently affected by the inertia of the baryons. This point is explained in detail by Wayne Hu. Here, I wish simply to draw out the conclusion: the ratio of the overall amplitudes of odd- and even-numbered peaks provides a tight constraint on the density of baryons, $\Omega_b$. It is worth remarking explicitly that this constraint on $\Omega_b$ arises from entirely different physics than the constraint that is set by primordial nucleosynthesis (see the review by Schramm and Turner, 1998).

It is also worth remarking that the relative amplitude of odd- and even-numbered peaks is affected to some degree by the overall tilt of the power spectrum, which in turn depends on the quantity $n_S - 1$. This brings up the general issue of the degeneracy of cosmic parameters. Changes in $n_S - 1$ can compensate for changes in $\Omega_b$, leaving the power spectrum essentially unaffected. To constrain $n_S$ and $\Omega_b$ separately, some additional observations are required to break the degeneracy. We have already looked briefly at one such case: those of CMB and supernovae results to constrain $\Omega_m$ and $\Omega_\Lambda$ separately. We will return to the issue of breaking degeneracy when we come to talk about polarization of the CMB.

### 1.4.1.2 Secondary fluctuations

Since these will be fully discussed by Matthias Bartelmann, I'll simply list some of the mechanisms that can induce secondary fluctuations. I'll also give an approximate range of $\ell$ over which these mechanisms operate.

The dominant mechanism is the Sunyaev–Zeldovich effect, the transfer of energy to some CMB photons by interactions with hot plasma. As we have seen in Section 1.2, if the plasma is generally distributed, a spectral distortion results. If the plasma is localized, as is true for the hot intergalactic medium in clusters of galaxies, anisotropies are introduced into the CMB. At frequencies below $\sim$220 GHz, where most anisotropy measurements are made, the Sunyaev–Zeldovich (SZ) effect produces small decrements in the temperature, and the typical range of angular scales corresponds to $\ell > 2000$. On the one hand SZ fluctuations can be viewed as a foreground contamination of the primary fluctuations at small angular scales; on the other hand, SZ observations are a rich source of science in their own right (again see Bartelmann's Chapter 4).

Next, there is the so-called integrated Sachs–Wolfe effect, the result of CMB photons falling into and crawling out of expanding density perturbations. This effect produces fluctuations primarily at large scales (low $\ell$).

Finally, there is a gravitational lensing. The brightness theorem can be used to show that the gravitational lensing of a purely isotropic CMB would produce no effect. Since the CMB does contain temperature fluctuations, however, we need to consider the effect of gravitational lensing by clusters of galaxies and other large-scale aggregates of matter. The effect is to distort the high-$\ell$ end of the CMB power spectrum, increasing the high-$\ell$ tail.

### 1.4.2 Foregrounds, systematics and techniques to control them

Now for some of the dirty problems encountered in measurements of the CMB anisotropy on small angular scales.

#### 1.4.2.1 Foregrounds

As noted in Section 1.3.3 above, both atmospheric and Galactic foregrounds are a smaller problem than at larger angular scales. The former permits small-scale anisotropy searches to be done from the ground, though working from the ground favors low frequencies. Nevertheless, the atmospheric emission restricts observations to selected frequency ranges to avoid atmospheric lines, and favors observations from high, dry sites. Hence the number of experiments employing mountaintops (as on Teide), or high plateaus such as the Atacama desert or the south pole. Galactic emission can be reduced by working in known areas of low Galactic synchrotron and/or dust emission.

While Galactic and atmospheric emission present smaller problems at high $\ell$, the opposite is true for foreground radio sources and, at high frequencies, dusty galaxies. If such sources are randomly distributed, it is easy to show that their contribution to the power spectrum of the sky expressed in $(\mu K)^2$ is proportional to $\theta^{-2}$ or $\ell^2$. At wavelengths longer than $\sim 1$ cm, the emission of most extragalactic sources is completely dominated by synchrotron emission (see Fig. 1.6). The temperature spectrum of synchrotron radiation drops sharply with increasing frequency; hence to minimize the effect of extragalactic sources (and also Galactic emission) high-frequency observations are favored. There are three limitations to this simple solution. First, at frequencies of roughly 100 GHz or higher, re-emission from dust enters (again, see Fig. 1.6). Second, observations at high frequencies except in the 90 GHz atmospheric window are difficult from the ground. Finally, and perhaps more fundamentally, the high-frequency properties of radio galaxies are poorly studied. There is abundant evidence that the high-frequency emission of radio galaxies cannot simply be extrapolated from their low-frequency fluxes and spectral indices (see Sadler *et al.*, 2006 and Lin *et al.*, 2009).

There are two complementary approaches taken to the problem of foreground radio sources. The first is careful modeling, based on extrapolation of radio properties from low-frequency surveys (de Zotti *et al.*, 2005, and Toffolatti *et al.*, 1998, for instance). The second is observational – either follow known sources to higher frequencies, or, even better, conduct a high-frequency survey of the sky. The former approach was taken by López-Caniego in a poster presented at the Winter School. The latter approach was taken by the Australia Telescope team, who have surveyed the entire southern sky at 20 GHz (preliminary results in Sadler *et al.*, 2006; full data release expected soon). Obviously the high-frequency observations link back to and improve the modeling. Now, here's a final technical remark. In most searches for small-scale anisotropies, the bright sources are masked out; as shown in the papers by de Zotti and Toffolatti referred to above, this substantially reduces the statistical fluctuation level in CMB images produced by the remaining foreground sources.

Despite all these precautions, the subtraction of foreground radio sources is a crucial feature of sensitive searches for CMB anisotropies on small angular scales. Virtually all of the ground-based experiments to be discussed below have had to devise techniques for locating, characterizing and subtracting (or correcting for) foreground sources. The WMAP team masked out several hundred sources in obtaining the results shown in Fig. 1.10. And the Planck team (see below) is actively trying to characterize the high frequency properties of radio sources before the launch of the satellite.

### 1.4.2.2 Instrumental systematics

As was the case for anisotropy measurements on larger scales, care must be taken to reduce any asymmetry in the beam switch if that technique is used. It is also important to reduce differential ground pick-up; hence all ground-based experiments use extensive ground shields.

It is also important to model any asymmetry in the shape of the beams on the sky. A beam with elliptical contours, rather than circular symmetry, samples and therefore mixes different values of $\ell$. Beam asymmetries thus affect the window function of an experiment. These considerations are particularly important when large focal plane arrays are used, since some detectors are necessarily located well off axis. This was the case for WMAP (see Fig. 1.11) and for Planck as well (see Fig. 1.12, showing Planck's focal plane array).

### 1.4.2.3 Interferometric measurements

Many of the systematics discussed above atmospheric emission, beam asymmetry, and ground pick-up may be sharply reduced by making interferometric measurements as in

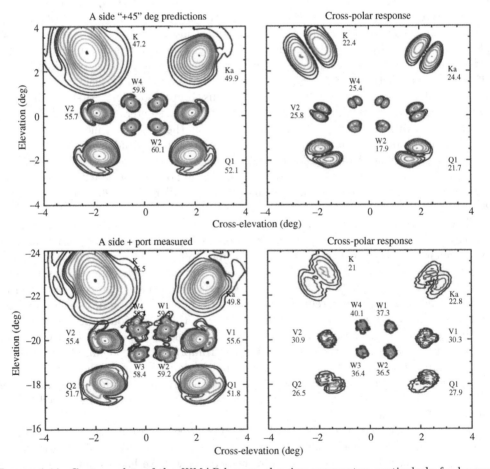

FIGURE 1.11. Contour plots of the *WMAP* beams, showing asymmetry, particularly for beams well displaced from the optical axis. Top panels: calculated beam shapes for both total power (left) and cross polarization (right). Bottom panels: measured beams. Note the accuracy with which the beam shapes can be calculated/modeled. Figure from Jarosik *et al.* (2007). NASA and the *WMAP* science team. Reproduced by permission of the AAS.

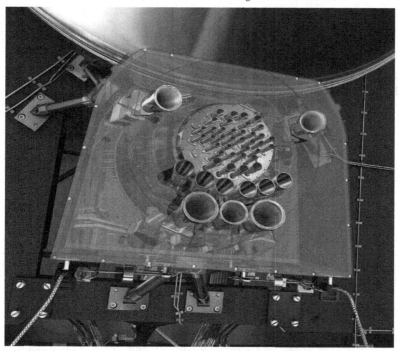

FIGURE 1.12. The focal plane array of the Planck mission. The central cluster of smaller horns contains the feed horns for the high frequency instrument (HFI). Arrayed around it are feed horns for the 70, 44 and 30 GHz receivers of the low frequency instrument (LFI) (ESA image).

the Very Small Array (VSA) at the Teide observatory. Ground pick-up and atmospheric emission are essentially common mode signals, and cancel out.

Since the angular resolution of an interferometric array is set by the size of array, not the diameter of the individual elements of the array, it is easy to obtain high resolution. On the other hand, a price is paid in sensitivity. An array of diameter $D$ consisting of $N$ elements each of diameter $d$ can be thought of as a single antenna of diameter $D$ and efficiency $N(d/D)^2$. These sensitivity considerations have led to the general use of close-packed arrays, so that $D$ is not radically different from $d$. These in turn have relatively narrow window functions since the range of angular scales sampled in observations at wavelength $\lambda$ run roughly from $\lambda/D$ to $\lambda/d$. Finally, interferometers in which the size of the array, $D$, can be adjusted allow the resolution to be changed, and thus allow the window function to be shifted.

The versatility and freedom from systematic effects of interferometers has made them a valuable addition to CMB studies; many of the groups listed in Table 1.1 below employ this technique.

### 1.4.2.4 Detectors

Coherent microwave detectors, generally based on high electron mobility transistor (HEMT) technology (see chapter 7 of Weiss *et al.*, 2006) are the frequent choice for observations at frequencies $\lesssim 90$ GHz. They are strongly favored for interferometry, since they can preserve the phase of incoming signals. At higher frequencies, on the other hand, bolometers are generally favored because of their much greater bandwidth. In the case of bolometric receivers, care must be taken to exclude incoming radiation at unwanted

TABLE 1.1.  Summary of basic characteristics of some CMB experiments cited in the text

| Project | Institutions | Sample paper | Features |
|---|---|---|---|
| 1. BOOMERanG | Caltech; Rome "La Sapienza" | Ruhl *et al.*, 2003 *Astrophys. J.* **599**, 786; Jones *et al.*, *Astrophys. J.* **647**, 823 | Antarctic balloon flight; spider-web bolometers at $\nu = 150$ GHz; $\ell \sim$ 50–1000; first consistent detection of second acoustic peak; third peak marginal |
| 2. CBI | Caltech | Mason *et al.*, 2003, *Astrophys. J.* **591**, 540 | Ground-based interferometer; $\nu =$ 26–36 GHz; range of $\ell \sim$ 400–3500; high $\ell$ excess (SZ?) |
| 3. DASI | Chicago | Halverson *et al.* 2002, *Astrophys. J.* **568**, 38; Pryke *et al.*, *Astrophys. J.* **568**, 46 | Compact interferometer; also at 26–36 GHz; lower $\ell$ probed; first detection of *polarization* |
| 4. MAXIMA | Berkeley | Abroe *et al.* 2004, *Astrophys. J.* **605**, 607 | Balloon-borne; 16 bolometers at 150, 240, 410 GHz; high sensitivity – 75 µK s$^{\frac{1}{2}}$ |
| 5. Archeops | IAS, IAP, etc. Paris | Benoit *et al.* 2003, *Astron. Astrophys.* **399**, L19 and L25 | Test-bed for Planck HFI; key observations of Galactic emission and CMB at $\ell = 30$–300 |
| 6. VSA | IAC; Cambridge; Jodrell | Scott *et al.*, 2003, *Mon. Not. R. Astron. Soc.* ***341***, *1076* | Interferometer with movable elements; 26–36 GHz; range of $\ell \sim$ 100–2000. |
| 7. COSMOSOMAS | IAC | Hildebrandt *et al.*, 2007 *Mon. Not. R. Astron. Soc.* **382**, 594 | 11–17 GHz total power experiment on Teide; detected anomalous dust |
| 8. ACBAR | Case Western; Berkeley | Kuo *et al.*, 2004, *Astrophys. J.* **600**, 32; Reichardt *et al.*, 2008 | Bolometers; 150 GHz; crucial measurements at high $\ell$ |

frequencies. This is the "filter problem," generally solved by having both high-pass and low-pass filters in the optical train, both at cryogenic temperatures to avoid re-emission from the filters themselves. Whatever the choice of detecting element, all recent CMB experiments have employed *arrays* of detectors, either in the form of several receivers in the focal plane of a single antenna or the multiple antennae used in interferometers. The reason for moving to arrays is simple. The sensitivity of individual detectors is nearing the quantum limit given by $T_{\rm rec} \sim h\nu/k$. The Planck 44 GHz receivers, for instance, are within a factor of $\sim$7 of this fundamental limit. Thus the only way to increase sensitivity is to use arrays of detectors. The Planck focal plane is shown in Fig. 1.12 as one example. The importance of progress in detector arrays is discussed in detail in chapter 7 of the useful summary of the current state of the art in CMB studies, the Weiss *et al.* (2006) report. There is growing interest in the "industrial" development of bolometric arrays.

The Atacama Cosmology Telescope, for instance, uses $32 \times 32$ arrays of TES (transition edge sensor) bolometers to "pave" the focal plane.

### 1.4.3 Observational programs

Since searches for small-scale anisotropies can be done from the ground (or balloon altitudes), and do not require satellites, these searches are being actively pursued all over the world. One site is here in the Canary Islands, but CMB observations are also being carried out in Chile, at the South Pole, in balloon flights, and elsewhere. The most recent – and stunning – results are from the Arcminute Cosmology Bolometer Array Receiver (ACBAR) collaboration. These appeared after the Winter School, and just before the final drafting of this article. In Fig. 1.13, I reproduce the ACBAR power spectrum.

There is beautiful complementarity between the ground-based experiments, including ACBAR, the VSA, Degree Angular Scale Interferometer (DASI), BOOMERanG and many others, on the one hand, and the WMAP observations from space on the other. Because of the limited size of its antenna, WMAP could push down in angular scale to roughly $1/4°$ (see Fig. 1.8). On the other hand, the ground-based observations have no trouble with resolution; but, because of the atmosphere, have difficulty in measuring fluctuations on large angular scale. The complementarity greatly increases the power of both sets of observations, and I will make use of observations of the entire power spectrum in the discussion in Section 1.4.4. It is, however, worth adding one word of caution. Comparing the results of two experiments – let us say WMAP and ACBAR – can introduce systematic error unless the calibrations of the two instruments are carefully tied together. Overlap of the window functions in $\ell$-space helps. But there is an advantage in making observations over a very extended range of $\ell$ with a *single* instrument, and that is one of the positive advantages of the Planck satellite to be discussed below. Planck's higher-frequency channels also allow for both higher sensitivity and higher resolution than obtained in WMAP; that will allow Planck to push out to $\ell \sim 3000$, allowing us to detect at least five acoustic peaks with a single experiment.

Because there are so many small-angular-scale anisotropy searches, and because this area of CMB studies is evolving so rapidly, I will not attempt to summarize all of the results. Instead, I provide Table 1.1 with brief remarks and indicative references of *some* of

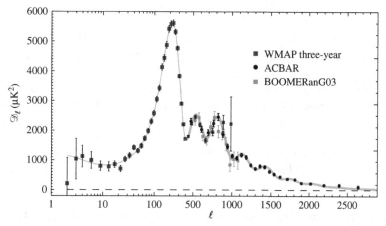

FIGURE 1.13. (From Reichardt *et al.*, 2008, with permission): The ACBAR power spectrum at $\ell \gtrsim 500$ joined to the WMAP power spectrum shown in Fig. 1.8. Notice the good agreement with earlier BOOMERanG results, as shown.

the ongoing programs. I have consciously included ground-based, balloon-borne, HEMT, bolometer and interferometry experiments. Now I will move directly to the constraints on the cosmic parameters set by these experiments in combination with WMAP, and in some cases with other astronomical results.

### 1.4.4 Quantitative results

The results I present below are based on the three-year WMAP data release and analysis, and on recent ground-based and balloon-borne isotropy measurements. Although I will emphasize observational results on small angular scales ($\ell > 300$), virtually all the results I report are affected by the WMAP measurements already discussed in Section 1.3. I will in general follow Spergel *et al.* (2007) and Reichardt *et al.* (2008), but I will simplify the analyses to get at the crucial cosmological results.

#### 1.4.4.1 Constraints on space curvature, k, and the overall density parameter, $\Omega_{TOT}$

As noted in Section 1.3.4.4, WMAP and all other relevant experiments point to a value of $\Omega_{TOT}$ very close to 1. It follows that the curvature term is extremely small, so we may confidently work with a cosmological model with $k = 0$. In Section 1.3.5.2, we gave the WMAP three-year results for $\Omega_{TOT}$: 0.989 ± 0.012. This value, I believe, is based on some prior constraints on both the value of the Hubble constant and the nature of the $\Lambda$ term. The less constrained evaluation made by the ACBAR team (Reichardt *et al.*, 2008), combining their results and the *WMAP* three-year results, gives $\Omega_{TOT} = 0.97 \pm 0.05$, entirely consistent with the earlier result as well as with flat geometry ($k = 0$).

#### 1.4.4.2 The Hubble constant, $H_0$

The Hubble constant, $H_0$, can be independently determined from the CMB results alone assuming reasonable values for $\Omega_m$: the value from *WMAP* is $73 \pm 3$ km s$^{-1}$ Mpc$^{-1}$. Adding the ACBAR results lowers the value to 72 in the same units. This is exactly the value obtained from the Hubble key project: $72 \pm 8$ km s$^{-1}$ Mpc$^{-1}$ (Freedman *et al.*, 2001). As a hint of things to come, however, the value derived for $H_0$ drifts down to more like 68 in the same units if we allow for the possibility of a running spectral index, that is a gradual change in the index $n_S$ as $\ell$ varies. This is an example of the degeneracy between the cosmic parameters that we will encounter in much sharper form below.

#### 1.4.4.3 The Baryon density

As was the case for Hubble constant, the CMB anisotropy results alone provide a precise measure of the baryon density. The value from WMAP (Spergel *et al.*, 2007) is $\Omega_b h^2 = 0.02229 \pm 0.00073$; here, $h = H_0/100$ km s$^{-1}$ Mpc$^{-1}$, so we use 0.72. With this value of $h$, we find $\Omega_b = 0.043 \pm 0.0013$. This result is, to my mind, a superb example of what is now being called "precision cosmology." As is well known, there is an entirely independent means of determining the baryon density, based on the abundance of light elements generated at an epoch of ∼3 min. That the value of $\Omega_b$ derived from primordial nucleosynthesis arguments ($\Omega_b h^2 = 0.0214 \pm 0.0020$, or $\Omega_b = 0.041 \pm 0.004$; see Kirkman *et al.*, 2003) agrees so well with the CMB results provides a strong basis for the belief that we do truly understand both the contents and the dynamics of the Universe back to a very early time in its history.

#### 1.4.4.4 The matter density, $\Omega_m$

Here, I will quote the result from the ACBAR analysis of the combined ACBAR/*WMAP* data: $\Omega_m = 0.26 \pm 0.03$. The WMAP value by itself was slightly smaller, but still consistent.

### 1.4.4.5 And thus $\Omega_\Lambda$

Since $\Omega_{\rm TOT}$ is simply the sum of the matter and the cosmological constant densities, we have $\Omega_\Lambda = 0.97 - 0.26 = 0.71$ with an uncertainty of roughly 10%. Reichardt *et al.* (2008) find that including a running index in $n_{\rm S}$ increases $\Omega_{\rm m}$ slightly, thus reducing $\Omega_\Lambda$.

As noted earlier (Section 1.3.5.2), combining the CMB results with measurements based on Type Ia supernovae substantially improves values for both $\Omega_{\rm m}$ and $\Omega_\Lambda$. The CMB results based on the location of the first acoustic peak give very precise measures of $\Omega_{\rm TOT} = \Omega_{\rm m} + \Omega_\Lambda$. On the other hand, the supernova results constrain something close to the difference between the matter and cosmological constant terms: $0.8\Omega_{\rm m} - 0.6\Omega_\Lambda$. When the two results are combined, we sharpen the precision of both quantities. Perhaps more important, the fact that the supernova results and the CMB results, independently, show that $\Omega_{\rm m}/\Omega_\Lambda \sim 1/3$ and $\Omega_{\rm m} + \Omega_\Lambda \approx 1$ bolsters our faith in both results.

### 1.4.4.6 A pause to reflect

Before continuing to the constraints on other crucial cosmological parameters, it is worth highlighting some conclusions based on the results so far. These are obvious, but are also important. First, the fact that $\Omega_{\rm b} \lesssim \frac{1}{4}\Omega_{\rm m}$ provides independent confirmation of the existence of non-baryonic dark matter. The fact that $\Omega_{\rm m}$ is so much less than $\Omega_{\rm TOT}$ from the CMB measurements alone tells us that dark energy must not only exist, but be the dominant contributor to the dynamics of the Universe today.

And, as I've remarked a couple of times, the agreement between the CMB results and results obtained from quite different observations – supernovae or light elements – gives us good confidence in the overall picture of the Universe based on cold dark matter, the existence of a cosmological constant and the primacy of gravity.

### 1.4.4.7 The amplitude of density and temperature fluctuations

We turn now to the overall amplitude of density and temperature fluctuations. As pointed out earlier, even the rough upper limits of $\Delta T/T$ available in the late 1980s ruled out pure baryonic models. Now let us examine constraints set by more recent measurements.

Let us characterize the amplitude of density fluctuations using the quantity $\sigma_8$. As a reasonable approximation, take $\sigma_8$ to represent the amplitude of density fluctuations on a scale of 8 Mpc, a distance scale chosen since early work by Peebles (reviewed in his 1980 book) showed that the two-point correlation function had roughly unit value at that length scale.

Here is the first case where there is some discordance. The WMAP three-year results alone favor a value of $\sigma_8 = 0.76 \pm 0.05$. This value is a bit low compared to values derived from other astronomical measurements. This apparent disagreement is widely discussed in the literature; both the papers by Spergel *et al.* (2007) and Reichardt *et al.* (2008) deal with it. Naïvely, using the results of Peebles (1980), we would expect $\sigma_8 \sim 1$. Other astronomical observations (such as weak lensing) favor $\sigma_8 \sim 0.9$–$1.0$.

A powerful way to resolve this apparent disagreement is through the use of Sunyaev–Zeldovich observations. The SZ signals are produced by clusters of galaxies, and the formation and properties of clusters is exquisitely sensitive to $\sigma_8$: the SZ contribution to the power spectrum is proportional to $\sigma_8^7$. The ACBAR instrument as well as some of the ground-based interferometers are capable of sampling the SZ fluctuations, and a careful characterization of their contribution to the power spectrum is a primary goal of both the South Pole Telescope (SPT) and the Atacama Cosmology Telescope (ACT). Using the results already available, the ACBAR team shows that the Sunyaev–Zel'dovich observations suggest a much higher value for $\sigma_8$ than WMAP, one consistent with 1.0. We can hope that SPT and ACT will soon resolve this question.

### 1.4.4.8 The spectral index of scalar fluctuations, $n_S$

The definition and role of $n_S$ are discussed in detail in the following chapter by Matarrese. Here, I simply mention again that $n_S = 1$ is the Harrison–Zel'dovich spectrum, and that $n_S \sim 0.9$–$1.0$ is expected in realistic inflationary models. So what do the observations tell us? First, since $n_S$ is an index in a power-law dependence, it is best constrained by observations stretching over a large range of $\ell$. Hence the value of combining WMAP and ground-based observations like those of ACBAR. Keep in mind, however, the danger of combining results at different scales from two different instruments; hence the attention in Reichardt *et al.* (2008) to calibration with WMAP and the importance of a third data set, that derived from BOOMERanG (Jones *et al.*, 2006) in tying the two results together. Given this concern, it is also easy to see why the results of Planck's all-sky survey to $\ell \sim 3000$ are eagerly awaited.

Nevertheless, WMAP alone constrains $n_S$ to $0.951 \pm 0.016$. Adding in the ACBAR results changes this value to $0.968 \pm 0.015$. The numbers are consistent, and both exclude 1.0. And of course both are fully consistent with a range of inflationary models.

Needless to say, if a change in the spectral index is allowed, expressed as $\mathrm{d}n_S/\mathrm{d}\ln k$, with $k$ the spatial frequency, $n_S$ changes substantially, dropping to more like $0.92 \pm 0.03$. Again, this value is consistent with inflationary models. However, the relatively large change in $n_S$ carries important consequences for the study of tensor fluctuations, discussed in Section 1.5.1 below.

The value of $n_S$ is also degenerate with another important cosmological parameter, $\tau$, the optical depth of the Universe introduced by reionization at redshifts $z = 5$–$10$. Reionization reintroduces free electrons, which scatter radiation coming from the surface of last scattering. So, in effect, some photons suffer a last scattering event at much lower redshifts. The effect (see Partridge 1995, chapter 8) is to damp out CMB fluctuations, to some degree, but only on small and medium scales. This clearly affects the overall "tilt" of the power spectrum, and hence both $n_S$ and measures of its change with $k$, and hence with $\ell$.

Given the interplay between $n_S$ and $\tau$, there was considerable interest as well as surprise when the WMAP team in 2003 reported a high value for $\tau$, $0.17 \pm 0.04$ (Bennett *et al.*, 2003). The analysis of the three-year WMAP data described in Hinshaw *et al.* (2007), Page *et al.* (2007) and Spergel *et al.* (2007) reduced the value to $0.088 \pm 0.030$, in better accord with expectations. Adding in the ACBAR results raises the value slightly, and decreases its uncertainty: $\tau = 0.095 \pm 0.014$. Not surprisingly, given its effect on $n_S$ and the degeneracy between $n_S$ and $\tau$, allowing a running spectral index changes $\tau$, increasing it to 0.105, with comparable errors. If values of $n_S$ and $\tau$ are degenerate, how do we obtain such precise results for both? What breaks the degeneracy? The answer is observations of polarized fluctuations, the topic I turn to next, in Section 1.5.

### 1.4.4.9 Summary

To summarize (and extend) the results in this section, I reproduce two figures. Figure 1.14 is taken from Spergel *et al.*, and shows the joint two-dimensional marginalized contours, at 68% and 95% confidence, for various combinations of the parameters discussed here and some additional ones. The solid lines are based entirely on WMAP three-year data; the smaller red contours include additional CMB observations and other astronomical data in the constraints.

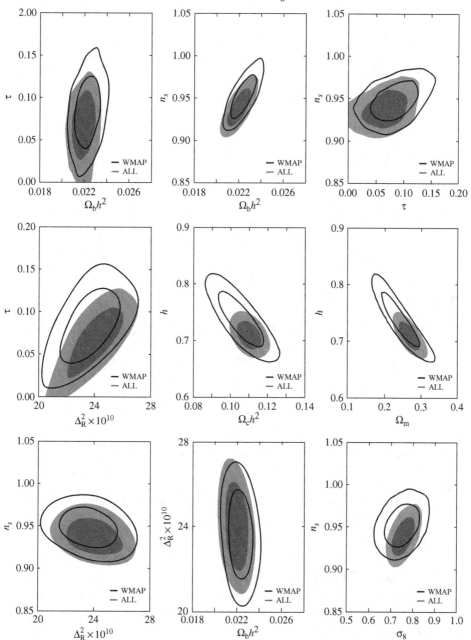

FIGURE 1.14. Constraints set on a variety of cosmological parameters on the basis of WMAP observations alone (dark contour lines), and by combining WMAP data with other astronomical and CMB measurements (light shaded areas). From Spergel *et al.* (2007); see paper for more details. NASA and the WMAP science team. Reproduced by permission of the AAS.

Figure 1.15 is taken from the ACBAR results (Reichardt *et al.*, 2008) and shows the likelihood distributions for a number of the parameters discussed here. This figure is a powerful reminder of how well we now know most of the fundamental constants in cosmology.

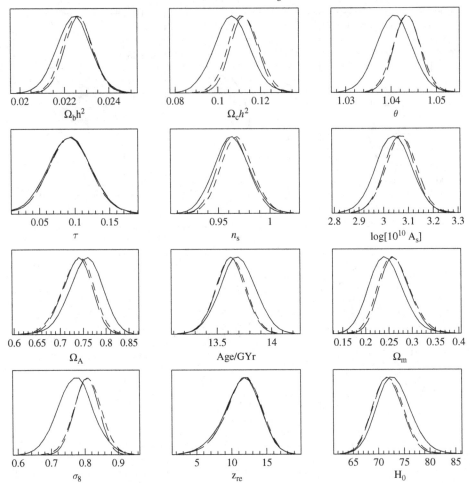

FIGURE 1.15. Likelihood distributions for key cosmological parameters from the ACBAR group (Reichardt *et al.*, 2008) with permission. Solid lines: WMAP three-year results; dashed lines: adding ACBAR (short-dashed) and ACBAR plus other CMB data (long-dashed). For future reference, note the close agreement in values for $n_S$.

## 1.5 Polarization, Planck and future prospects

As noted just above, the detection of polarized fluctuations in the CMB helps remove some of its degeneracies in values of parameters based solely on the temperature power spectrum. Polarized fluctuations were first detected in 2002 by the DASI team (Kovac *et al.*, 2002). Since then, these results have been confirmed by many other groups, and I think it fair to say that our knowledge of the polarization power spectrum of fluctuations in 2008 is roughly comparable to our knowledge of the temperature power spectrum in 1996 (as shown in Fig. 1.7). There is the expectation that the Planck satellite soon will do for polarized fluctuations what the first-year WMAP data did for temperature fluctuations – provide precise measurements of the details of the power spectrum.

### 1.5.1 Polarized fluctuations

Polarization of CMB photons can be produced by Thomson scattering from free electrons. If the radiation field is completely isotropic, no net polarization results. The

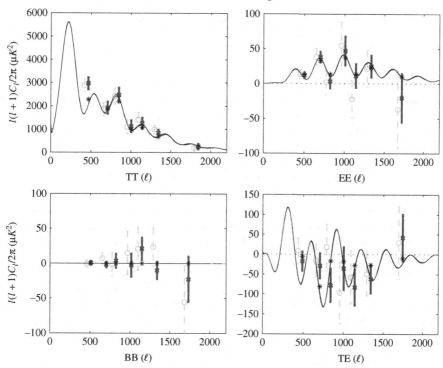

FIGURE 1.16. Temperature, polarization and TE correlation measurements made by the CBI team (from Sievers *et al.*, 2007) – these are shown by the heavy blue bars. The curves are *not* fitted to the data, but are drawn from the cosmological model that best fits all CMB temperature power spectrum measurements, including early results from WMAP. Reproduced by permission of the AAS.

observed polarization thus depends on the existence of a quadrupole moment in the radiation field. Large-scale polarization is expected from the overall quadrupole moment, discussed in Section 1.3; this small signal was actively sought early in the history of CMB studies (see Partridge, 1995, sections 6.9 and 8.3.3, and, for a recent upper limit; O'Dell *et al.*, 2003). The smaller-scale temperature fluctuations discussed in Sections 1.3 and 1.4 also induce polarization on comparable angular scales. It follows that the temperature and the polarization of the CMB are correlated; thus we expect to see features in the polarization power spectrum similar to those seen in the power spectrum shown in Fig. 1.8. The direct correlations between temperature and polarization fluctuations are expressed in the TE curve shown in Fig. 1.16.

The pure polarization fluctuations may be decomposed into two signals with different symmetry, the so-called *E*-modes and *B*-modes. The former have zero curl, and thus resemble the properties of an electric field; the latter, like a magnetic field, have non-zero curl. Wayne Hu provides more details; see also his fine pedagogical Website.[3] *E*-mode polarization is produced by both scalar and tensor perturbations. The *B*-mode, on the other hand, is produced uniquely by tensor perturbations. In turn, tensor perturbations are produced only by gravitational waves, perturbations in the metric rather than in the distribution of matter. Gravitational wave, tensor fluctuations are naturally expected in most inflationary models. Thus the most direct method now available to us to constrain

---

[3] http://background.uchicago.edu/~whu/

the properties of inflation is through the characterization of $B$-mode polarization. Hence the great interest in their possible detection (see the useful pedagogical summary in chapters 2 and 3 of Weiss et al., 2006).

### 1.5.1.1 Polarized foregrounds

Given their importance in breaking degeneracies and more particularly in providing clues to the properties of inflation, a great deal of effort has gone into the detection of polarized fluctuations. The problem is that the amplitude of the polarized fluctuations is small (see below) and our understanding of the polarization properties of astronomical foregrounds is limited. If the polarization percentage of foregrounds were the same or smaller than the expected polarization percentage of the CMB, then we could as conveniently correct for it in polarization images as we do in temperature images of the CMB. That is we could do so if we understood the polarization of the foregrounds as well. As it happens, the polarization percentages are roughly comparable; the difficulty is that we know less about the polarization properties of Galactic emission or even extragalactic sources than we do about the total power properties of either.

This was recognized by an interagency Task Force on CMB research (Weiss et al., 2006). Chapter 4 of that report provides a good summary of our present understanding of polarized foregrounds, but I would refer readers also to the work of the Archeops team (Ponthieu et al., 2005) and the long and detailed paper on the three-year WMAP results by Page et al. (2007).

### 1.5.1.2 The EE (E-mode) and TE signals

For reasonable values of the cosmological parameters, the amplitude of the $E$-mode polarization signals is predicted to be several percent of the amplitude of the unpolarized, temperature fluctuations. The details of that calculation need not concern us, since the polarized fluctuations now have been robustly detected. Indeed, now that we have detected the $E$-mode signals, we fold the argument backwards and use the result to better constrain values of certain cosmological parameters. As can be seen from Figs. 1.16 and 1.17, the overall amplitude of the $E$-mode signals is $\sim$1–10% of the temperature fluctuations. Note that the relative amplitude of the $E$-mode signal increases gradually with $\ell$; not surprisingly the first detection was at $\ell \sim 300$–800 (Kovac et al., 2002). Since 2002, both WMAP and ground-based programs have refined measurements of both the EE and TE power spectra. The results (Figs. 1.16 and 1.17) are good enough to reveal the polarization analog of the acoustic peaks in the CMB temperature power spectrum, and to pin down the phase relation between the $E$-mode and temperature fluctuations, as shown in Figure 1.17.

Polarization of the CMB relies on Thomson scattering, and hence is induced only when free electrons are present. As a consequence, polarization is imprinted only at the epoch of last scattering (on scales $\lesssim 2°$) and again at reionization (on larger scales, or $\ell \lesssim 10$). Zaldarriaga (1997) and Kaplinghat et al. (2003) provide useful discussions.

The amplitude of $E$-mode polarization at low $\ell$ thus offers an independent measure of the optical depth $\tau$ of reionization.[4] This breaks the degeneracy between $\tau$ and $n_S$ referred to in Section 1.4.4.8. The large change in the reported value of $\tau$ between the first-year WMAP data release (Spergel et al., 2003) and the more complete three-year release (Page et al., 2007; Spergel et al., 2007) was a consequence of much better constraints on E-mode polarization at low $\ell$ (Fig. 1.17). With $\tau$ well determined at $\sim$0.09, limits on the

---

[4] Strictly, the issue is the relative amplitude of temperature and E-mode polarization signals at low $\ell$.

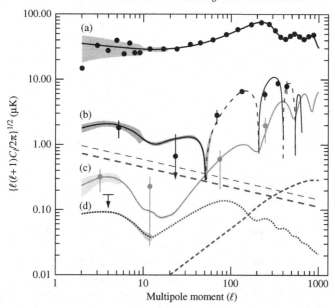

FIGURE 1.17. Power spectra for CMB fluctuations. From top to bottom: (a) power spectra for temperature (total intensity) CMB fluctuations (dark top line); (b) TE correlated fluctuations; grey solid part of the line indicates a positive correlation, while the dashed part indicates $\ell$-ranges of anticorrelation; (c) E-mode polarized fluctuations (grey solid line). Superimposed (filled dots) are the WMAP three-year measurements (from Page *et al.*, 2007). The bottom dotted line is the theoretical curve for B-mode polarization drawn for $r = 0.3$. The bottom dashed line shows B-mode fluctuations induced by gravitational lensing (not discussed here). The diagonal dashed lines are suggestions of the foreground polarization at 50–100 GHz, showing the need for careful foreground subtraction to detect polarization at low $\ell$. NASA and the WMAP science team. Reproduced by permission of the AAS.

scalar spectral index, $n_S$, could be tightened. Recall also from Fig. 1.15 that other CMB measurements are consistent with the WMAP value $n_S = 0.95 \pm 0.02$.

### 1.5.1.3 B-Modes

So far, there are only upper limits on B-mode polarization. As a consequence, we do need to pay attention to predictions of the amplitude of these fluctuations and that in turn depends crucially on the ratio of the amplitude of tensor perturbations to scalar perturbations. Both classes of perturbations are predicted by most theories for the rapid expansion ("inflation") of the early Universe. There is, however, a huge range in the predicted ratio of tensor/scalar amplitudes, $r$. A recent and useful summary of classes of inflation models, and their predictions of $r$, is given in chapters 2 and 3 of Weiss *et al.* (2006); see also Matarrese's chapter 2, this volume. At some risk of oversimplifying, let me assert that standard "vanilla" models predict $r$ to lie very roughly in the range 0.01–1.0. We can sharpen this estimate by making use of links between $r$, the properties of the inflationary potential, and the small departure of $n_S$ from the exact Harrison–Zel'dovich value of 1.00. I adopt the notation of Leach and Liddle (2003); see also Chapter 2.

For slow-roll inflation,

$$r = 16\varepsilon_1, \tag{1.8a}$$

$$1 - n_S = 2\varepsilon_1 + \varepsilon_2 \tag{1.8b}$$

where $\varepsilon_1$ and $\varepsilon_2$ relate to the shape of the inflationary potential:

$$\varepsilon_1 = \frac{M_{\mathrm{Pl}}^2}{16\pi}\left(\frac{V'}{V}\right)^2, \tag{1.9a}$$

$$\varepsilon_2 = \frac{M_{\mathrm{Pl}}^2}{4\pi}\left[\left(\frac{V'}{V}\right)^2 - \frac{V''}{V}\right]. \tag{1.9b}$$

Here, $M_{Pl}$ is the Planck mass. The quantity $(1 - n_S)$ can also be expressed as

$$1 - n_S = \frac{M_{\mathrm{Pl}}^2}{16\pi}\left[6\left(\frac{V'}{V}\right)^2 - 2\left(\frac{V''}{V}\right)\right]. \tag{1.10}$$

To relate $r$ and $1 - n_S$, we need to know how $V''$ and $V'$ relate. It is often assumed that $V'' \ll V'$, or $\varepsilon_2 \ll \varepsilon_1$, in which case $r \sim 8(1 - n_S)$. Now we see the importance of a precise value of $n_S$. Using the best available estimate from WMAP, $n_S = 0.95 \pm 0.02$ (see Section 1.4.4.8), we find $r \sim 0.4 \pm 0.2$ in the simplest case.

The value of $r$ translates directly into an expected amplitude of $B$-mode fluctuations. For instance, the $B$-mode power spectrum sketched in Fig. 1.17 assumes $r = 0.3$. The WMAP (and other) experiments have placed upper limits on $B$-mode fluctuations; these constrain $r$ to be $\lesssim 0.3$.

Thus CMB observations are already beginning to eliminate some models of inflation. They are also helping to constrain the energy scale associated with inflation. If we express that energy $V$ in GeV,

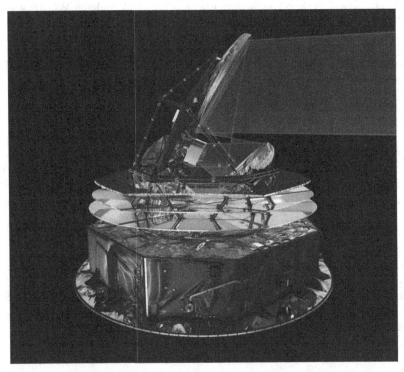

FIGURE 1.18.   Artist's impression of the Planck satellite. (ESA. Image by AOES Medialab).

TABLE 1.2. Summary of Planck instrument characteristics. Values from the Planck team(2005)

| Instrument | LFI | | | HFI | | | | | |
|---|---|---|---|---|---|---|---|---|---|
| Center frequency (GHz) | 30 | 44 | 70 | 100 | 143 | 217 | 353 | 545 | 857 |
| Detector technology | HEMT LNA arrays | | | Bolometer arrays | | | | | |
| Detector temperature | $\sim 20$ K | | | 0.1 K | | | | | |
| Cooling requirements | $H_2$ sorption cooler | | | $H_2$ sorption + 4 K J–T stage + dilution cooler | | | | | |
| Number of unpolarized detectors | 0 | 0 | 0 | 0 | 4 | 4 | 4 | 4 | 4 |
| Number of linearly polarized detectors | 4 | 6 | 12 | 8 | 8 | 8 | 8 | 0 | 0 |
| Angular resolution (FWHM, arcmin) | 33 | 24 | 14 | 9.5 | 7.1 | 5 | 5 | 5 | 5 |
| Bandwidth (GHz) | 6 | 8.8 | 14 | 33 | 47 | 72 | 116 | 180 | 283 |
| Average $\Delta T/T$ per pixel $\times 10^{-6}$ | 2.0 | 2.7 | 4.7 | 2.5 | 2.2 | 4.8 | 14.7 | 147 | 6700 |

$$r = 0.001 \left( \frac{V}{10^{16} \text{ GeV}} \right)^4 \qquad (1.11)$$

and we can conclude $V \lesssim 4 \times 10^{16}$ GeV.

Given the links between $B$-mode polarization, tensor fluctuations, and the energy scale and properties of inflation, we would very much like to *detect* B-mode fluctuations, not just set upper limits on their amplitude. This is "the next big thing" in CMB studies, the subject of sustained analysis in the report of a Task Force on CMB Research chaired by Rai Weiss (2006). The authors of that report concur that reaching $r \sim 0.01$ is possible – but only with arrays of $\sim 10^3$ detectors, and only with an improved understanding of polarized foregrounds. In the meantime, both the Planck mission and other sensitive polarization experiments have a chance to detect the B modes if $r \sim 0.2$–0.6, as our current value of $n_S$ suggests for simple inflation models.

### 1.5.2 The Planck mission

Planck, scheduled for launch in early 2009, is the European Space Agency's first CMB satellite (Planck Team, 2005). It is the fourth CMB satellite, following RELIKT, COBE and WMAP. Like WMAP, it will orbit the second Lagrange point, about 1.5 million km further from the Sun than the Earth. The spin axis of the satellite will be kept close to the Earth–Sun line, thus reducing and keeping relatively constant the stray light from these two bright astronomical sources. As the satellite spins, it surveys the entire sky each six months, with a higher density of observations at the ecliptic poles.

Planck carries two instruments: the Low Frequency Instrument (LFI) has polarization-sensitive detectors based on HEMT receivers operating at 30, 44 and 70 GHz; the High Frequency Instrument (HFI) employs spider-web bolometers at six frequencies from 100 to 857 GHz. Their properties and average sensitivity are listed in Table 1.2. The WMAP results suggest that the best measurements of CMB fluctuations will come from the 44, 70, 100 and 143 GHz receivers, but Planck's wide frequency range will permit better characterization and hence removal of astronomical foregrounds. Planck is also more

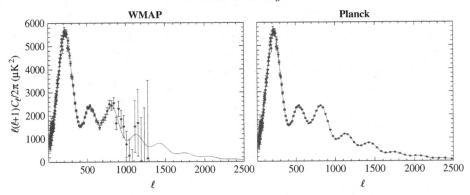

FIGURE 1.19. Comparing the sensitivity and angular resolution of the WMAP and Planck missions. (From Figure 2.8 in the Planck Team, 2005, with permission).

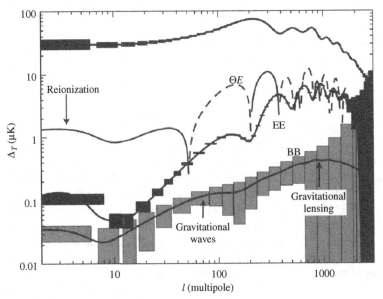

FIGURE 1.20. Hoped-for sensitivities for the full Planck mission (shaded bars, Hu & Dodelson 2002). The $E$-mode polarization (EE) will be precisely characterized (to something like the precision of WMAP's first-year plot of the *temperature* power spectrum). The $B$-mode (BB) signals at $\ell \lesssim 10$ and $\ell \sim 50$–$100$ should be detected if $r$ is as large as 0.1, assumed in the plot. Planck may also detect the B-mode signal at $\ell > 300$ generated from E-mode fluctuations by gravitational lensing (not discussed here). Reprinted, with permission, from the *Annual Review of Astronomy and Astrophysics*, Volume 40 © 2002 by Annual Reviews (www.annualreviews.org).

sensitive than WMAP (in part because it employs actively cooled receivers) and has better angular resolution. Figure 1.19 shows a comparison between WMAP (three-year) and Planck measurements of the temperature power spectrum. Planck should be able to measure precisely the scale and amplitude of at least five of the acoustic peaks. Planck may also (barely) detect $B$-mode polarization, as Fig. 1.20 suggests. If it does, it will play the same pioneering role for $B$-mode polarization that COBE did 16 years ago for temperature fluctuations.

### 1.5.3 Some future observational goals

Clearly, a major goal of CMB studies is the detection, and eventual characterization, of B-mode polarization fluctuations. This will allow astronomers to contribute to particle physics – by limiting $V$ – just as they did 25 years ago by constraining the number of neutrino families.

Another major goal is constraining the evolution of structure in the Universe by a careful study of clusters of galaxies using the Sunyaev–Zeldovich effect. This is the aim of both the South Pole Telescope (SPT) team (Ruhl *et al.*, 2004) and the Atacama cosmology telescope (ACT) project (Kosowsky, 2003). Studies of the numbers and properties of clusters as a function of cosmic epoch will in turn sharpen our knowledge of both $\sigma_8$ and the equation of state for dark energy. If we write $P = w\rho c^2$, $w = -1$ for a classical cosmological constant, but other values of $w$ are allowed by other, more esoteric forms of Dark Energy. Cosmic Microwave Background observations (see Fig. 1.14) already point to a value of $w$ close to $-1$; SPT and ACT should provide a definitive value.

### Acknowledgments

I am deeply grateful to the organizers of this Winter School, José Alberto Rubiño-Martin and Rafa Rebolo, for inviting me to the school, and to return to Tenerife. My CMB research and the preparation of this manuscript are supported in part by National Science Foundation grant AST-0606975 to Haverford College. I would also like to thank Lillian Dietrich for her help throughout the preparation of my talks and this paper – and for typing all 47 drafts of it. Finally I would like thank Edmond Rodriguez for guiding me through the maze of LaTeX.

## REFERENCES

ASTIER, P., GUY, J., REGNAULT, N. *et al.* (2006). *Astron. Astrophys.* **447**, 31.

BENNETT, C. L., HILL, R. S., HINSHAW, G. *et al.* (2003). *Astrophys. J. Suppl. Ser.* **148**, 97.

BIANCHI, L. (1898). *Mem. Soc. Ital.* **11**, 267.

BOUGHN, S. P. and POBER, J. C. (2007). *Astrophys. J.* **661**, 938.

BURIGANA, C., DANESE, L. and DE ZOTTI, G. (1991). *Astron. Astrophys.* **246**, 49.

CONDON, J. J. (1992). *Ann. Rev. Astron. and Astrophys.* **30**, 575.

CRUZ, M., TUROK, N., VIELVA, P., MARTÍNEZ-GONZÁLEZ, E. and HOBSON, M. (2007). *Science* **318**, 1612.

DANESE, L. and PARTRIDGE, R. B. (1989). *Astrophys. J.* **342**, 604.

DE ZOTTI, G., RICCI, R., MESA, D. *et al.* (2005). *Astron. Astrophys.* **431**, 893.

DICKE, R. H., PEEBLES, P. J. E., ROLL, P. G. and WILKINSON, D. T. (1965). *Astrophys. J.* **142**, 414.

DRAINE, B. T. and LAZARIAN, A. (1998). *Astrophys. J.* **508**, 157.

DRAINE, B. T. and LAZARIAN, A. (1999). *Astrophys. J.* **512**, 740.

EISENSTEIN, D. J., HU, W. and TEGMARK, M. (1999). *Astrophys. J.* **518**, 2.

ERIKSEN, H. K., HANSEN, F. K., BANDAY, A. J. GORSKI, K. M. and LILJE, P. B. (2004). *Astrophys. J.* **605**, 14.

FIXSEN, D. J., CHENG, E. S., GALES, J. M. *et al.* (1996). *Astrophys. J.* **473**, 576.

FIXSEN, D. J. and MATHER, J. (2002). *Astrophys. J.* **581**, 817.

FREEDMAN, W. L., MADORE, B. F., GIBSON, B. K. *et al.* (2001). *Astrophys. J.* **553**, 47.

GERVASI, M., ZANNONI, M., TARTARI, A., BOELLA, G. and SIRONI, G. (2008) *Astrophys. J.* **688**, 24.

GUSH, H. P., HALPERN, M. and WISHNOW, E. H. (1990). *Phys. Rev. Lett.* **65**, 537.

HU, W. and DODELSON, S. (2002). *Ann. Rev. Astron. and Astrophys.* **40**, 171.

HINSHAW, G. *et al.* (2007). *Astrophys. J. Suppl. Ser.* **170**, 288 (WMAP 3 year results).

ILLARIONOV, A. F. and SUNYAEV, R. A. (1975). *Sov. A. J.* **18**, 691.

JAROSIK, N., BARNES, C., GREASON, M. R. *et al.* (2007). *Astrophys. J. Suppl. Ser.* **170**, 263.

JONES, W. C., ADE, P. A. R., BOCK, J. J. *et al.* (2006). *Astrophys. J.* **647**, 823.

KAPLINGHAT, M., CHU, M., HAIMAN, Z. *et al.* (2003). *Astrophys. J.* **583**, 24.

KIRKMAN, D., TYTLER, D., SUZUKI, N., O'MEARA, J. M. and LUBIN, D. (2003). *Astrophys. J. Suppl.* **149**, 1.

KOGUT, A., BANDAY, A. J., BENNETT, C. L. *et al.* (1996). *Astrophys. J.* **460**, 1.

KOGUT, A., FIXSEN, D., FIXSEN, S. *et al.* (2006). *New Astronomy* **50**, 925 (see also Singal, J. *et al.* arXiv: 0901.0546).

KOMPANEETS, A. S. (1957). Soviet Physics – *J.E.T.P.*, **4**, 730 (*Zh. Eks. Teor. Fys.*, **31**, 876 [1956]).

KOSOWSKY, A. (2003). *New Astron.* **47**, 939 (the Atacama Cosmology Telescope); see also http://www.physics.princeton.edu/act/.

KOVAC, J. M. and the DASI team (2002). *Nature* **420**, 772.

LEACH, S. M. and LIDDLE, A. R. (2003). *Mon. Not. R. Astron. Soc.* **341**, 1151.

LIFSCHITZ, E. M. (1946). *Zh. Eksper. Teor. Fiz.* **16**, 587.

LIN, Y.-T., PARTRIDGE, B., POBER, J. C. *et al.* (2009). *Astrophys. J.* **694**, 992.

MATHER, J. C., CHENG, E. S., EPLEE, R. E., JR *et al.* (1990). *Astrophys. J. Lett.* **354**, L37.

O'DELL, C. W., KEATING, B. G., DE OLIVEIRA-COSTA, A., TEGMARK, M., and TIMBIE, P. T. (2003). *Phys. Rev. D* **68**, 042002.

PAGE, L., NOLTA, M. R., BARNES, C. *et al.* (2003). *Astrophys. J. Suppl. Ser.* **148**, 233.

PAGE, L. A., HINSHAW, G., KOMATSU, E. *et al.* (2007). *Astrophys. J. Suppl. Ser.* **170**, 335 (WMAP three-year results).

PARTRIDGE, R. B. (1995). *3K: The Cosmic Microwave Background Radiation*, Cambridge University Press.

PEACOCK, J. A. (1999). *Cosmological Physics*, Cambridge University Press.

PEEBLES, P. J. E. (1980). *The Large Scale Structure of the Universe*, Princeton University Press.

PEEBLES, P. J. E., PAGE, L. A. and PARTRIDGE, R. B. (2009). *Finding the Big Bang*, Cambridge University Press.

PEN, U.-L., SELJAK, U. and TUROK, N. (1997). *Phys. Rev. Lett.* **79**, 1611.

PENZIAS, A. A. and WILSON, R. W. (1965). *Astrophys. J.* **142**, 419.

Planck Team (2005). http://www.rssd.esa.int/index.php?project=Planck&page=pubdocs_top.

PONTHIEU, N., MACÍAS-PÉREZ, J. F., TRISTRAM, M. *et al.* (2005). *Astron. Astrophys.* **444**, 327.

REICHARDT, C. L., ADE, P. A. R., BOCK, J. J. *et al.* (2008). astro-ph/0801.1491 (newest ACBAR results).

RIESS, A. G., STROLGER, L. G., TONRY, J. *et al.* (2004). *Astrophys. J.* **607**, 665.

RUHL, J. E. (2004). *Proc. SPIE*, **5498**, 11 (the South Pole Telescope); see also http://pole.uchicago.edu/spt/.

SADLER, E. M., RICCI, R., EKERS, R. D. *et al.* (2006). *Mon. Not. R. Astron. Soc.* **371**, 898.

SCHRAMM, D. N. and TURNER, M. S. (1998). *Rev. Mod. Phys.* **70**, 303.

SIEVERS, J. L. *et al.* (2007). *Astrophys. J.* **660**, 976.

SMOOT, G. F., DE AMICI, G., FRIEDMAN, S., *et al.* (1985). *Astrophys. J. Lett.* **291**, L23.

SMOOT, G. F., LEVIN, S. M., KOGUT, A., DE AMICI, G. and WITEBSKY, C. (1987). *Radio Science* **22**, 521.

SMOOT, G. F., BENNETT, C. L., KOGUT, A. *et al.* (1992). *Astrophys. J. Lett.* **396**, L1.

SPERGEL, D. N., VERDE, L., PEIRIS, H. V. *et al.* (2003). *Astrophys. J. Suppl. Ser.* **148**, 175, and other papers in that issue (WMAP first-year results).

SPERGEL, D. N., BEAN, R., DORÉ, O. *et al.* (2007). *Astrophys. J. Suppl. Ser.* **170**, 377 (WMAP three-year results).

STAGGS, S. T., JAROSIK, N. C., MEYER, S. S. and WILKINSON, D. T. (1996). *Astrophys. J. Lett.* **473**, L1

STOKES, R. A., PARTRIDGE, R. B. and WILKINSON, D. T. (1967). *Phys. Rev. Lett.* **19**, 1199.

SUNYAEV, R. A. and ZELDOVICH, YA. B. (1969). *Astrophys. Space Sci.* **4**, 301.

SUNYAEV, R. A. and ZELDOVICH, YA. B. (1980a). *Ann. Rev. Astron. and Astrophys.* **18**, 537.

SUNYAEV, R. A. and ZELDOVICH, YA. B. (1980b). *Mon. Not. R. Astron. Soc.* **190**, 413.

TEGMARK, M., EISENSTEIN, D. J., HU, W. and KRON, R. (1998). arXiv:astro-ph/9805117

THADDEUS, P. (1972). *Ann. Rev. Astron. Astrophys.* **10**, 305.

TOFFOLATTI, L., ARGUESO GOMEZ, F., DE ZOTTI, G. *et al.* (1998). *Mon. Not. R. Astron. Soc.* **297**, 117.

WEINBERG, S. (1972). *Gravitation and Cosmology*, J. Wiley and Sons.

WEISS, R. (1980). *Ann. Rev. Astron. Astrophys.* **18**, 489.

WEISS, R., BOCK, J., CHURCH, S. *et al.* (2006). *Task Force Report on CMB Research.* astro-ph/0604101.

WILKINSON, D. T. (1967). *Phys. Rev. Lett.* **19**, 1195.

WRIGHT, E. L., MATHER, J. C., FIXSEN, D. J. *et al.* (1994). *Astrophys. J.* **420**, 450.

ZALDARRIAGA, M. (1997). *Phys. Rev. D* **55**, 1822.

# 2. The inflationary universe

## SABINO MATARRESE

### Abstract

These lecture notes aim at providing a concise and self-contained introduction to the theory of inflation in the early Universe and its predictions in terms of cosmological observables. After a review of the most important dynamical properties of the general model, the focus is on its predictions in terms of the generation, by quantum vacuum fluctuations, and subsequent classical evolution, of scalar and tensor (gravitational-wave) perturbations. Particular emphasis is given both to the spectral properties and the possible non-Gaussian features of perturbations.

## 2.1 Introduction

The inflationary scenario, proposed in the early 1980s (Kazanas, 1980; Starobinsky, 1980; Guth, 1981; Sato, 1981) has revolutionized our view of the physics of the early Universe and established a coherent framework to explain the large-scale smoothness of the Universe as well as the origin of the seeds of the cosmological perturbations, which gave rise to all the structures we see today. Moreover, it has modified our description of the "initial conditions" out of which the Universe evolved to its present state (see, e.g. Linde, 1983), as well as the way we speculate on its global structure (see, e.g. Linde *et al.*, 1994; Linde, 2005).

This chapter aims at providing a concise and self-contained introduction to the inflationary model and to its most important observational predictions. After reviewing the main dynamical properties of the model, we will focus on its consequences for the generation and subsequent evolution of scalar and tensor (gravitational-wave) perturbations.

We will start by discussing the dynamics of inflation. Next, we will introduce the fundamental concepts of cosmological perturbation theory and discuss the generation of scalar and tensor modes from quantum vacuum oscillations, the subsequent classical evolution and their spectral properties. Finally, we will synthetically describe some aspects of perturbation theory beyond linear order and their consequences for statistics beyond the power spectrum, aiming at describing non-Gaussian features of cosmological perturbations.

The interested reader can find a more comprehensive introduction to the topics covered by this chapter in, e.g. Mukhanov (1992), Liddle and Lyth (1993); Kolb (1999), Lyth and Riotto (1999), Riotto (2002), Kinney (2003), and Bartolo *et al.* (2004b).

## 2.2 Basics of the inflationary model

According to the inflationary paradigm, during an early period in the history of the Universe, well before the epoch of primordial nucleosynthesis, the Universe expansion was accelerated. This simple kinematical feature, which can be taken as being the definition of inflation, allows us to solve some fundamental shortcomings of the Hot Big Bang model, such as the horizon and flatness problems. But inflation has another attractive feature, which today is regarded as its most important one: it can generate the primordial fluctuations which are the seeds for the inhomogeneities in the large scale structure (LSS) of the Universe and for the temperature anisotropies and polarization of the cosmic microwave background (CMB) that we observe today. In the inflationary picture primordial density and gravitational-wave perturbations were created from quantum fluctuations "redshifted" out of the Hubble radius, where they remain "frozen".

### 2.2.1 The horizon and flatness problems

Massless particles like photons travel along null geodesics such that $dr = dt$, where $r$ is the comoving radial coordinate, $a(t)$ is the scale-factor of the Universe, $t$ is the cosmic time, i.e. the proper time of comoving observers; the physical distance a photon could have traveled from the beginning of the Universe until time $t$ is called the particle horizon, and is given by $d_H(t) = a(t) \int_0^t dt'/a(t')$ (throughout these lectures we use units such that $c = \hbar = k_B = 1$). An observer at time $t$ is able to receive signals from all other observers in the Universe only if the above integral diverges. If, on the other hand, the integral converges not all the observers are in causal contact with one another and we define $d_H(t)$ as the *particle horizon* of the comoving observer at time $t$. Now, if the dominant cosmic fluid has equation of state $P = w\rho$, it easily follows by integrating the Friedmann equations (with negligible spatial curvature), that $a(t) \propto t^\alpha$ with $\alpha = 2/3(1 + w)$, and $d_H(t) = t^{1-\alpha} \propto H^{-1}$, where $H(t) \equiv \dot{a}/a$ is the Hubble parameter and dots denote differentiation with respect to proper time $t$. So the distance to the horizon is finite in the standard cosmology when considering matter or radiation, and in general a perfect fluid with $w > -1/3$. Note that in this case $d_H(t) \sim H^{-1}$, up to numerical factors: this is the reason why cosmologists often use the terms horizon and Hubble radius interchangeably. In inflationary models, as we will shortly see, the cosmic equation of state parameter $w < -1/3$ and the horizon grows with respect to the Hubble radius: in a de Sitter space, for instance, such a growth is exponential. A given physical length scale $\lambda$ is said to be within the horizon if $\lambda < H^{-1}$, and it is outside the horizon if $\lambda > H^{-1}$. One can also define a *comoving* Hubble radius, $r_H(t) \equiv 1/(aH) = 1/\dot{a}$. The horizon problem can be stated as follows [see, e.g. Lucchin and Matarrese (1985b)]: in a universe dominated by a standard fluid with $w > -1/3$, the comoving Hubble radius – setting the effective comoving scale of causal correlation – grows with time; hence, larger and larger wavelengths get in causal contact with increasing time as soon as they cross the Hubble radius ("enter the horizon"). However, the Universe appears extremely smooth over very large distance scales, even though – according to the standard Big Bang model – such distant regions had not been able to establish mutual correlation by exchanging any causal signal. The inflationary solution to this problem is extremely simple: if $r_H(t)$ had decreased for some time in the early Universe, those scales which now appear as establishing causal contact for the very first time when entering the horizon may have actually been in causal contact in the past: e.g. a wavelength which enters the horizon now has been inside the horizon during inflation. A decrease of the Hubble radius with time, i.e. $\dot{r}_H < 0$, corresponds to $\ddot{a} > 0$, *i.e.* acceleration: this is the characteristic feature which defines the inflationary process.

Another well-known shortcoming of the standard Hot Big Bang model is the so-called flatness (or oldness) problem. To understand where the problem comes from, let us extrapolate the validity of Einstein's equations back to the Planck era, when the temperature of the Universe was $T_{\text{Planck}} \sim M_P \equiv (8\pi G)^{-1/2} \sim 10^{19}$ GeV. If the Universe is perfectly spatially flat, then its global density parameter $\Omega \equiv 8\pi G\rho/(3H^2)$ is identically one at all times. On the other hand, if there is even a small amount of spatial curvature, the time dependence of $(\Omega - 1)$ is quite different. Indeed, in whole generality, using the first Friedmann equation [Eq. (2.6) below] one can write

$$\Omega - 1 = \kappa r_H^2, \tag{2.1}$$

where the constant $\kappa$, which represents the curvature of constant time hypersurfaces, is 1 ($-1$) for a closed (open) Universe and 0 for the flat Einstein–de Sitter case.

During the radiation-dominated period, we have $H^2 \propto \rho_R \propto a^{-4}$ and $\Omega - 1 \propto a^2$, while during the matter-dominated era, $\rho_M \propto a^{-3}$ and $\Omega - 1 \propto a$. In both cases $(\Omega - 1)$ decreases going backwards in time. Since we know that $(\Omega_0 - 1)$ is of order unity at present, we can deduce its value at $t_{\text{Planck}}$ (the time at which the temperature of the Universe is $T_{\text{Planck}} \sim 10^{19}$ GeV)

$$\frac{|\Omega - 1|_{T = T_{\text{Planck}}}}{|\Omega - 1|_{T = T_0}} \sim \left(\frac{a_{\text{Planck}}^2}{a_0^2}\right) \sim \left(\frac{T_0^2}{T_{\text{Planck}}^2}\right) \sim \mathcal{O}(10^{-64}). \tag{2.2}$$

where 0 stands for the present epoch, and $T_0 \sim 10^{-13}$ GeV is the present-day temperature of the CMB radiation. In order to get the correct value of $(\Omega_0 - 1) \sim 1$ at present, the value of $(\Omega - 1)$ at early times has to be "fine-tuned" to values amazingly close to zero, but without being exactly zero.

### 2.2.2 Slow-roll dynamics of the inflaton field

Let us start by considering a homogeneous and isotropic Friedmann–Robertson–Walker (FRW) universe model, with line-element

$$ds^2 = g_{\mu\nu}dx^\mu dx^\nu = dt^2 + a^2(t)\left[\frac{dr^2}{1 - \kappa r^2} + r^2\left(d\theta^2 + \sin^2\theta d\phi^2\right)\right], \tag{2.3}$$

where $g_{\mu\nu}$ is the metric tensor, $t$ is the cosmic time, $r$, $\theta$ and $\phi$ are comoving coordinates, $a(t)$ is the Universe scale factor and $\kappa$ is the curvature constant of constant time hypersurfaces. Hereafter greek indices will be taken to run from 0 to 3, while latin indices, labeling spatial coordinates, will run from 1 to 3. The dynamics is governed by Einstein's field equations

$$G_{\mu\nu} = 8\pi G T_{\mu\nu}, \tag{2.4}$$

where $G_{\mu\nu}$ is Einstein's tensor, $G$ Newton's constant and $T_{\mu\nu}$ the energy-momentum tensor. In the case of the metric above, $T_{\mu\nu}$ describes a perfect fluid with energy density $\rho$ and isotropic pressure $P$; in such a case

$$T_{\mu\nu} = (\rho + P)u_\mu u_\nu + P g_{\mu\nu}, \tag{2.5}$$

where $u^\mu$ is the four-velocity of the observer (corresponding to a fluid-element, in our case). The Einstein equations give the Friedmann equations

$$H^2 = \frac{8\pi G}{3}\rho - \frac{\kappa}{a^2} \tag{2.6}$$

$$\frac{\ddot{a}}{a} = \frac{4\pi G}{3}(\rho + 3P), \tag{2.7}$$

where $H$ is the Hubble expansion rate. The latter equation immediately implies that a period of accelerated expansion, $\ddot{a} > 0$, takes place only if $P < -\rho/3$. In particular, a period in which $P = -\rho$ is called a *de Sitter* phase. The continuity equation $\nabla^\mu T_{\mu\nu} = 0$ leads to

$$\dot{\rho} = -3H(\rho + P) \tag{2.8}$$

and from the above Friedmann equations, neglecting the spatial curvature term (which would be soon redshifted away as $a^{-2}$), we see that in a de Sitter stage $H = \text{const}$, and the scale factor grows exponentially, $a(t) \propto e^{Ht}$.

The condition $P < -\rho/3$ can be satisfied by a neutral scalar field, the so-called *inflaton* $\varphi$.

The action for a minimally coupled scalar field $\varphi$ is given by

$$S = \int d^4x \sqrt{-g}\mathcal{L} = \int d^4x \sqrt{-g}\left[-\frac{1}{2}g^{\mu\nu}\partial_\mu\varphi\partial_\nu\varphi - V(\varphi)\right], \qquad (2.9)$$

where $g$ is the metric determinant, and $g_{\mu\nu}$ is the contravariant metric tensor ($g_{\mu\nu}g^{\nu\lambda} = \delta_\mu^\lambda$). Finally $V(\varphi)$ specifies the scalar-field potential. By varying the action with respect to $\varphi$ one obtains the Klein–Gordon equation

$$\nabla^\mu\nabla_\mu\varphi = \frac{\partial V}{\partial\varphi}, \qquad (2.10)$$

where

$$\nabla^\mu\nabla_\mu\varphi = \frac{1}{\sqrt{-g}}\partial_\nu\left(\sqrt{-g}\,g^{\mu\nu}\,\partial_\mu\varphi\right). \qquad (2.11)$$

In a FRW universe described by the metric (2.3), the evolution equation for $\varphi$ becomes

$$\ddot{\varphi} + 3H\dot{\varphi} - \frac{\nabla^2\varphi}{a^2} + V'(\varphi) = 0, \qquad (2.12)$$

where $V'(\varphi) = (dV(\varphi)/d\varphi)$. Note, in particular, that a friction term $3H\dot{\varphi}$ appears in the equation: a scalar field rolling down its potential suffers a friction due to the expansion of the Universe. The energy-momentum tensor for a minimally coupled scalar field $\varphi$ is given by

$$T_{\mu\nu} = -2\frac{\partial\mathcal{L}}{\partial g^{\mu\nu}} + g_{\mu\nu}\mathcal{L} = \partial_\mu\varphi\partial_\nu\varphi + g_{\mu\nu}\left[-\frac{1}{2}g^{\alpha\beta}\partial_\alpha\varphi\partial_\beta\varphi - V(\varphi)\right]. \qquad (2.13)$$

We can now split the inflaton field as $\varphi(t,\mathbf{x}) = \varphi_0(t) + \delta\varphi(t,\mathbf{x})$, where $\varphi_0$ is the "classical" (infinite wavelength) field, that is the expectation value of the inflaton field on the initial isotropic and homogeneous state, while $\delta\varphi(t,\mathbf{x})$ represents the quantum fluctuations around $\varphi_0$. In this section, we will be only concerned with the evolution of the classical field $\varphi_0$. In order not be overwhelmed by the notation, we will keep indicating from now on the classical value of the inflaton field by $\varphi$. A homogeneous scalar field $\varphi(t)$ behaves like a perfect fluid with background energy density and pressure

$$\rho_\varphi = \frac{\dot{\varphi}^2}{2} + V(\varphi), \qquad (2.14)$$

$$P_\varphi = \frac{\dot{\varphi}^2}{2} - V(\varphi). \qquad (2.15)$$

Therefore if $V(\varphi) \gg \dot{\varphi}^2$ we obtain $P_\varphi \simeq -\rho_\varphi$.

From this simple calculation, we immediately realize that a scalar field whose energy is dominant in the Universe and whose potential energy dominates over the kinetic term gives rise to accelerated expansion, i.e. inflation. Inflation is thus driven by the vacuum energy of the inflaton field. Ordinary matter fields, in the form of a radiation fluid, and the spatial curvature $\kappa$ are usually neglected during inflation because their contribution to the energy density is redshifted away during the accelerated expansion. For the very same reason also any small inhomogeneities are wiped out as soon as inflation sets in, thus justifying the use of the background FRW metric. Moreover the basic picture we have discussed here refers only to the simplest models of inflation, where only a single scalar field is present. Non-standard models of inflation involving more than one scalar field have also been largely considered in the literature [see, e.g. Bartolo *et al.* (2004b) and references therein].

Let us now quantify under which circumstances a scalar field may give rise to a period of inflation. The equation of motion of the homogeneous scalar field is

$$\ddot{\varphi} + 3H\dot{\varphi} + V'(\varphi) = 0. \tag{2.16}$$

If we require that $\dot{\varphi}^2 \ll V(\varphi)$, the scalar field slowly rolls down its potential. Such a *slow-roll* period can be achieved if the inflaton field $\varphi$ is in a region where the potential is sufficiently flat. We may also expect that – if the potential is flat – $\ddot{\varphi}$ is negligible as well. We will assume that this is true and we will quantify this condition soon. The Friedmann equation (2.6) becomes

$$H^2 \simeq \frac{8\pi G}{3} V(\varphi), \tag{2.17}$$

where we have assumed that the inflaton field dominates the energy density of the Universe. The new equation of motion becomes

$$3H\dot{\varphi} \approx -V'(\varphi), \tag{2.18}$$

which gives $\dot{\varphi}$ as a function of $V'(\varphi)$. Using Eq. (2.18) the slow-roll conditions then require

$$\dot{\varphi}^2 \ll V(\varphi) \implies \frac{(V')^2}{V} \ll H^2 \tag{2.19}$$

and

$$\ddot{\varphi} \ll 3H\dot{\varphi} \implies V'' \ll H^2. \tag{2.20}$$

Eqs. (2.19) and (2.20) represent the flatness conditions on the potential which are conveniently parametrized in terms of the so-called *slow-roll parameters*, which are built from $V$ and its derivatives $V'$, $V''$, $V'''$, $V^{(n)}$, with respect to $\varphi$. In particular, one defines the two slow-roll parameters

$$\epsilon \equiv \frac{M_P^2}{2} \left(\frac{V'}{V}\right)^2,$$

$$\eta \equiv M_P^2 \left(\frac{V''}{V}\right). \tag{2.21}$$

Achieving a successful period of inflation requires the slow-roll parameters to be $\epsilon, |\eta| \ll 1$. Indeed there exists a full hierarchy of slow-roll parameters [see e.g. Liddle and Lyth (1993)]. For example one can define the slow-roll parameter related to the third derivative of the potential

$$\xi = M_P^4 \left(\frac{V'V'''}{V^2}\right), \tag{2.22}$$

which is a second-order slow-roll parameter. The parameter $\epsilon$ can also be written as $\epsilon = -\dot{H}/H^2$, thus it quantifies how much the Hubble rate $H$ changes with time during inflation. In particular notice that, since

$$\frac{\ddot{a}}{a} = \dot{H} + H^2 = (1 - \epsilon) H^2, \tag{2.23}$$

inflation can be attained only if $\epsilon < 1$; as soon as this condition fails, inflation ends. At first order in the slow-roll parameters, $\epsilon$ and $\eta$ can be considered constant, since the potential is very flat. In fact it is easy to see that $\dot{\epsilon}, \dot{\eta} = \mathcal{O}\left(\epsilon^2, \eta^2\right)$ [with $\mathcal{O}(\epsilon, \eta)$ and $\mathcal{O}(\epsilon^2, \eta^2)$ we indicate general combinations of the slow-roll parameters of lowest order and next order, respectively].

Despite the simplicity of the inflationary paradigm, the number of inflationary models that have been proposed so far is enormous, differing for the kind of potential and for the underlying particle physics theory [see, e.g. the review by Lyth and Riotto (1999)]. It may be worth mentioning a useful classification in connection with the observations, in which single-field inflationary models are divided into three broad groups: "small field," "large field" (or chaotic), and "hybrid" type, according to the region occupied in the $(\epsilon - \eta)$ space by a given inflationary potential (Dodelson et al., 1997). Typical examples of the large-field models $(0 < \eta < 2\epsilon)$ are polynomial potentials $V\left(\varphi\right) = \Lambda^4 \left(\varphi/\mu\right)^p$, and exponential potentials (Lucchin and Matarrese, 1985a), $V\left(\varphi\right) = \Lambda^4 \exp\left(\varphi/\mu\right)$. The small-field potentials $(\eta < -\epsilon)$ are typically of the form $V\left(\varphi\right) = \Lambda^4\left[1 - \left(\varphi/\mu\right)^p\right]$, while generic hybrid potentials $(0 < 2\epsilon < \eta)$ are of the form $V\left(\varphi\right) = \Lambda^4\left[1 + \left(\varphi/\mu\right)^p\right]$.

A crucial quantity for the inflationary dynamics and for understanding the generation of the primordial perturbations during inflation is the Hubble radius (also called the Hubble horizon size) $R_H = H^{-1}$. The Hubble radius represents a characteristic length scale beyond which coherent causal processes cannot operate. A key point is that during inflation the comoving Hubble horizon, $(aH)^{-1}$, decreases with time as the scale factor, $a$, grows quasi-exponentially, and the Hubble radius remains almost constant [indeed the decrease of $(aH)^{-1}$ is a consequence of the accelerated expansion, $\ddot{a} > 0$, characterizing inflation]. Therefore, a given comoving length scale will become larger than the Hubble radius and *leaves the Hubble horizon*. On the other hand, the comoving Hubble radius increases as $(aH)^{-1} \propto a^{1/2}$ and $a$ during radiation and matter dominated era, respectively.

We have defined inflation as a period of accelerated expansion of the Universe; however, this is actually not sufficient. A successful inflation must last for a long enough period to solve the horizon and flatness problems. By "long enough period" we mean a period of accelerated expansion of the Universe long enough that a small, smooth patch whose size is smaller than the Hubble radius can grow to encompass *at least* the entire observable Universe. Typically the amount of inflation is measured in terms of the number of e-foldings, defined as

$$N_{\mathrm{TOT}} = \int_{t_i}^{t_f} H \, dt, \qquad (2.24)$$

where $t_i$ and $t_f$ are the time inflation starts and ends respectively. To explain the smoothness of the observable Universe, we impose that the largest scale we observe today, the present horizon $H_0^{-1}$ ($\sim 4000$ Mpc), was reduced during inflation to a value $\lambda_{H_0}$ at $t_i$, smaller than $H_I^{-1}$ during inflation. Then, it follows that we must have $N_{\mathrm{TOT}} > N_{\mathrm{min}}$, where $N_{\mathrm{min}} \approx 60$ is the number of e-foldings before the end of inflation when the present Hubble radius leaves the horizon. A very useful quantity is the number of e-foldings from the time when a given wavelength $\lambda$ leaves the horizon during inflation to the end of inflation,

$$N_\lambda = \int_{t(\lambda)}^{t_f} H \, dt = \ln\left(\frac{a_f}{a_\lambda}\right), \qquad (2.25)$$

where $t(\lambda)$ is the time when $\lambda$ leaves the horizon during inflation and $a_\lambda = a(t(\lambda))$. The cosmologically interesting scales probed by CMB anisotropies correspond to $N_\lambda \simeq 40$–$60$.

One of the attractive features of the inflationary scenario is that the solution of the flatness problem actually comes as an "extra bonus" of having solved the horizon problem. Indeed, since $r_H$ decreases because of the accelerated expansion, from Eq. (2.1) we immediately deduce that $|\Omega - 1|$ is pushed towards such a tiny value at the end of inflation (the exact value depending on the number of inflation e-folding, on the exact equation of state during inflation and on its value at the "onset" of inflation), that it can easily accommodate for the present observation of an almost flat Universe.

Inflation ends when the inflaton field starts to roll fast along its potential. During this regime $V'' > H^2$ (or $\eta > 1$). The scalar field will reach the minimum of its potential and will start to oscillate around it. By this time any other contribution to the energy density and entropy of the Universe has been redshifted away by the inflationary expansion. However we know that the Universe must be repopulated by a hot radiation fluid in order for the standard Big Bang cosmology to set in. This is achieved through a process, called *reheating*, by which the energy of the inflaton field is transferred to radiation during the oscillating phase. In the ordinary scenario of reheating (Abbott *et al.*, 1982; Albrecht *et al.*, 1982; Dolgov and Linde, 1982; Linde, 1982) such a transfer corresponds to the decay of the inflaton field into other lighter particles to which it couples through a decay rate $\Gamma_\varphi$. Such a decay damps the inflaton oscillations and when the decay products thermalize and form a thermal background the Universe is finally reheated. Alternatively, reheating may occur through the so-called preheating (Kofman *et al.*, 1994).

## 2.3 Generation of perturbations during inflation

We will now discuss the generation and subsequent evolution of quantum fluctuations of the inflaton field $\delta\varphi(t, \mathbf{x})$. These vacuum fluctuations are associated with primordial energy density perturbations, which survive after inflation and are the origin of all the structures in the Universe. Our current understanding of the origin of structure in the Universe is that once the Universe became matter-dominated, primeval density inhomogeneities were amplified by gravity and grew into the structure we see today. In this section we will summarize the process by which these "seed" perturbations are generated during inflation.

In order for structure formation to occur via gravitational instability, there must have been small preexisting fluctuations on relevant scales, which entered the Hubble radius during the radiation and matter-dominated eras. However in the standard Big Bang model these small perturbations have to be put in by hand, because it is impossible to produce fluctuations on any length scales larger than the horizon size. Inflation provides a causal mechanism to generate both density perturbations and gravitational waves. A key ingredient of this mechanism is the fact that during inflation the comoving Hubble radius $(aH)^{-1}$ decreases with time. Consequently, the wavelength of a quantum fluctuation in the scalar field whose potential energy drives inflation soon exceeds the Hubble radius. The quantum fluctuations arise on scales which are much smaller than the comoving Hubble radius $(aH)^{-1}$, which is the scale beyond which causal processes cannot operate. On such small scales one can use the usual flat space-time quantum field theory to describe the scalar-field vacuum fluctuations. The inflationary expansion then stretches the wavelength of quantum fluctuations to outside the horizon; thus, gravitational effects become more and more important and amplify the quantum fluctuations, the result being that a net number of scalar-field particles are created by the changing cosmological background (Mukhanov and Chibisov, 1981; Guth and Pi, 1982; Hawking, 1982; Linde, 1982b; Starobinsky, 1982). On large scales the perturbations follow a classical evolution. Since

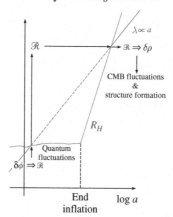

FIGURE 2.1. Quantum perturbations in the curvature, $\mathcal{R}$, are created during inflation. Their scale $\lambda$ is stretched from microscopic to cosmological scales during inflation. From Kolb (1999).

microscopic physics does not affect the evolution of fluctuations when their wavelength is outside the horizon, the amplitude of fluctuations is "frozen" and fixed at some non-zero value $\delta\varphi$ at the horizon crossing, because of a large friction term $3H\dot{\varphi}$ in the equation of motion of the field $\varphi$. The amplitude of the fluctuations on superhorizon scales then remains almost unchanged for a very long time, whereas their wavelength grows exponentially. Therefore, the appearance of such frozen fluctuations is equivalent to the appearance of a classical field $\delta\varphi$ that does not vanish after having averaged over some macroscopic interval of time. Moreover, the same mechanism also generates stochastic gravitational waves (Starobinsky, 1979; Abbott and Wise, 1984).

The fluctuations of the scalar field produce primordial perturbations in the energy density, $\rho_\varphi$, which are then inherited by the radiation and matter to which the inflaton field decays during reheating after inflation. Once inflation has ended, however, the Hubble radius increases faster than the scale factor, so the fluctuations eventually reenter the Hubble radius during the radiation- or matter-dominated eras. The fluctuations that exit around 60 e-foldings or so before reheating reenter with physical wavelengths in the range accessible to cosmological observations. These spectra, therefore, preserve the signature of inflation, giving us a direct observational connection to the physics of inflation.

The physical inflationary processes that give rise to the structures we observe today are illustrated in Fig. 2.1.

### 2.3.1 Quantum fluctuations of a scalar field during inflation

Let us first consider the case of a scalar field $\chi$ with an effective potential $V(\chi)$ in a pure de Sitter stage, during which $H$ is constant. Let us take $\chi$ as a scalar field different from the inflaton driving the accelerated expansion.

We first split the scalar field $\chi(\tau, \mathbf{x})$ as

$$\chi(\tau, \mathbf{x}) = \chi(\tau) + \delta\chi(\tau, \mathbf{x}), \tag{2.26}$$

where $\chi(\tau)$ is the homogeneous classical value of the scalar field, $\delta\chi$ its fluctuations and $\tau$ the conformal time, related to the cosmic time $t$ through $\mathrm{d}\tau = \mathrm{d}t/a(t)$. The scalar field $\chi$ is quantized by implementing the standard technique of second quantization. To proceed we first make the following field redefinition

$$\widetilde{\delta\chi} = a\delta\chi. \tag{2.27}$$

Introducing the creation and annihilation operators $a_{\mathbf{k}}$ and $a_{\mathbf{k}}^{\dagger}$ we promote $\widetilde{\delta\chi}$ to an operator which can be decomposed as

$$\widetilde{\delta\chi}(\tau, \mathbf{x}) = \int \frac{d^3\mathbf{k}}{(2\pi)^{3/2}} \left[ u_k(\tau) a_{\mathbf{k}} e^{i\mathbf{k}\cdot\mathbf{x}} + u_k^*(\tau) a_{\mathbf{k}}^{\dagger} e^{-i\mathbf{k}\cdot\mathbf{x}} \right]. \tag{2.28}$$

The creation and annihilation operators for $\widetilde{\delta\chi}$ satisfy the standard commutation relations

$$[a_{\mathbf{k}}, a_{\mathbf{k}'}] = 0, \quad [a_{\mathbf{k}}, a_{\mathbf{k}'}^{\dagger}] = \delta^{(3)}(\mathbf{k} - \mathbf{k}'), \tag{2.29}$$

and the modes $u_k(\tau)$ are normalized so that they satisfy the condition

$$u_k^* u_k' - u_k u_k^{*'} = -i, \tag{2.30}$$

deriving from the usual canonical commutation relations between the operators $\widetilde{\delta\chi}$ and its conjugate momentum $\Pi = \widetilde{\delta\chi}'$. Here primes denote derivatives with respect to the conformal time $\tau$. The evolution equation for the scalar field $\chi(\tau, \mathbf{x})$ is given by the Klein–Gordon equation

$$\nabla^{\mu}\nabla_{\mu}\chi = \frac{\partial V}{\partial\chi}. \tag{2.31}$$

The Klein–Gordon equation, in an unperturbed FRW universe, reads

$$\chi'' + 2\mathcal{H}\chi' = -a^2\frac{\partial V}{\partial\chi}, \tag{2.32}$$

where $\mathcal{H} \equiv a'/a$ is the Hubble expansion rate in conformal time. Now, we perturb the scalar field but neglect the metric perturbations in the Klein–Gordon equation (2.31), the eigenfunctions $u_k(\tau)$ obey the equation of motion

$$u_k'' + \left( k^2 - \frac{a''}{a} + m_\chi^2 a^2 \right) u_k = 0, \tag{2.33}$$

where $m_\chi^2 = \partial^2 V/\partial\chi^2$ is the effective mass of the scalar field. The modes $u_k(\tau)$ at very short distances must reproduce the form for the ordinary flat space-time quantum field theory. Thus, well within the horizon, in the limit $k/aH \to \infty$, the modes should approach plane waves of the form

$$u_k(\tau) \to \frac{1}{\sqrt{2k}} e^{-ik\tau}. \tag{2.34}$$

Equation (2.33) has an exact solution in the case of a de Sitter stage. Before recovering it, let us study the limiting behavior of Eq. (2.33) on subhorizon and superhorizon scales. On subhorizon scales $k^2 \gg a''/a$, the mass term is negligible so that Eq (2.33) reduces to

$$u_k'' + k^2 u_k = 0, \tag{2.35}$$

whose solution is a plane wave $u_k \propto \exp(-ik\tau)$.

Thus, fluctuations with wavelength within the cosmological horizon oscillate as in Eq. (2.34). This is what we expect since in the ultraviolet limit, i.e. for wavelengths much smaller than the horizon scales, we are approximating the space-time as flat. On superhorizon scales $k^2 \ll a''/a$, Eq. (2.33) reduces to

$$u_k'' - \left( \frac{a''}{a} - m_\chi^2 a^2 \right) u_k = 0. \tag{2.36}$$

Let us see what happens in the case of a massless scalar field ($m_\chi^2 = 0$). There are two solutions of Eq. (2.36), a growing and a decaying mode:

$$u_k = B_+(k)a + B_-(k)a^{-2}. \tag{2.37}$$

We can fix the amplitude of the growing mode, $B_+$, by matching the (absolute value of the) solution (2.37) to the plane-wave solution (2.34) when the fluctuation with wavenumber $k$ leaves the horizon ($k = aH$)

$$|B_+(k)| = \frac{1}{a\sqrt{2k}} = \frac{H}{\sqrt{2k^3}}, \tag{2.38}$$

so that the quantum fluctuations of the original scalar field $\chi$ on superhorizon scales are constant,

$$|\delta\chi_k| = \frac{|u_k|}{a} = \frac{H}{\sqrt{2k^3}}. \tag{2.39}$$

Now, let us derive the exact solution without any matching tricks. The exact solution to Eq. (2.33) introduces some corrections due to a non-vanishing mass of the scalar field. In a de Sitter stage, as $a = -(H\tau)^{-1}$

$$\frac{a''}{a} - m_\chi^2 a^2 = \frac{2}{\tau^2}\left(1 - \frac{1}{2}\frac{m_\chi^2}{H^2}\right), \tag{2.40}$$

so that Eq. (2.33) can be recast in the form

$$u_k'' + \left(k^2 - \frac{\nu_\chi^2 - \frac{1}{4}}{\tau^2}\right)u_k = 0, \tag{2.41}$$

where

$$\nu_\chi^2 = \left(\frac{9}{4} - \frac{m_\chi^2}{H^2}\right). \tag{2.42}$$

When the mass $m_\chi^2$ is constant in time, Eq. (2.41) is a Bessel equation whose general solution for *real* $\nu_\chi$ reads

$$u_k(\tau) = \sqrt{-\tau}\left[c_1(k)H_{\nu_\chi}^{(1)}(-k\tau) + c_2(k)H_{\nu_\chi}^{(2)}(-k\tau)\right], \tag{2.43}$$

where $H_{\nu_\chi}^{(1)}$ and $H_{\nu_\chi}^{(2)}$ are the Hankel functions of first and second kind, respectively. This result actually coincides with the solution found by Bunch and Davies (1978) for a free massive scalar field in de Sitter space. If we impose the condition that in the ultraviolet regime $k \gg aH$ ($-k\tau \gg 1$) the solution matches the plane-wave solution $e^{-ik\tau}/\sqrt{2k}$ that we expect in flat space-time, and knowing that

$$H_{\nu_\chi}^{(1)}(x \gg 1) \sim \sqrt{\frac{2}{\pi x}}e^{i\left(x - \frac{\pi}{2}\nu_\chi - \frac{\pi}{4}\right)}, \quad H_{\nu_\chi}^{(2)}(x \gg 1) \sim \sqrt{\frac{2}{\pi x}}e^{-i\left(x - \frac{\pi}{2}\nu_\chi - \frac{\pi}{4}\right)},$$

we set $c_2(k) = 0$ and $c_1(k) = \frac{\sqrt{\pi}}{2}e^{i\left(\nu_\chi + \frac{1}{2}\right)\frac{\pi}{2}}$, which also satisfy the normalization condition (2.30). The exact solution becomes

$$u_k(\tau) = \frac{\sqrt{\pi}}{2}e^{i\left(\nu_\chi + \frac{1}{2}\right)\frac{\pi}{2}}\sqrt{-\tau}H_{\nu_\chi}^{(1)}(-k\tau). \tag{2.44}$$

We are particularly interested in the asymptotic behavior of the solution when the mode is well outside the horizon. On superhorizon scales, since $H_{\nu_\chi}^{(1)}(x \ll 1) \sim \sqrt{2/\pi}\, e^{-i\frac{\pi}{2}}\, 2^{\nu_\chi - \frac{3}{2}}\, (\Gamma(\nu_\chi)/\Gamma(3/2))\, x^{-\nu_\chi}$, the fluctuation (2.44) becomes

$$u_k(\tau) = e^{i\left(\nu_\chi - \frac{1}{2}\right)\frac{\pi}{2}} 2^{\left(\nu_\chi - \frac{3}{2}\right)} \frac{\Gamma(\nu_\chi)}{\Gamma(3/2)} \frac{1}{\sqrt{2k}} (-k\tau)^{\frac{1}{2} - \nu_\chi}. \tag{2.45}$$

Thus we find that on superhorizon scales, the fluctuation of the scalar field $\delta\chi_k \equiv u_k/a$ with a non-vanishing mass is not exactly constant, but it acquires a dependence upon time

$$|\delta\chi_k| = 2^{\left(\nu_\chi - \frac{3}{2}\right)} \frac{\Gamma(\nu_\chi)}{\Gamma(3/2)} \frac{H}{\sqrt{2k^3}} \left(\frac{k}{aH}\right)^{\frac{3}{2} - \nu_\chi}. \tag{2.46}$$

Notice that the solution (2.46) is valid for values of the scalar-field mass $m_\chi \leq \frac{3}{2}H$. If the scalar field is very light, $m_\chi \ll 3/2H$, we can introduce the parameter $\eta_\chi = (m_\chi^2/3H^2)$ in analogy with the slow-roll parameters $\epsilon$ and $\eta$ for the inflaton field, and make an expansion of the solution in Eq. (2.46) to lowest order in $\eta_\chi = (m_\chi^2/3H^2) \ll 1$ to find

$$|\delta\chi_k| = \frac{H}{\sqrt{2k^3}} \left(\frac{k}{aH}\right)^{\frac{3}{2} - \nu_\chi}, \tag{2.47}$$

with $\frac{3}{2} - \nu_\chi \simeq \eta_\chi$. Eq. (2.47) shows that when the scalar field $\chi$ is light, its quantum fluctuations, first generated on subhorizon scales, are gravitationally amplified and stretched to superhorizon scales because of the accelerated expansion of the Universe during inflation.

### 2.3.2 Power spectrum

A useful quantity to characterize the properties of a perturbation field is the *power spectrum*. For a given random field $f(t, \mathbf{x})$ which can be Fourier transformed (since we work in flat space) as

$$f(t, \mathbf{x}) = \int \frac{\mathrm{d}^3 k}{(2\pi)^{3/2}} e^{i\mathbf{k}\cdot\mathbf{x}} f_\mathbf{k}(t), \tag{2.48}$$

the ("dimensionless") power spectrum $\mathcal{P}_f(k)$ can be defined through

$$\langle f_{\mathbf{k}_1} f_{\mathbf{k}_2}^* \rangle \equiv \frac{2\pi^2}{k^3} \mathcal{P}_f(k)\, \delta^{(3)}\left(\mathbf{k}_1 - \mathbf{k}_2\right), \tag{2.49}$$

where the angular brackets denote ensemble averaging. Notice that the alternative Fourier-transform definition $f(t, \mathbf{x}) = (2\pi)^{-3} \int \mathrm{d}^3 k\, e^{i\mathbf{k}\cdot\mathbf{x}} f_\mathbf{k}(t)$ will also be used.

The power spectrum measures the amplitude of the fluctuations at a given scale $k$; indeed from the definition (2.49) the mean-square value of $f(t, \mathbf{x})$ in real space is

$$\langle f^2(t, \mathbf{x}) \rangle = \int \frac{\mathrm{d}k}{k} \mathcal{P}_f(k). \tag{2.50}$$

Thus, according to our definition the power spectrum, $\mathcal{P}_f(k)$ is the contribution to the variance per unit logarithmic interval in the wavenumber $k$. This is standard notation in the literature for the inflationary power spectrum. However, another definition of power spectrum, given by the quantity $P_f(k)$ related to $\mathcal{P}_f(k)$ by the relation $P_f(k) = 2\pi^2 \mathcal{P}_f(k)/k^3$, or $\langle f_{\mathbf{k}_1} f_{\mathbf{k}_2}^* \rangle = P_f(k)\delta^{(3)}\left(\mathbf{k}_1 - \mathbf{k}_2\right)$, will be also used.

To describe the slope of the power spectrum it is standard practice to define a *spectral index* $n_f(k)$, through

$$n_f(k) - 1 \equiv \frac{d \ln \mathcal{P}_f}{d \ln k}. \tag{2.51}$$

In the case of a scalar-field $\chi$ the power-spectrum $\mathcal{P}_{\delta\chi}(k)$ can be evaluated by combining Eqs. (2.27), (2.28), and (2.29)

$$\langle \delta\chi_{\mathbf{k}_1} \delta\chi_{\mathbf{k}_2}^* \rangle = \frac{|u_k|^2}{a^2} \delta^{(3)}(\mathbf{k}_1 - \mathbf{k}_2), \tag{2.52}$$

yielding

$$\mathcal{P}_{\delta\chi}(k) = \frac{k^3}{2\pi^2} |\delta\chi_k|^2, \tag{2.53}$$

where, as usual, $\delta\chi_k \equiv u_k/a$.

The expression in Eq. (2.53) is completely general. In the case of a de Sitter phase and a very light scalar field $\chi$, with $m_\chi \ll \frac{3}{2}H$ we find from Eq. (2.47) that the power spectrum on superhorizon scales is given by

$$\mathcal{P}_{\delta\chi}(k) = \left( \frac{H}{2\pi} \right)^2 \left( \frac{k}{aH} \right)^{3-2\nu_\chi}. \tag{2.54}$$

Thus in this case the dependence on time is tiny, and the spectral index slightly deviates from unity,

$$n_{\delta\chi} - 1 = 3 - 2\nu_\chi = 2\eta_\chi. \tag{2.55}$$

A useful expression to keep in mind is that of a massless free scalar field in de Sitter space. In this case from Eqs. (2.39) and (2.44) with $\nu_\chi = \frac{3}{2}$ we obtain

$$\delta\chi_k = (-H\tau) \left( 1 - \frac{i}{k\tau} \right) \frac{e^{-ik\tau}}{\sqrt{2k}}. \tag{2.56}$$

The corresponding two-point correlation function for the Fourier modes is

$$\langle \delta\chi(\mathbf{k}_1) \delta^*\chi(\mathbf{k}_2) \rangle = \delta^{(3)}(\mathbf{k}_1 - \mathbf{k}_2) \frac{H^2\tau^2}{2k_1} \left( 1 + \frac{1}{k^2\tau^2} \right), \tag{2.57}$$

$$\approx \delta^{(3)}(\mathbf{k}_1 - \mathbf{k}_2) \frac{H^2}{2k_1^3} \quad \text{(for } k_1\tau \ll 1\text{)}, \tag{2.58}$$

with a power spectrum which, on superhorizon scales, is given by

$$\mathcal{P}_{\delta\chi}(k) = \left( \frac{H}{2\pi} \right)^2, \tag{2.59}$$

which is exactly scale-invariant.

So far, we have analyzed the time evolution and computed the spectrum of the quantum fluctuations of a generic scalar field $\chi$ assuming that the scale factor evolves like in a pure de Sitter expansion, $a(\tau) = -1/(H\tau)$. However, during inflation the Hubble rate is not exactly constant, but changes with time as $\dot{H} = -\epsilon H^2$ (quasi-de Sitter expansion). In this section we will solve for the perturbations in a quasi-de Sitter expansion. Let us now consider a scalar field $\chi$ with a very small effective mass, $\eta_\chi = (m_\chi^2/3H^2) \ll 1$, and proceed by making an expansion to lowest order in $\eta_\chi$ and the inflationary parameter $|\epsilon| \ll 1$. Thus from the definition of conformal time

$$a(\tau) \simeq -\frac{1}{H}\frac{1}{\tau(1-\epsilon)},\qquad(2.60)$$

so that

$$\frac{a''}{a} = a^2 H^2 \left(2 + \frac{\dot{H}}{H^2}\right) \simeq \frac{2}{\tau^2}\left(1 + \frac{3}{2}\epsilon\right).\qquad(2.61)$$

In this way we again obtain the Bessel equation (2.41) where now $\nu_\chi$ is given by $\nu_\chi \simeq \frac{3}{2} + \epsilon - \eta_\chi$, to lowest order in $\eta_\chi$ and $\epsilon$. Notice that the time derivatives of the slow-roll parameters are next order in the slow-roll parameters themselves, i.e. $\dot{\epsilon}, \dot{\eta} \sim \mathcal{O}(\epsilon^2, \eta^2)$, so we can safely treat $\nu_\chi$ as being constant to our order of approximation. Thus, the solution is given by Eq. (2.44) with the new expression of $\nu_\chi$. On large scales and to lowest order in the slow-roll parameters we find

$$|\delta\chi_k| = \frac{H}{\sqrt{2k^3}}\left(\frac{k}{aH}\right)^{\frac{3}{2}-\nu_\chi}.\qquad(2.62)$$

Notice that the quasi-de Sitter expansion yields a correction of order $\epsilon$ in comparison with Eq. (2.47). Since on superhorizon scales, from Eq. (2.62)

$$\delta\chi_{\mathbf{k}} \simeq \frac{H}{\sqrt{2k^3}}\left(\frac{k}{aH}\right)^{\eta_\chi-\epsilon} \simeq \frac{H}{\sqrt{2k^3}}\left[1 + (\eta_\chi - \epsilon)\ln\left(\frac{k}{aH}\right)\right],\qquad(2.63)$$

we get

$$|\delta\dot{\chi}_{\mathbf{k}}| \simeq |H\,\eta_\chi\,\delta\chi_{\mathbf{k}}| \ll |H\,\delta\chi_{\mathbf{k}}|,\qquad(2.64)$$

which shows that the fluctuations are (nearly) frozen on superhorizon scales. Therefore, a way to characterize the perturbations is to compute their power spectrum on scales larger than the horizon, where one finds

$$\mathcal{P}_{\delta\chi}(k) \simeq \left(\frac{H}{2\pi}\right)^2\left(\frac{k}{aH}\right)^{3-2\nu_\chi}.\qquad(2.65)$$

## 2.4 Cosmological perturbations

In order to study the perturbed Einstein's equations, we first write down the perturbations on a spatially flat FRW background. We shall first consider the fluctuations of the metric and then those of the energy-momentum tensor. Unless otherwise specified, we will work with conformal time $\tau$, and primes will denote differentiation with respect to $\tau$.

### 2.4.1 The metric tensor

The components of a spatially flat FRW metric perturbed to first order can be written as

$$g_{00} = -a^2(\tau)\left(1 + 2\phi\right),$$
$$g_{0i} = a^2(\tau)\hat{\omega}_i,$$
$$g_{ij} = a^2(\tau)\left[(1 - 2\psi)\delta_{ij} + \hat{\chi}_{ij}\right].\qquad(2.66)$$

The functions $\phi, \hat{\omega}_i, \psi$ and $\hat{\chi}_{ij}$ are the general perturbations of the metric. Notice that such an expansion could a priori include terms of arbitrary order in perturbation theory (Matarrese *et al.*, 1998); here we will only include first-order terms; but second-order (and higher-order) contributions should be taken into account for a quantitative evaluation of deviations from the Gaussian statistics. It is standard use to split the perturbations into the so-called scalar, vector and tensor parts according to their transformation properties with respect to the three-dimensional space with metric $\delta_{ij}$; here the scalar parts are related to a scalar potential, the vector parts to transverse (divergence-free) vectors and the tensor parts to transverse trace-free tensors. Thus, in our case

$$\hat{\omega}_i = \partial_i \omega + \omega_i, \qquad (2.67)$$

$$\hat{\chi}_{ij} = D_{ij}\chi + \partial_i \chi_j + \partial_j \chi_i + \chi_{ij}, \qquad (2.68)$$

where $\omega$ and $\chi$ are scalar modes (just like $\phi$ and $\psi$ defined above), $\omega_i$ and $\chi_i$ are transverse vectors, *i.e.* $\partial^i \omega_i = \partial^i \chi_i = 0$, $\chi_{ij}$ is a symmetric transverse and trace-free tensor, *i.e.* $\partial^i \chi_{ij} = 0$, $\chi^i_{\ i} = 0$ and the differential operator $D_{ij} = \partial_i \partial_j - (1/3)\,\delta_{ij}\,\nabla^2$ is a trace-free operator. Note that here and in the following latin indices are raised and lowered using $\delta^{ij}$ and $\delta_{ij}$, respectively.

The reason why such a splitting has been introduced is that, in linear theory, these different modes are decoupled from each other in the perturbed evolution equations, so that they can be studied separately. This nice property does not hold anymore beyond the linear regime, where e.g. second-order perturbations are coupled – sourced – by linear perturbations.

For our purposes the metric in Eq. (2.66) can be simplified. Indeed, we will neglect the vector modes $\omega_i$, and $\chi_i$, as they cannot be generated by scalar fields. Note that the same reasoning does not apply to higher-order perturbations: in the nonlinear case scalar, vector and tensor modes are dynamically coupled, the second- and higher-order vector and tensor contributions are generated by first-order scalar perturbations even if they were initially zero (Matarrese *et al.*, 1998).

The contravariant metric tensor can be obtained by requiring that $g_{\mu\nu}g^{\nu\lambda} = \delta^\lambda_\mu$. Next, using $g_{\mu\nu}$ and $g^{\mu\nu}$ one can calculate the connection coefficients $\Gamma^\alpha_{\beta\gamma}$ and the Einstein tensor components $G^\mu_\nu$ in the metric fluctuations. Their complete expressions up to second order can be found in Bartolo *et al.* (2004b).

Let us conclude this section by noting that in the following we will sometimes refer to the *Poisson gauge* which is defined by the conditions $\omega = \chi = \chi_i = 0$. Then, one scalar degree of freedom is eliminated from $g_{0i}$ and one scalar and two vector degrees of freedom from $g_{ij}$. This gauge generalizes the so-called longitudinal gauge to include vector and tensor modes (where needed).

### 2.4.2 The energy-momentum tensor

Let us consider a fluid characterized by the energy-momentum tensor

$$T^\mu_{\ \nu} = (\rho + P)\,u^\mu u_\nu + P\delta^\mu_\nu, \qquad (2.69)$$

where $\rho$ is the energy density, $P$ the pressure, and $u^\mu$ is the fluid four-velocity subject to the constraint $g_{\mu\nu}u^\mu u^\nu = -1$. Notice that we do not include any anisotropic stress term in our energy-momentum tensor, i.e. we make the perfect fluid hypothesis. Indeed, we will devote a specific analysis to the energy-momentum tensor of a scalar field, given its importance for the standard scenario of inflation. We also restrict ourselves to the case where the equation of state of the cosmic fluid $w = P/\rho$ is constant, with $w = 1/3$ for a radiation fluid and $w = 0$ for collisionless matter (dust). We now expand the basic matter variables $u^\mu$, $\rho$ and $P$ around the homogeneous background. For the velocity we write

$$u^\mu = \frac{1}{a}\left(\delta^\mu_0 + v^\mu\right). \tag{2.70}$$

From the normalization condition we obtain $v^0 = -\phi$. Notice that the velocity perturbation $v^i$ also splits into a scalar (irrotational) and a vector (solenoidal) part, as

$$v^i = \partial^i v + v^i_S, \tag{2.71}$$

with $\partial_i v^i_S = 0$. According to what we said, we can neglect the linear vector velocity perturbation. However, in the following expression for the perturbed energy-momentum tensor we are still completely general by including also linear vector and tensor perturbation modes.

Using the perturbed metric we find for $u_\mu = g_{\mu\nu}u^\nu$

$$u_0 = -a\left(1 + \phi\right),$$
$$u_i = a\left(v_i + \partial_i\omega\right). \tag{2.72}$$

The energy density $\rho$ can be split into a homogeneous background term $\rho_0(\tau)$ and a perturbation $\delta\rho(\tau, x^i)$ as follows

$$\rho(\tau, x^i) = \rho_0(\tau) + \delta\rho(\tau, x^i). \tag{2.73}$$

The same decomposition can be adopted for the pressure $P$, where in our case $\delta P = w\delta\rho$.

Using the expression (2.73) for the energy density and the expressions for the velocity into Eq. (2.69) we can calculate $T^\mu_\nu = T^{\mu(0)}_\nu + \delta T^\mu_\nu$ where $T^{\mu(0)}_\nu$ corresponds to the background value, and

$$T^{0(0)}_0 + \delta T^0_0 = -\rho_0 - \delta\rho, \tag{2.74}$$
$$T^{i(0)}_0 + \delta T^i_0 = -\left(1 + w\right)\rho_0 v^i, \tag{2.75}$$
$$T^{i(0)}_j + \delta T^i_j = w\rho_0\left(1 + \frac{\delta\rho}{\rho_0}\right)\delta^i_j, \tag{2.76}$$

### 2.4.3 Gauge transformations

Let us consider an infinitesimal coordinate transformation up to first order

$$\tilde{x}^\mu(\lambda) = x^\mu - \xi^\mu, \tag{2.77}$$

where $\xi^\mu(\tau, x^i)$ is a vector field defining the gauge transformation. Specifying its time and space components, one can write

$$\xi^0 = \alpha, \tag{2.78}$$

and

$$\xi^i = \partial^i\beta + d^i, \tag{2.79}$$

where we have split the space component into a scalar and a vector part with $\partial_i d^i = 0$. From a practical point of view fixing a gauge is equivalent to fixing a coordinate system. In particular the function $\xi^0$ selects constant-$\tau$ hypersurfaces, *i.e.* a time-slicing, while $\xi^i$ selects the spatial coordinates within those hypersurfaces.

If we now expand a generic tensor $T(\tau, x^i)$ defined in the perturbed, i.e. physical world as

$$T(\tau, x^i) = T_0 + \delta T(\tau, x^i), \tag{2.80}$$

where $T_0$ is the background value, its perturbation transforms as

$$\widetilde{\delta T} = \delta T + \pounds_\xi T_0, \tag{2.81}$$

where $\pounds_\xi$ is the Lie derivative along the vector $\xi^\mu$.

Thus, for example, the energy density perturbation transforms as

$$\widetilde{\delta\rho} = \delta\rho + \rho_0'\alpha. \tag{2.82}$$

By transforming the metric tensor perturbations $\delta g_{\mu\nu}$ and $\delta^{(2)}g_{\mu\nu}$ according to Eq. (2.81) one finds that $\psi$ transforms as

$$\widetilde{\psi} = \psi - \frac{1}{3}\nabla^2\beta - \frac{a'}{a}\alpha. \tag{2.83}$$

A generalization of these transformation rules to second order can be found in Matarrese *et al.* (1998).

### 2.4.4 Curvature perturbations on spatial slices of uniform density

At linear order the intrinsic spatial curvature on hypersurfaces of constant conformal time $\tau$ and for a flat Universe is given by Bardeen (1980).

$$^{(3)}R = \frac{4}{a^2}\nabla^2\,\hat{\psi}, \tag{2.84}$$

where for simplicity of notation we have indicated

$$\hat{\psi} \equiv \psi + \frac{1}{6}\nabla^2\chi. \tag{2.85}$$

The quantity $\hat{\psi}$ is usually referred to as *curvature perturbation*. However, the curvature perturbation $\psi$ is *not* gauge invariant, but is defined only on a given slicing. In fact, under a transformation on constant time hypersurfaces $\tau \to \tau + \alpha$ [change of the slicing in Eq. (2.83)]

$$\hat{\psi} \to \hat{\psi} - \mathcal{H}\alpha, \tag{2.86}$$

where we have used Eq. (2.83) and the transformation $\widetilde{\chi} = \chi + 2\beta$ (Matarrese *et al.*, 1998). If we consider the *slicing of uniform energy density* which is defined to be the slicing where there is no perturbation in the energy density, $\delta\rho = 0$, from Eq. (2.82) we have $\alpha = \delta\rho/\rho_0'$ and the curvature perturbation $\hat{\psi}$ on uniform density perturbation slices – usually indicated by $-\zeta$ – is given by

$$-\zeta \equiv \widetilde{\hat{\psi}}|_\rho = \hat{\psi} + \mathcal{H}\frac{\delta\rho}{\rho_0'}. \tag{2.87}$$

This quantity is gauge-invariant and it is a clear example of how to find a gauge-invariant quantity by selecting in an unambiguous way a proper time slicing. It was first introduced in Bardeen *et al.* (1983) as a conserved quantity on large scales for purely adiabatic perturbations.

Notice that such a combination can be regarded also as the density perturbation on uniform curvature slices, where $\psi = \chi = 0$, the so-called *spatially flat gauge* (Kodama and Sasaki, 1984). The energy density $\rho$ here has to be regarded as the total energy density. If the matter content of a system is made of several fluids it is possible to define similarly the curvature perturbations associated with each individual energy density components $\rho_i$, which to linear order are given by (Lyth *et al.*, 2003; Malik *et al.*, 2003)

$$\zeta_i = -\hat{\psi} - \mathcal{H}\left(\frac{\delta\rho_i}{\rho_i'}\right). \tag{2.88}$$

Here and in the following, if not specified, we drop the subscript '0' referring to the background quantities for simplicity of notation. Notice that the total curvature perturbation in Eq. (2.87) is given in terms of the individual curvature perturbations as

$$\zeta = \sum_i \frac{\rho_i'}{\rho} \zeta_i. \tag{2.89}$$

### 2.4.5 Adiabatic and entropy perturbations

The gauge-invariant curvature perturbations previously introduced are usually adopted to characterize the so-called adiabatic perturbations. In fact adiabatic perturbations are such that a net perturbation in the total energy density and – via the Einstein equations – in the intrinsic spatial curvature are produced. However, as we have seen, neither the energy density nor the curvature perturbations are gauge-invariant, hence the utility of using the variable $\zeta$ to define such perturbations. Thus the notion of adiabaticity applies when the properties of a fluid, in the physical perturbed space-time, can be described *uniquely* in terms of its energy density $\rho$. For example, the pressure perturbation will be adiabatic if the pressure is a unique function of the energy density $P = P(\rho)$ [see Lyth and Wands (2003) for an exhaustive discussion on this point].

On the other hand, by the same token, to define a non-adiabatic (or entropy) perturbation of a given quantity $X$ it is necessary to "extract" that part of the perturbation which does not depend on the energy density. A very general prescription to do that is to consider the value of the perturbation $\delta X$ on the hypersurfaces of uniform energy density

$$\delta X_{\text{nad}} \equiv \widetilde{\delta X}|_\rho, \tag{2.90}$$

since this quantity will vanish for adiabatic perturbations when $X = X(\rho)$. Specifically, the non-adiabatic pressure perturbation will be given by the pressure perturbation on slices of uniform energy density $\widetilde{\delta P}|_\rho$. Being specified in a non-ambiguous slicing, the entropy perturbations, as defined in Eq. (2.90), turn out to be gauge-invariant. Notice that such a definition holds true both when considering quantities on uniform total energy-density hypersurfaces and when considering hypersurfaces of uniform energy density relative to each individual component when more than one fluid is present.

Before moving to the explicit expressions for the adiabatic and entropy perturbations an important remark is in order. In general the perturbations will not be exclusively of adiabatic or of entropy type, but both perturbation modes will be present. Indeed, the non-adiabatic pressure perturbation $\widetilde{\delta P}|_\rho$ sources the total curvature perturbation $\zeta$ on large scales. Such a coupling is the mechanism responsible for the generation of cosmological perturbations in the curvaton and in the inhomogeneous reheating scenarios, contrary to the standard single-field inflationary scenario where only adiabatic perturbations are involved.

At first order the non-adiabatic pressure perturbation is given by (Bardeen, 1980; Kodama and Sasaki, 1984)

$$\delta P_{\text{nad}} \equiv \widetilde{\delta P}|_\rho = \delta P - c_s^2 \delta \rho, \tag{2.91}$$

where $c_s^2 = P_0'/\rho_0'$ is the adiabatic sound speed of the fluid. As a check of what we said above, notice that this quantity is indeed gauge-invariant.

It can be shown that in the presence of more than one fluid the total non-adiabatic pressure perturbation can be split into two parts

$$\delta P_{\text{nad}} = \delta P_{\text{int}} + \delta P_{\text{rel}}. \tag{2.92}$$

The first part is given by the sum of the intrinsic entropy perturbation of each fluid

$$\delta P_{\text{int}} = \sum_i \delta P_{\text{intr},i}, \tag{2.93}$$

where

$$\delta P_{\text{intr},i} = \delta P_i - c_i^2 \delta \rho_i \tag{2.94}$$

is the intrinsic non-adiabatic pressure perturbation of that fluid (which is a gauge-invariant quantity) with $c_i^2 = p_i'/\rho_i'$ the adiabatic sound speed of the individual fluid. The second part is given by the relative entropy perturbation between different fluids (Malik *et al.*, 2003)

$$\delta P_{\text{rel}} = \frac{1}{6\mathcal{H}\rho'} \sum_{ij} \rho_i'\rho_j' \left(c_i^2 - c_j^2\right) \mathcal{S}_{ij}, \tag{2.95}$$

where $\mathcal{S}_{ij}$ is the relative energy-density perturbation whose gauge-invariant definition is expressed in terms of the curvature perturbations $\zeta_i$ of Eq. (2.88) as (Wands *et al.*, 2000; Malik *et al.*, 2003)

$$\mathcal{S}_{ij} = 3\left(\zeta_i - \zeta_j\right). \tag{2.96}$$

Notice that for fluids with no intrinsic entropy perturbations, the pressure perturbation will be adiabatic if the relative entropy perturbations vanish

$$\zeta_i = \zeta_j. \tag{2.97}$$

In such a case this is the condition to have *pure* adiabatic perturbations. As a consequence, from Eq. (2.89) we see that the total curvature is equally shared by the different components $\zeta = \zeta_i$.

On the other hand a *pure* isocurvature perturbation is such that the individual components compensate with each other in order to leave the curvature perturbation unperturbed. This is the reason why these are also referred to as *isocurvature* perturbations.

In Wands *et al.* (2000) it has been shown how to derive the evolution equation for the curvature perturbation $\zeta$ simply from the continuity equation for the energy density, without making any use of Einstein's equations. The result is that even on large scales the curvature perturbation can evolve being sourced by the non-adiabatic pressure of the system according to Garcia-Bellido and Wands (1996), Lyth and Riotto (1999) and Wands *et al.* (2000):

$$\zeta' = -\frac{\mathcal{H}}{\rho + P} \delta P_{\text{nad}}. \tag{2.98}$$

For purely adiabatic perturbations the curvature perturbation is conserved on large scales, thus making $\zeta$ the proper quantity to characterize the amplitude of adiabatic perturbations. Eq. (2.98) shows in particular that the notion of isocurvature perturbation is valid only at some initial epoch. Indeed, the fact that the non-adiabatic pressure perturbation sources the curvature perturbation on large scales was already known in the literature (Bardeen, 1980; Mollerach, 1990; Mukhanov *et al.*, 1992) However, it was only recently that this issue has received renewed attention, being applied in the context of the curvaton scenario as an alternative way to produce adiabatic density perturbation starting from an initial entropy mode.

It may be worth noticing that, as far as the evolution of the entropy perturbation itself is concerned, it has been shown that the non-adiabatic part of a perturbation is sourced on large scales only by other entropy perturbations, and that there is no source term coming from the overall curvature perturbation (Lyth *et al.*, 2003). If we indicate generically an entropy perturbation as $\mathcal{S}$ then its equation of motion on large scales is $\mathcal{S}' = \beta \mathcal{H} \mathcal{S}$, where $\beta$ is a time-dependent function which depends on the particular system under study. This result has also been obtained on very general grounds within the "separate universe approach" of Wands *et al.* (2000).

### *2.4.6 Evolution of cosmological perturbations*

Our starting point is given by the perturbed Einstein equations $\delta G^\mu_\nu = 8\pi G \, \delta T^\mu_\nu$. As the matter content we take the generic fluid defined by the energy-momentum tensor given in Section 2.4.2. Here we only report those equations which we shall use to derive our main results. Specifically, in the Poisson gauge the first-order $(0-0)$- and the $(i-0)$-components of Einstein's equations read

$$\frac{1}{a^2}\left[6\,\mathcal{H}^2\phi + 6\,\mathcal{H}\psi' - 2\nabla^2\psi\right] = -8\pi G\delta\rho, \qquad (2.99)$$

$$\frac{2}{a^2}\left(\mathcal{H}\partial^i\phi + \partial^i\psi'\right) = -8\pi G\left(1+w\right)\rho_0 v^i, \qquad (2.100)$$

where $w \equiv P/\rho$ is the equation of state of the fluid.

In the Poisson gauge the non-diagonal part of the $(i-j)$-component of Einstein equations, gives

$$\psi = \phi, \qquad (2.101)$$

and, on superhorizon scales, Eq. (2.99) gives

$$\psi = -\frac{1}{2}\frac{\delta\rho}{\rho_0} = \frac{3(1+w)}{2}\,\mathcal{H}\frac{\delta\rho}{\rho'}, \qquad (2.102)$$

where in the last step we have used the background continuity equation $\rho' = -3\mathcal{H}\rho\left(1+w\right)$.

Using the spatially flat gauge $\psi = \chi = 0$, from the $(0-0)$-component of the Einstein equation one gets a similar result for the gravitational potential $\phi$.

$$\phi = -\frac{1}{2}\frac{\delta\rho}{\rho_0}. \qquad (2.103)$$

Notice that Eqs. (2.99), (2.100), (2.102) and (2.103) hold also when referring to the total energy density $\rho$ and the total velocity perturbation and equation of state $w$ in the case of a multiple-component system.

From the definition of the curvature perturbation at linear order

$$\zeta = -\hat{\psi} - \delta\rho/\rho' \qquad (2.104)$$

and using Eq. (2.102) in the Poisson gauge we determine

$$\psi = -\frac{3(1+w)}{5+3w}\,\zeta. \qquad (2.105)$$

Such a relation is very useful to relate the gravitational potential $\psi$ during either the radiation- or the matter-dominated epoch to the gauge-invariant curvature perturbation $\zeta$ at the end of the "reheating" phase. In fact in the case of standard single-field inflation the perturbations are always adiabatic through the different phases and thus the curvature perturbation $\zeta$ remains always constant on superhorizon scales, so that we can write $\zeta = \zeta_I$, where the subscript "$I$" means that $\zeta$ is evaluated during the inflationary stage.

### 2.4.7 Curvature perturbation in the standard scenario

The *standard scenario* is associated to single-field models of inflation, and the observed density perturbations are due to fluctuations of the inflaton field itself. When inflation ends, the inflaton oscillates about the minimum of its potential and decays, thereby reheating the Universe. The initial inflaton fluctuations are adiabatic on large scales and are transferred to the radiation fluid during reheating. In such a standard scenario the inflaton decay rate has no spatial fluctuations.

During inflation the inflaton field dominates the energy density of the Universe and the energy-density perturbations produced by its quantum fluctuations generate an adiabatic curvature perturbation. Let us consider a (minimally coupled with gravity) inflaton field $\varphi(\tau, \mathbf{x})$ with a potential $V(\varphi)$. The evolution equation is the Klein–Gordon equation

$$\nabla^\mu \nabla_\mu \varphi = \frac{\partial V}{\partial \varphi}. \tag{2.106}$$

Perturbing Eq. (2.106) we obtain that the inflaton fluctuations at first order obey

$$\delta\varphi'' + 2\,\mathcal{H}\delta\varphi' - \nabla^2\delta\varphi + a^2\delta\varphi \frac{\partial^2 V}{\partial\varphi^2}\,a^2 + 2\,\phi\,\frac{\partial V}{\partial\varphi}$$

$$- \varphi_0'\left[\phi' + 3\,\psi' + \nabla^2\omega\right] = 0. \tag{2.107}$$

A straightforward way to calculate the curvature perturbation generated on large scales is to solve the Klein–Gordon equation in the spatially flat gauge defined by the requirement $\psi = 0$ and $\chi = 0$. In fact, in this gauge the perturbations of the scalar field correspond to the so-called Sasaki–Mukhanov gauge-invariant variables (Sasaki, 1986; Mukhanov, 1998)

$$Q_\varphi = \delta\varphi + \frac{\varphi'}{\mathcal{H}}\hat\psi. \tag{2.108}$$

As usual we introduce the field $\widetilde{Q}_\varphi = aQ_\varphi$. The Klein–Gordon equation in the spatially flat gauge now reads (in Fourier space; Taruya and Nambu, 1998)

$$\widetilde{Q}_\varphi'' + \left(k^2 - \frac{a''}{a} + \mathcal{M}_\varphi^2 a^2\right)\widetilde{Q}_\varphi = 0, \tag{2.109}$$

where

$$\mathcal{M}_\varphi^2 = V_{\varphi\varphi} - \frac{8\pi G}{a^3}\left(\frac{a^3}{H}\dot\varphi^2\right) \tag{2.110}$$

is an effective mass of the inflaton in this gauge To lowest order in the slow-roll parameters this is given by

$$\frac{\mathcal{M}_\varphi^2}{H^2} = 3\eta - 6\epsilon, \tag{2.111}$$

where $\epsilon = (1/16\pi G)\left(V_\varphi/V\right)^2$ and $\eta = (1/8\pi G)\left(V_{\varphi\varphi}/V\right)$ are the inflaton slow-roll parameters. Equation 2.109 has the same form as Eq. (2.33) and thus we can just follow the same procedure described in detail in the previous sections, by simply replacing $m_\chi^2$ with $\mathcal{M}_\varphi^2$. The equation of motion for $\widetilde{Q}_\varphi$ or for the corresponding eigenvalues $u_k(\tau)$ thus becomes

$$u_k'' + \left(k^2 - \frac{\nu_\varphi^2 - \frac{1}{4}}{\tau^2}\right)u_k = 0, \tag{2.112}$$

with $\nu_\varphi \simeq 3/2 + 3\epsilon - \eta$.

We can then conclude that on superhorizon scales and to lowest order in the slow-roll parameters the inflaton fluctuations are

$$|Q_\varphi(k)| = \frac{H}{\sqrt{2k^3}} \left(\frac{k}{aH}\right)^{\frac{3}{2} - \nu_\varphi}.$$  (2.113)

In order to calculate the curvature perturbation on large scales we can consider it on comoving hypersurfaces, which in the case of a single scalar field reads (Lukash, 1980; Lyth, 1985; Lidsey *et al.*, 1997; Lyth and Riotto, 1999)

$$\mathcal{R} = \hat{\psi} + \frac{\mathcal{H}}{\varphi'}\delta\varphi.$$  (2.114)

Notice that the comoving curvature perturbation $\mathcal{R}$ and the uniform energy density curvature perturbation $\zeta$ are simply related by (see e.g. Gordon *et al.*, 2001)

$$-\zeta = \mathcal{R} + \frac{2\rho}{9(\rho + p)} \left(\frac{k}{aH}\right)^2 \phi$$  (2.115)

where $\phi$ is the gravitational potential in the longitudinal gauge. Therefore, on large scales $\mathcal{R} \simeq -\zeta$. From Eq. (2.108) it is evident that

$$\mathcal{R} = \frac{\mathcal{H}}{\varphi'}Q_\varphi.$$  (2.116)

Thus we obtain the power spectrum of the curvature perturbation on large scales

$$\mathcal{P}_\mathcal{R} = \left(\frac{H^2}{2\pi\dot{\varphi}}\right)^2 \left(\frac{k}{aH}\right)^{3 - 2\nu_\varphi} \simeq \left(\frac{H^2}{2\pi\dot{\varphi}}\right)^2_*,$$  (2.117)

where the asterisk stands for the epoch when a given mode leaves the horizon during inflation. From Eq. (2.117) one immediately reads the spectral index of the curvature perturbation to lowest order in the slow-roll parameters

$$n_\mathcal{R} - 1 \equiv \frac{\mathrm{d}\ln\mathcal{P}_\mathcal{R}}{\mathrm{d}\ln k} = 3 - 2\nu_\varphi = -6\epsilon + 2\eta.$$  (2.118)

From this result one can check that during inflation the curvature perturbation mode is constant on superhorizon scales $\mathcal{R}' \simeq -\zeta' \simeq 0$ [from which the last equality in Eq. (2.117) follows]. This is a well-known result: the curvature mode is the quantity which allows observable perturbations to be connected to primordial perturbations produced during inflation (Bardeen *et al.*, 1983; Liddle and Lyth, 1993; Lyth and Riotto, 1999). This result comes from the fact that in single-field slow-roll models of inflation the intrinsic entropy perturbation of the inflaton field is negligible on large scales (Bassett *et al.*, 1999; Lyth and Riotto, 1999; Gordon *et al.*, 2001; Bartolo *et al.*, 2004a). It can be shown that the above result also holds during the reheating phase on large scales (see, e.g. Bartolo *et al.*, 2004b).

Analyzing the possible variation of the spectral index with wavenumber, i.e. the so-called *running spectral index*, requires going to next-to-leading order in the slow-roll parameters. One can easily show that (see e.g. Liddle and Lyth, 1993)

$$\frac{\mathrm{d}n_\mathcal{R}}{\mathrm{d}\ln k} = -2\xi + 16\epsilon\eta - 24\epsilon^2.$$  (2.119)

## 2.4.8 Gravitational waves

Quantum fluctuations in the gravitational field are generated in a similar way as the scalar perturbations discussed so far. Gravitational waves may be viewed as ripples of space-time in our FRW background and in general the linear tensor perturbations may be written as $h_{ij} = \chi_{ij}$. A generic symmetric three-tensor $h_{ij}$ would generally have six degrees of freedom, but, as we already said, tensor perturbations are traceless, $\delta^{ij} h_{ij} = 0$, and transverse $\partial^i h_{ij} = 0$ ($i = 1, 2, 3$). With these four constraints, there remain two physical degrees of freedom, or polarization states, which are usually indicated $\lambda = +, \times$. More precisely, we can write $h_{ij} = h_+ e^+_{ij} + h_\times e^\times_{ij}$, where $e^+$ and $e^\times$ are the polarization tensors, which have the following properties: $e_{ij} = e_{ji}$, $k^i e_{ij} = 0$, and $e^i_i = 0$

It is important to stress that linear tensor modes $h_{ij}$ are gauge-invariant and therefore represent physical degrees of freedom.

If the stress-energy-momentum tensor is that of a perfect fluid, or the one provided by the inflaton $T_{\mu\nu} = \partial_\mu \phi \partial_\nu \phi - g_{\mu\nu} \mathcal{L}$, the tensor modes do not have any source and the gauge-invariant tensor amplitude $v_{\mathbf{k}} = a M_{\mathrm{P}} \frac{1}{\sqrt{2}} h_{\mathbf{k}}$, satisfies the equation

$$v''_{\mathbf{k}} + \left( k^2 - \frac{a''}{a} \right) v_{\mathbf{k}} = 0, \qquad (2.120)$$

which is the equation of motion for a massless scalar field in a FRW background. We can therefore conclude that on superhorizon scales tensor modes scale like

$$|v_{\mathbf{k}}| = \left( \frac{H}{2\pi} \right) \left( \frac{k}{aH} \right)^{\frac{3}{2} - \nu_T}, \qquad (2.121)$$

where $\nu_T \simeq 3/2 - \epsilon$. Since fluctuations are frozen on superhorizon scales, a way of characterizing tensor perturbations is to compute their power spectrum on scales larger than the horizon

$$\mathcal{P}_T(k) = \frac{k^3}{2\pi^2} \sum_\lambda |h_{\mathbf{k}}|^2 = 4 \times 2 \frac{k^3}{2\pi^2} |v_{\mathbf{k}}|^2. \qquad (2.122)$$

This gives the tensor power spectrum on superhorizon scales

$$\mathcal{P}_T(k) = \frac{8}{M_P^2} \left( \frac{H}{2\pi} \right)^2 \left( \frac{k}{aH} \right)^{n_T} \equiv A_T^2 \left( \frac{k}{aH} \right)^{n_T}, \qquad (2.123)$$

where we have defined the *tensor spectral index*

$$n_T = \frac{\mathrm{d} \ln \mathcal{P}_T}{\mathrm{d} \ln k} = 3 - 2\nu_T = -2\epsilon. \qquad (2.124)$$

The tensor perturbation spectrum is almost scale-invariant. Notice that the amplitude of the tensor modes depends only on the value of the Hubble rate during inflation. This amounts to saying that it depends only on the energy scale $V^{\frac{1}{4}}$ associated to the inflaton potential. A detection of gravitational waves from inflation will therefore be a direct measurement of the energy scale associated to inflation. The present prospects for detecting tensor modes are mostly based on the (so-called $B$-mode) polarization of the CMB (see chapter 3 by Wayne Hu, this volume).

The results obtained so far for the scalar and tensor perturbations allow us to predict an important *consistency relation*, which holds for models of inflation driven by one single field $\phi$. Defining the tensor-to-scalar amplitude ratio as

$$r = \frac{A_T^2}{A_{\mathcal{R}}^2} \approx \frac{8 \left(\frac{H}{2\pi M_{\mathrm{P}}}\right)^2}{(2\epsilon)^{-1} \left(\frac{H}{2\pi M_{\mathrm{P}}}\right)^2} \approx 16\,\epsilon, \qquad (2.125)$$

which means that

$$r \approx -8n_T. \qquad (2.126)$$

It should be stressed that this consistency relation is specific of single-field slow-roll models of inflation. For a generalization of this relation to the case of multifield inflation, see e.g. Wands *et al.* (2002).

## 2.5 Inflation and non-Gaussianity

Let us now briefly review some recent results about the possibility that the primordial density perturbations generated during or immediately after inflation have an appreciable level of non-Gaussianity.

Non-Gaussianity should be considered as an important observable, which will allow us to provide information about the mechanism that leads to the production of the structures we see today. To characterize it one can consider the n-point correlation functions

$$\langle \delta(\mathbf{x}_1)\delta(\mathbf{x}_2)...\delta(\mathbf{x}_n) \rangle \qquad (2.127)$$

of a given perturbation field $\delta(\mathbf{x})$. If this random field is Gaussian then all its statistical properties are characterized by the two-point correlation function (or its Fourier transform, the power spectrum) $\langle \delta(\mathbf{x})\delta(\mathbf{x}') \rangle$ (all even n-point correlation functions can be expressed in terms of $\langle \delta(\mathbf{x})\delta(\mathbf{x}') \rangle$, while odd ones vanish identically). Thus, a non-vanishing three-point function, or its Fourier transform, the *bispectrum*, is often used as an indicator of non-Gaussianity. Its importance comes from the fact that it is the lowest-order statistics able to distinguish non-Gaussian from Gaussian perturbations (for more details, see chapter 6 by Enrique Martínez-González, this volume).

As a consequence of the assumed flatness of the inflaton potential any intrinsic non-Gaussianity (NG) during standard single-field slow-roll inflation is generally small, hence adiabatic perturbations originated by the quantum fluctuations of the inflaton field during standard inflation are nearly Gaussian distributed. Despite the simplicity of the inflationary paradigm, the mechanism by which the cosmological perturbations are generated cannot be considered as fully established yet. Besides the standard single-field slow-roll model of inflation, during the last few years many alternative scenarios for the generation of cosmological perturbations have been proposed, such as the curvaton (Mollerach, 1990; Moroi and Takahashi, 2001; Enqvist and Sloth, 2002; Lyth and Wands, 2002) and the inhomogeneous reheating scenario (Kofman, 2003; Dvali *et al.*, 2004). In both these mechanisms the curvature fluctuations from the inflaton are usually assumed to be small and the density perturbations can be traced back to the quantum fluctuations of a scalar field different from the inflaton. During the inflationary expansion fluctuations of this field are produced on superhorizon scales. Since during inflation the additional field(s) gives a negligible contribution to the total energy density its initial fluctuations are of isocurvature type and they are subsequently transformed into adiabatic ones, well after the end of inflation.

A key point is that primordial NG is *model-dependent*. For this reason detecting or even just constraining primordial NG signals is one of the most promising ways to shed light on the physics of inflation. Let us now make these statements more quantitative. Bardeen's gauge-invariant gravitational potential $\Phi$, produced by curvature fluctuations at the end of inflation, can be written as (see, e.g. Falk *et al.*, 1993; Gangui *et al.*, 1994; Komatsu and Spergel, 2001)

$$\Phi = \Phi_L + f_{NL} \left( \Phi_L^2 - \langle \Phi_L^2 \rangle \right), \qquad (2.128)$$

where $\Phi_L$ is a Gaussian random field, $\Phi_L^2$ and $f_{NL}$ is a dimensionless parameter measuring the expected level of quadratic NG. In more general terms, the parameter $f_{NL}$ should be replaced by a suitable function, and the product by a (double) convolution. Standard single-field slow-roll inflation produces $f_{NL}$ ($f_{NL} \ll 1$; much larger values of $|f_{NL}|$ are allowed by non-standard inflationary models). Higher (e.g. cubic) levels of NG are of course also possible.

The bispectrum implied by such a quadratic non-linearity is proportional to $f_{NL}$, namely

$$\langle \Phi(\mathbf{k}_1)\Phi(\mathbf{k}_2)\Phi(\mathbf{k}_3) \rangle = (2\pi)^3 \delta^{(3)} \left( \sum_i \mathbf{k}_i \right) [2 f_{NL} \, P_\Phi(k_1) P_\Phi(k_2) + \text{cycl.}] \qquad (2.129)$$

where $\Phi(\mathbf{k})$ is the gravitational potential Fourier transform, and $P_\Phi(k)$ is the linear power spectrum defined by

$$\langle \Phi_L(\mathbf{k}_1)\Phi_L(\mathbf{k}_2) \rangle = (2\pi)^3 \delta^{(3)}(\mathbf{k}_1 + \mathbf{k}_2) P_\Phi(k). \qquad (2.130)$$

The general recipe to evaluate the level of non-Gaussianity in a given model is as follows:

- Evaluate the non-linearities generated during the primordial epoch (i.e. the value of $\zeta$, the non-linear generalization of the linear curvature perturbation introduced so far).
- Evolve the gravitational potentials to (at least!) second order in the radiation- and matter-dominated epochs by matching the conserved and gauge-invariant variable $\zeta$ to its initial value.
- Evaluate the additional second- and higher-order corrections that arise when the non-Gaussianity produced in the gravitational potential is transferred to CMB anisotropies and matter perturbations.[1]

Let us now briefly list the various classes of inflationary models on the basis of their prediction for NG. Tables 2.1 and 2.2 summarize the main predictions on the non-linearity parameter $f_{NL}$ of various inflationary models (see Bartolo *et al.*, 2004b, for more details on the various models).

- *Standard single-field slow-roll inflation.* A firm prediction of single-field slow-roll models of inflation (Falk *et al.*, 1993; Gangui *et al.*, 1994; Acquaviva *et al.*, 2003; Maldacena, 2003) is that the intrinsic inflationary perturbations have an NG level of the order of the slow-roll parameters, which are $\ll 1$ (unless the potential has some specific features). In this case the large-scale NG level is dominated by second-order gravitational corrections, leading to an effective scale and angle-dependent $f_{NL} \sim \mathcal{O}(1)$ (see Bartolo *et al.*, 2004b,c,e, 2005; D'Amico *et al.*, 2008).
- *Multifield inflation.* In this general class a larger NG level is allowed, in that the NG in one of the fields can be transferred to inflaton perturbations through a cross-correlation mechanism between entropic and adiabatic modes (Bartolo *et al.*, 2002).

---

[1] The potential $\Phi$ adopted to quantify NG in a given model should be related to the curvature perturbation $\zeta$. If the NG is large (say $|f_{NL}| > 10$) then the linear relation $\Phi = 3\zeta/5$ (during matter dominance) can be safely adopted. For smaller levels of NG, a more refined, second-order relation is needed (Bartolo *et al.*, 2004f). Note that in NG analyses of CMB data, $\Phi$ is conventionally normalized so that $\Delta T/T = -\Phi/3$, if temperature anisotropies were computed in the pure Sachs–Wolfe limit.

TABLE 2.1.  Predictions for $f_{\mathrm{NL}}$ from different scenarios

| Model | $f_{\mathrm{NL}}$ | Comments |
|---|---|---|
| Single-field inflation | $\mathcal{O}(\epsilon, \eta)$ | $\epsilon, \eta$ slow-roll parameters |
| Curvaton | $\frac{5}{4r} - \frac{5}{6}r - \frac{5}{3}$ | $r \approx \left(\frac{\rho_\sigma}{\rho}\right)_{\mathrm{decay}}$ |
| Inhomogeneous reheating | $-\frac{5}{4} - I$ | $I = -\frac{5}{2} + \frac{5}{12}\frac{\bar{\Gamma}}{\alpha \Gamma_1}$ "minimal case" $I = 0$ ($\alpha = \frac{1}{6}, \Gamma_1 = \bar{\Gamma}$) |
| Multiple scalar fields | $\frac{\mathcal{P}_S}{\mathcal{P}_{\mathcal{R}}} \cos^2 \Delta \left(4 \cdot 10^3 \cdot \frac{V_{\chi\chi}}{3H^2}\right) \cdot 60\frac{H}{\chi}$ | Order of magnitude estimate of absolute value |
| Warm inflation | $-\frac{5}{6}\left(\frac{\dot{\varphi}_0}{H^2}\right)\left[\ln\left(\frac{\Gamma}{H}\right)\frac{V'''}{\Gamma}\right]$ | $\Gamma$: inflaton decay rate |
| Ghost inflation | $-85 \cdot \beta \cdot \alpha^{-8/5}$ | Equilateral configuration |
| DBI | $-0.2\gamma^2$ | Equilateral configuration |
| Preheating scenarios | e.g. $\frac{M_{\mathrm{Pl}}}{\varphi_0}e^{Nq/2} \sim 50$ | $N$: number of inflaton oscillations |
| Inhomogeneous preheating and inhomogeneous hybrid inflation | e.g. $\frac{5}{6}\lambda_\varphi\left(\frac{M_{\mathrm{Pl}}}{m_\chi}\right)^2 \sim 100$ | $\lambda_\varphi$: inflaton coupling to waterfall field $\chi$ |
| Generalized single-field (non-standard kinetic terms: including k- and brane inflation) | $-\frac{35}{108}\left(\frac{1}{c_s^2}-1\right) + \frac{5}{81}\left(\frac{1}{c_s^2}-1-2\frac{\lambda}{\Sigma}\right)$ | High when sound speed $c_s \ll 1$ or $\lambda/\Sigma \gg 1$ (equilateral configuration) |

TABLE 2.2. Predictions for $f_{\rm NL}$ from some unconventional scenarios

| Models | $f_{\rm NL}$ | Comments |
| --- | --- | --- |
| Warm inflation | $-15L(r) < f_{\rm NL} < (33/2)L(r)$ | $L(r) \simeq \ln(1 + r/14)$; $r = \frac{\Gamma}{3H} \gg 1$ |
| Generalized slow-roll/ higher-order kinetic terms | $f_{\rm NL} \gg +1$ | Equilateral configuration |
| Excited in. states + interact. | $\sim \left(6.3 \times 10^{-4} \frac{M_{\rm P}}{M}\right)^5 \sim (1 - 100)$ | Flatten configuration M: cutoff scale |
| Ekpyrotic models | $-50 \leq f_{\rm NL} \leq +200$ | Depends on sharpness of conversion from isocurvature to curvature modes |

- *Curvaton* perturbations are produced from an initial isocurvature mode associated with fluctuations of a light scalar, the "curvaton," whose energy density is negligible during inflation. These NG isocurvature perturbations are then transformed into adiabatic ones when the curvaton decays into radiation long after inflation; $f_{\mathrm{NL}}$ can be large enough to already be constrained by observational data. In particular notice that in this scenario $f_{\mathrm{NL}} \gg 1$ when $r \ll 1$, yielding $f_{\mathrm{NL}} \sim 5/(4r)$ which is the limit studied in (Lyth *et al.*, 2003). However, second-order corrections become important in the range $f_{\mathrm{NL}} \sim 1$ (Bartolo *et al.*, 2004d).

- *Inhomogeneous reheating.* If superhorizon spatial fluctuations in the decay rate of the inflaton are induced, NG adiabatic perturbations are produced in the final reheating temperature in different regions of the Universe. Typical values of $f_{\mathrm{NL}}$ can be within reach of the Planck satellite sensitivity.

- *DBI, ghost and generalized single-field inflation.* In DBI inflation (Alishahiha *et al.*, 2004; Silverstein and Tong, 2004) the inflaton dynamics is described by a Dirac–Born–Infeld action coupled to gravity, and the inflaton can also roll relatively fast. An interesting feature is that NG is bounded from below: $|f_{\mathrm{NL}}| > 5$. Ghost inflation (Arkani-Hamed *et al.*, 2004a,b) also leads to high values of $|f_{\mathrm{NL}}|$. A characteristic property of these models is that $f_{\mathrm{NL}} \sim c_{\mathrm{s}}^{-2}$ (as is the case also for inflation driven by non-standard inflaton kinetic terms, k-inflation), where $c_s$ is an effective sound speed. Moreover, it has been shown that the NG signal of these models is stronger for equilateral configurations of the bispectrum (Babich *et al.*, 2004).

- *Other non-standard inflation models.* Sizeable levels of NG can be produced during a preheating stage after inflation. Also density perturbations generated at the end of inflation can display high NG: as in inhomogeneous reheating, the post-inflation evolution starts from different values in different regions. This case includes inhomogeneous preheating models from a global broken symmetry or some hybrid inflation models. All of these scenarios can lead to a NG level such that $|f_{\mathrm{NL}}| \sim 100$, depending on the model parameters.

Putting constraints on the parameter $f_{\mathrm{NL}}$ allows us either to rule out or confirm the predictions of specific models of inflation. As an example, a positive detection of $|f_{\mathrm{NL}}| \sim 10$ by the Planck satellite would imply that all standard single-field slow-roll models of inflation are ruled out. On the contrary, any improvement of the limits on the amplitude of $f_{\mathrm{NL}}$ will allow us to strongly reduce the class of viable non-standard inflationary models, thus providing a unique clue on the fluctuation generation mechanism.

## 2.6 Conclusions

As we saw in this chapter, the inflationary model for the early Universe provides us with a causal mechanism for the generation of perturbations. These perturbations can be characterized by a number of parameters that allow direct contact with observations. In particular, the scalar spectral index and its running, and the tensor-to-scalar ratio are currently being constrained by observations of CMB temperature anisotropies and polarization as well as data sets relative to the LSS of the Universe (see, e.g., for some recent analyses Kinney *et al.*, 2008; Komatsu *et al.*, 2009; Verde and Peiris, 2008). From the analysis of the WMAP five-year data, combined with other datasets, Komatsu *et al.* (2009) find $n_{\mathcal{R}} = 0.960 \pm 0.014$ (68% confidence limits, CL), $-0.0728 < \mathrm{d}n_{\mathcal{R}}/\mathrm{d}\ln k < 0.0087$ and $r < 0.20$ (95% CL).

The level of primordial non-Gaussianity also starts to be severely constrained by CMB (and LSS) observations through the parameter $f_{\mathrm{NL}}$. The WMAP team (Komatsu *et al.*,

2009), from an analysis of the five-year data, find $-9 < f_{NL} < 111$ (95% CL), for the "local" model of Eq. (2.128).

Given the precision level achievable by future detectors, such as the Planck Surveyor satellite, it is important to account accurately for all second-order corrections to compute the level of non-Gaussianity. This implies, for instance, a calculation of the second-order radiation transfer function, as initiated in Bartolo *et al.* (2006, 2007).

It is also crucial to exploit the information content of the scale- and shape-dependence of non-Gaussianity, as the NG signal of different scenarios is expected to peak at some specific configurations (Babich *et al.*, 2004; LoVerde *et al.*, 2008). The analysis of the five-year WMAP data in Komatsu *et al.* (2009), for instance, yields the 95% confidence range $-151 < f_{NL} < 253$, for the so-called "equilateral" models.

Testing the level and the shape of non-Gaussianity of the primordial fluctuations will more and more become one of the most powerful probes of the inflationary paradigm and will enable us to discriminate among different models that would otherwise be indistinguishable.

## Acknowledgments

I would like to thank the organizers, R. Rebolo and J.A. Rubiño-Martín, as well as the students of the XIX Canary Island Winter School of Astrophysics. I would also like to thank N. Bartolo and A. Riotto for many useful discussions and N. Bartolo for providing me with the two updated tables on inflationary NG.

## REFERENCES

ABBOTT, L. F., FARHI, E. and WISE, M. B. (1982). *Phys. Lett.* **B117**, 29.

ABBOTT, L. F. and WISE, M. B. (1984). *Nucl. Phys.* **B244**, 541.

ACQUAVIVA, V., BARTOLO, N., MATARRESE, S. and RIOTTO, A. (2003). *Nucl. Phys.* **B667**, 119.

ALISHAHIHA, M., SILVERSTEIN, E. and TONG, D. (2004). *Phys. Rev. D* **70**, 123505-1.

ARKANI-HAMED, N., CHENG, H. C., LUTY, M. A. and MUKOHYAMA, S. (2004a). *J. High Energy Phys.* **0405**, 074-1–36.

ARKANI-HAMED, N., CREMINELLI, P., MUKOHYAMA, S. and ZALDARRIAGA, M. (2004b). *J. Cosmol. Astropart. Phys.* **0404**, 001-1.

ALBRECHT, A., STEINHARDT, P. J., TURNER, M. S. and WILCZEK, F. (1982). *Phys. Rev. Lett.* **48**, 1437.

BABICH, D., CREMINELLI, P. and ZALDARRIAGA, M. (2004). *J. Cosmol. Astroparticle Phys.* **0408**, 009-1.

BARDEEN, J. M. (1980). *Phys. Rev. D* **22**, 1882.

BARDEEN, J. M., STEINHARDT, P. J. and TURNER, M. S. (1983). *Phys. Rev. D* **28**, 679.

BARTOLO, N., MATARRESE, S. and RIOTTO, A. (2002). *Phys. Rev. D* **65**, 103505.

BARTOLO, N., CORASANITI, P. S., LIDDLE, A. R. and MALQUARTI, M. (2004a). *Phys. Rev. D* **70**, 043532-1.

BARTOLO, N., KOMATSU, E., MATARRESE S. and RIOTTO, A. (2004b). *Phys. Rept.* **402**, 103.

BARTOLO, N., MATARRESE, S. and RIOTTO, A. (2004c). *J. High Energy Phys.* **0404**, 006-1.

BARTOLO, N., MATARRESE, S. and RIOTTO, A. (2004d). *Phys. Rev. D* **69**, 043503-1.

BARTOLO, N., MATARRESE, S. and RIOTTO, A. (2004e). *J. Cosmol. Astroparticle Phys.* **0401**, 003-1.

BARTOLO, N., MATARRESE, S. and RIOTTO, A. (2004f). *Phys. Rev. Lett.* **93**, 231301-1.

BARTOLO, N., MATARRESE, S. and RIOTTO, A. (2005). *J. Cosmol. Astroparticle Phys.* **0508**, 010-1.

BARTOLO, N., MATARRESE, S. and RIOTTO, A. (2006). *J. Cosmol. Astroparticle Phys.* **0606**, 024-1.

BARTOLO, N., MATARRESE, S. and RIOTTO, A. (2007). *J. Cosmol. Astroparticle Phys.* **0701**, 019-1–53.

BASSETT, B. A., TAMBURINI, F., KAISER, D. I. and MARTEENS, R. (1999). *Nucl. Phys.* **B561**, 188.

BUNCH, T. S. and DAVIES, P. C. W. (1978). *Proc. R. Soc. Lond.* **A360**, 117.

D'AMICO, G., BARTOLO, N., MATARRESE, S. and RIOTTO, A. (2008). *J. Cosmol. Astroparticle Phys.* **0801**, 005-1.

DODELSON, S., KINNEY, W. H. and KOLB, E. W. (1997). *Phys. Rev. D* **56**, 3207.

DOLGOV, A. D. and LINDE, A. D. (1982). *Phys. Lett.* **B116**, 329.

DVALI, G., GRUZINOV, A. and ZALDARRIAGA, M. (2004). *Phys. Rev. D* **69**, 083505-1.

ENQVIST, K. and SLOTH, M. S. (2002). *Nucl. Phys.* **B626**, 395.

FALK, T., RANGARAJAN, R. and SREDNICKI, M. (1993). *Astrophys. J.* **403**, L1.

GANGUI, A., LUCCHIN, F., MATARRESE, S. and MOLLERACH, S. (1994). *Astrophys. J.* **430**, 447.

GARCIA-BELLIDO, J. and WANDS, D. (1996). *Phys. Rev. D* **53**, 5437.

GORDON, C., WANDS, D., BASSETT, B. A. and MAARTENS, R. (2001). *Phys. Rev. D* **63**, 023506-1.

GUTH, A. (1981). *Phys. Rev. D* **23**, 347.

GUTH, A. and PI, S. Y. (1982). *Phys. Rev. Lett.* **49**, 1110.

HAWKING, S. W. (1982). *Phys. Lett.* **B115**, 295.

KAZANAS, D. (1980). *Astrophys. J.* **241**, L59.

KINNEY, W. H. (2003). arXiv:astro-ph/0301448.

KINNEY, W. H., KOLB, E. W., MELCHIORRI, A. and RIOTTO A. (2008). arXiv:0805.2966

KODAMA, H. and SASAKI, M. (1984). *Prog. Theor. Phys. Suppl.* **78**, 1.

KOFMAN, L. (2003). arXiv:astro-ph/0303614.

KOFMAN, L., LINDE, A. D. and STAROBINSKY, A. A. (1994). *Phys. Rev. Lett.* **73**, 3195.

KOLB, E. W. (1999). arXiv:hep-ph/9910311.

KOMATSU, E. and SPERGEL, D. N. (2001). *Phys. Rev. D* **63**, 063002-1.

KOMATSU, E., DUNKLEY, J., NOLTA, M. R., *et al.* (2009). *Astrophys. J. Suppl.* **180**, 330.

LIDDLE, A. R. and LYTH, D. H. (1993). *Phys. Rept.* **231**, 1.

LIDSEY, J. E., LIDDLE, A.R, KOLB, E. W. *et al.* (1997). *Rev. Mod. Phys.* **69**, 373.

LINDE, A. D. (1982). *Phys. Lett.* **B108**, 389.

LINDE, A. D. (1982b). *Phys. Lett.* **B116**, 335.

LINDE, A. D. (1983). *Phys. Lett.* **B129**, 177.

LINDE, A. (2005). *Phys. Scripta* **T117**, 40.

LINDE, A. D., LINDE, D. and MEZHLUMIAN, A. (1994). *Phys. Rev. D* **49**, 1783.

LOVERDE, M., MILLER, A., SHANDERA, S. and VERDE, L. (2008). *J. Cosmol. Astropart. Phys.* **0804**, 014-1.

LUCCHIN, F. and MATARRESE, S. (1985a). *Phys. Rev. D* **32**, 1316.

LUCCHIN, F. and MATARRESE, S. (1985b). *Phys. Lett.* **B164**, 282.

LUKASH, V. (1980). Production of phonons in an isotropic universe, *Sov. Phys. JETP* **52**, 807 [*Zh. Eksp. Teor. Fiz.* **79**, 1601].

LYTH, D. H. (1985). *Phys. Rev. D* **31**, 1792.

LYTH, D. H. and RIOTTO, A. (1999). *Phys. Rept.* **314**, 1.

LYTH, D. H. and WANDS, D. (2002). *Phys. Lett.* **B524**, 5.

LYTH, D. H. and WANDS, D. (2003). *Phys. Rev. D* **68**, 103515-1.

LYTH, D. H., UNGARELLI, C. and WANDS, D. (2003). *Phys. Rev. D* **67**, 023503-1.

MALDACENA, J. M. (2003). *J. High Energy Phys.* **0305**, 013-1.

MALIK, K. A. and WANDS, D. (2004). *Class. Quant. Grav.* **21**, L65.

MALIK, K. A., WANDS, D. and UNGARELLI, C. (2003). *Phys. Rev. D* **67**, 063516-1.

MATARRESE, S., MOLLERACH, S. and BRUNI, M. (1998). *Phys. Rev. D* **58**, 043504-1.

MOLLERACH, S. (1990). *Phys. Rev. D* **42**, 313.

MOROI, T. and TAKAHASHI, T. (2001). *Phys. Lett.* **B522**, 215 [Erratum-ibid. (2002), **B539**, 303].

MUKHANOV, V. F. (1998). *Sov. Phys. JETP* **67**, 1297 [*Zh. Eksp. Teor. Fiz.* **94N7**, 1].

MUKHANOV, V. F. and CHIBISOV, G. V. (1981). *JETP Lett.* **33**, 532 [*Pisma Zh. Eksp. Teor. Fiz.* **33**, 549].

MUKHANOV, V. F., FELDMAN, H. A. and BRANDENBERGER, R. H. (1992). *Phys. Rept.* **215**, 203.

RIOTTO, A. (2002). arXiv:hep-ph/0210162

SASAKI, M. (1986). *Prog. Theor. Phys.* **76**, 1036.

SATO, K. (1981). *Mon. Not. R. Astron. Soc.* **195**, 467.

SILVERSTEIN, E. and TONG, D. (2004). *Phys. Rev. D* **70**, 103505-1.

STAROBINSKY, A. A. (1979). *JETP Lett.* **30**, 682 [*Pisma Zh. Eksp. Teor. Fiz.* **30**, 719].

STAROBINSKY, A. A. (1980). *Phys. Lett.* **B91**, 99.

STAROBINSKY, A. A. (1982). *Phys. Lett.* **B117**, 175.

TARUYA, A. and NAMBU, Y. (1998). *Phys. Lett.* **B428**, 37.

VERDE, L. and PEIRIS, H. V. (2008). *J. Cosmol. Astropart. Phys.* **07**, 009.

WANDS, D., BARTOLO, N., MATARRESE, S. and RIOTTO, A. (2002). *Phys. Rev. D* **66**, 043520-1.

WANDS, D., MALIK, K. A., LYTH, D. H. and LIDDLE, A. R. (2000). *Phys. Rev. D* **62**, 043527-1.

# 3. CMB theory from nucleosynthesis to recombination

## WAYNE HU

## Abstract

These lecture notes comprise an informal but pedagogical introduction to the well-established physics and phenomenology of the cosmic microwave background (CMB) between Big Bang nucleosynthesis and recombination. The dominant properties of the spectrum, temperature anisotropy and polarization anisotropy of the CMB all arise from this period. The physical processes involved are reviewed and it is shown how they are related to the observed phenomenology.

## 3.1 Introduction

These lecture notes comprise an introduction to the well-established physics and phenomenology of the cosmic microwave background (CMB) between Big Bang nucleosynthesis and recombination. We take our study through recombination since most of the temperature and polarization anisotropy observed in the CMB formed during the recombination epoch when free electrons became bound into hydrogen and helium.

The other reason for considering only this restricted range from nucleosynthesis to recombination is that these notes are meant to complement the other lectures of the XIX Canary Island Winter School of Astrophysics. While they are self-contained and complete in and of themselves, they omit important topics that will be covered elsewhere in this volume: namely inflation (Sabino Matarrese), observations (Bruce Partridge), statistical analysis (Licia Verde), secondary anisotropy (Matthias Bartelmann), and non-Gaussianity (Enrique Martínez-González).

Furthermore the approach taken here of introducing only as much detail as necessary to build physical intuition is more suitable as a general overview rather than a rigorous treatment for training specialists. As such these notes complement the more formal lecture notes for the Trieste school which may anachronistically be viewed as a continuation of these notes in the same notation (Hu, 2003).

The outline of these notes are as follows. We begin in Section 3.2 with a brief thermal history of the CMB. We discuss the temperature and polarization anisotropy and acoustic peaks from recombination in Sections 3.3 and 3.4 and conclude in Section 3.5. We take units throughout with $\hbar = c = k_B = 1$ and illustrate effects in the standard cosmological constant cold dark matter Universe with adiabatic inflationary initial conditions ($\Lambda$CDM).

## 3.2 Brief thermal history

In this section, we discuss the major events in the thermal history of the CMB (see Fig. 3.1). We begin in Section 3.2.1 with the formation of the light elements and the original prediction of relic radiation. We continue in Section 3.2.2 with the processes that thermalize the CMB into a blackbody. Finally in Section 3.2.3, we discuss the recombination epoch where the main sources of temperature and polarization anisotropy lie.

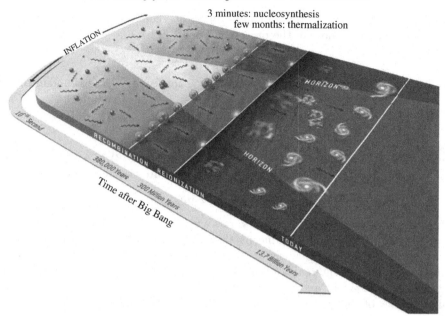

FIGURE 3.1. A brief thermal history: nucleosynthesis, thermalization, recombination and reionization. Adapted from Hu and White, 2004.

### 3.2.1 Nucleosynthesis and prediction of the CMB

Let us begin our brief thermal history with the relationship between the CMB and the abundance of light elements established at an energy scale of $10^2$ keV, timescale of a few minutes, temperature of $10^9$ K and redshift of $z \sim 10^8$–$10^9$. This is the epoch of nucleosynthesis, the formation of the light elements. The qualitative features of nucleosynthesis are set by the low baryon–photon number density of our Universe. Historically, the sensitivity to this ratio was used by Gamow and collaborators in the late 1940s to predict the existence and estimate the temperature of the CMB. Its modern use is the opposite: with the photon density well measured from the CMB spectrum, the abundance of light elements determines the baryon density.

At the high temperature and densities of nucleosynthesis, radiation is rapidly thermalized to a perfect blackbody and the photon number density is a fixed function of the temperature $n_\gamma \propto T^3$ (see below). Apart from epochs in which energy from particle annihilation or other processes is dumped into the radiation, the baryon–photon number density ratio remains constant.

Likewise nuclear statistical equilibrium, while satisfied, makes the abundance of the light elements of mass number $A$ follow the expectations of a Maxwell–Boltzmann distribution for the phase space occupation number

$$f_A = e^{-(m_A - \mu_A)/T} e^{-p_A^2/2m_A T}, \tag{3.1}$$

where $p_A$ is the particle momentum, $m_A$ is the rest mass, and $\mu_A$ is the chemical potential. Namely their number density

$$n_A \equiv g_A \int \frac{d^3 p_A}{(2\pi)^3} f_A$$

$$= g_A \left(\frac{m_A T}{2\pi}\right)^{3/2} e^{(\mu_A - m_A)/T}. \tag{3.2}$$

Here $g_A$ is the degeneracy factor. In equilibrium, the chemical potentials of the various elements are related to those of the proton and neutron by

$$\mu_A = Z\mu_p + (A - Z)\mu_n, \tag{3.3}$$

where $Z$ is the charge or number of protons. Using this relation, the abundance fraction

$$X_A \equiv A\frac{n_A}{n_b} = A^{5/2}g_A 2^{-A}\left[\left(\frac{2\pi T}{m_b}\right)^{3/2}\frac{2\zeta(3)\eta_{b\gamma}}{\pi^2}\right]^{A-1}e^{B_A/T}X_p^Z X_n^{A-Z}, \tag{3.4}$$

where $X_p$ and $X_n$ are the proton and neutron abundance, $\zeta(3) \approx 1.202$, and $n_b$ is the baryon number density. The two controlling quantities are the binding energy

$$B_A = Zm_p + (A - Z)m_n - m_A, \tag{3.5}$$

and the baryon–photon number density ratio $\eta_{b\gamma} = n_b/n_\gamma$. That the latter number is of order $10^{-9}$ in our Universe means that light elements form only well after the temperature has dropped below the binding energy of each species. Nuclear statistical equilibrium holds until the reaction rates drop below the expansion rate. At this point, the abundance freezes out and remains constant.

Gamow's back-of-the-envelope estimate (Gamow, 1948, refined by Alpher and Herman, 1948) was to consider the neutron capture reaction that forms deuterium

$$n + p \leftrightarrow d + \gamma \tag{3.6}$$

with a binding energy of $B_2 = 2.2\,\text{MeV}$. Given Eq. (3.4)

$$X_2 = \frac{3}{\pi^2}\left(\frac{4\pi T}{m_b}\right)^{3/2}\eta_{b\gamma}\zeta(3)e^{B_2/T}X_pX_n, \tag{3.7}$$

a low baryon–photon ratio, and $X_n \approx X_p \approx 1/2$ for estimation purposes, the critical temperature for deuterium formation is $T \sim 10^9\,\text{K}$. In other words the low baryon–photon ratio means that there are sufficient numbers of photons to dissociate deuterium until well below $T \sim B_2$. Note that this condition is only logarithmically sensitive to the exact value of the baryon–photon ratio chosen for the estimate and so the reasoning is not circular.

Furthermore, that we observe deuterium at all and not all helium and heavier elements means that the reaction must have frozen out at near this temperature. Given the thermally averaged cross section of

$$\langle\sigma v\rangle \approx 4.6 \times 10^{-20}\,\text{cm}^3\,\text{s}^{-1}, \tag{3.8}$$

the freezeout condition

$$n_b\langle\sigma v\rangle \approx H \approx t^{-1}, \tag{3.9}$$

and the time–temperature relation $t(T = 10^9\,\text{K}) \approx 180\,\text{s}$ from the radiation-dominated Friedmann equation, we obtain an estimate of the baryon number density

$$n_b(T = 10^9\,\text{K}) \sim 1.2 \times 10^{17}\,\text{cm}^{-3}. \tag{3.10}$$

Comparing this density to the an observed current baryon density and requiring that $n_b \propto a^{-3}$ and $T \propto a^{-1}$ yields the current temperature of the thermal background. For example, taking the modern value of $\Omega_b h^2 \approx 0.02$ and $n_b(a = 1) = 2.2 \times 10^{-7}\,\text{cm}^{-3}$ yields the rough estimate

$$T(a = 1) \approx 12\,\text{K}. \tag{3.11}$$

This value is of the same order of magnitude as the observed CMB temperature 2.725 K as well as the original estimates (Alpher and Herman, 1948; Dicke *et al.*, 1965). Modern-day estimates of the baryon–photon ratio also rely on deuterium (e.g. Tytler *et al.*, 2000). We shall see that the CMB has its own internal measure of this ratio from the acoustic peaks. Agreement between the nucleosynthesis and acoustic peak measurements argue that the baryon–photon ratio has not changed appreciably since $z \sim 10^8$.

### 3.2.2 Thermalization and spectral distortions

Between nucleosynthesis and recombination, processes that create and destroy photons and hence thermalize the CMB fall out of equilibrium. The lack of spectral distortions in the CMB thus constrains any process that injects energy or photons into the plasma after this epoch.

In a low baryon–photon ratio Universe, the main thermalization process is double, also known as radiative, Compton scattering

$$e^- + \gamma \leftrightarrow e^- + \gamma + \gamma. \tag{3.12}$$

The radiative Compton scattering rate becomes insufficient to maintain a blackbody at a redshift of (Danese and de Zotti, 1982)

$$z_{\text{therm}} = 2.0 \times 10^6 (1 - Y_{\text{p}}/2)^{-2/5} \left( \frac{\Omega_{\text{b}} h^2}{0.02} \right)^{-2/5}, \tag{3.13}$$

where $Y_{\text{p}}$ is the primordial mass function of helium, corresponding to a timescale of order a few months.

After this redshift, energy or photon injection appears as a spectral distortion in the spectrum of the CMB. The form of the distortion is determined by Compton scattering

$$e^- + \gamma \leftrightarrow e^- + \gamma, \tag{3.14}$$

since it is still sufficiently rapid compared with respect to the expansion while hydrogen remains ionized. Because a blackbody has a definite number density of photons at a given temperature, energy exchange via Compton scattering alone can only produce a Bose–Einstein spectrum for the photon distribution function

$$f = \frac{1}{e^{(E-\mu)/T} - 1} \tag{3.15}$$

with $\mu$ determined by the conserved number density and temperature. The evolution to a Bose–Einstein distribution is determined by solving the Kompaneets equation (Zeldovich and Sunyaev, 1969; Sunyaev and Zeldovich, 1970; Illarionov and Sunyaev, 1975). The Kompaneets equation is the Boltzmann equation for Compton scattering in a homogeneous medium. It includes the effects of electron recoil and the second-order Doppler shift that exchange energy between the photons and electrons. If energy is injected into the plasma to heat the electrons, Comptonization will try to redistribute the energy to the CMB. Consequently the first step in Comptonization is a so called "*y*-distortion" in the spectrum as low-energy photons in the Rayleigh Jeans tail of the CMB gain energy from the second-order Doppler shift (see Fig. 3.2).

After multiple scatterings, the spectrum settles into the Bose–Einstein form with a "$\mu$-distortion." The transition occurs when the energy-transfer weighted optical depth approaches unity, *i.e.* $\tau_K(z_K) = 4 \int dt n_e \sigma_T T_e / m_e = 1$ where $n_e$ is the electron density,

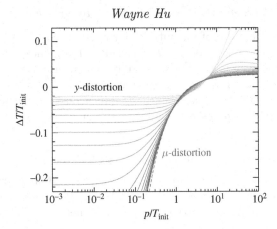

FIGURE 3.2. Comptonization process. Energy injected into the CMB through heating of the electrons is thermalized by Compton scattering. Under the Kompaneets equation, a $y$-distortion first forms as low frequency photons gain energy from the electrons. After multiple scatterings the distribution is thermalized to a chemical potential or $\mu$-distortion. Adapted from Hu, 1995.

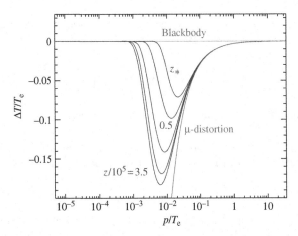

FIGURE 3.3. Low-frequency thermalization by bremsstrahlung. Creation and absorption of photons at low frequencies by bremsstrahlung bring a chemical potential distortion back to blackbody at the electron temperature $T_e$ between $z_K$ and recombination. Adapted from Hu and Silk, 1993.

$\sigma_T$ is the Thomson cross section, $T_e$ is the electron temperature and $m_e$ is the electron mass

$$z_K \approx 5.1 \times 10^4 (1 - Y_p/2)^{-1/2} \left( \frac{\Omega_b h^2}{0.02} \right)^{-1/2}. \tag{3.16}$$

Above this redshift energy injection creates a $\mu$-distortion, below a $y$-distortion.

At very low frequencies, bremsstrahlung,

$$e^- + p \leftrightarrow e^- + p + \gamma, \tag{3.17}$$

is still efficient in creating and absorbing photons. In Fig. 3.3 we show how a $\mu$-distortion continues to evolve until recombination which brings the low-frequency spectrum back to a blackbody but now at the electron temperature.

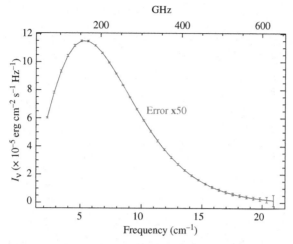

FIGURE 3.4. CMB spectrum from FIRAS. No spectral distortions from a blackbody have been discovered to date.

The best limits to date are from the Cosmic Background Explorer Far Infrared Absolute Spectrometer (COBE FIRAS) from intermediate to high frequencies: $|\mu| < 9 \times 10^{-5}$ and $|y| < 1.5 \times 10^{-5}$ at 95% confidence (see Fig 3.4 and Fixsen *et al.*, 1996). After subtracting out Galactic emission, no spectral distortions of any kind are detected and the spectrum appears to be a perfect blackbody of $\bar{T} = 2.725 \pm 0.002\,\mathrm{K}$ (Mather *et al.*, 1999).

### 3.2.3 Recombination

While the recombination process

$$\mathrm{p} + \mathrm{e}^- \leftrightarrow \mathrm{H} + \gamma, \tag{3.18}$$

is rapid compared to the expansion, the ionization fraction obeys an equilibrium distribution just like that considered for light elements for nucleosynthesis or the CMB spectrum thermalization. As in the former two processes, the qualitative behavior of recombination is determined by the low baryon–photon ratio of the Universe.

Taking number densities of the Maxwell–Boltzmann form of Eq. (3.2), we obtain

$$\frac{n_\mathrm{p} n_\mathrm{e}}{n_\mathrm{H}} \approx e^{-B/T} \left(\frac{m_\mathrm{e} T}{2\pi}\right)^{3/2} e^{(\mu_\mathrm{p} + \mu_\mathrm{e} - \mu_\mathrm{H})/T}, \tag{3.19}$$

where $B = m_\mathrm{p} + m_\mathrm{e} - m_\mathrm{H} = 13.6\,\mathrm{eV}$ is the binding energy and we have set $g_\mathrm{p} = g_\mathrm{e} = \frac{1}{2} g_\mathrm{H} = 2$. Given the vanishingly small chemical potential of the photons, $\mu_\mathrm{p} + \mu_\mathrm{e} = \mu_\mathrm{H}$ in equilibrium.

Next, defining the ionization fraction for a hydrogen-only plasma

$$n_\mathrm{p} = n_\mathrm{e} = x_\mathrm{e} n_\mathrm{b},$$
$$n_\mathrm{H} = n_\mathrm{b} - n_\mathrm{p} = (1 - x_\mathrm{e}) n_\mathrm{b}, \tag{3.20}$$

we can rewrite Eq. (3.19) as the Saha equation

$$\frac{n_\mathrm{e} n_\mathrm{p}}{n_\mathrm{H} n_\mathrm{b}} = \frac{x_\mathrm{e}^2}{1 - x_\mathrm{e}} = \frac{1}{n_\mathrm{b}} \left(\frac{m_\mathrm{e} T}{2\pi}\right)^{3/2} e^{-B/T}. \tag{3.21}$$

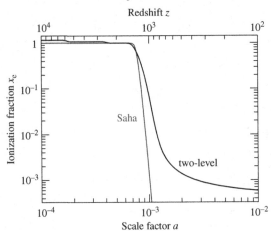

FIGURE 3.5. Hydrogen recombination in Saha equilibrium vs. the calibrated two-level calculation of RECFAST. The features before hydrogen recombination are due to helium recombination.

Recombination occurs at a temperature substantially lower than $T = B$ again because of the low baryon–photon ratio of the Universe. We can see this by rewriting the Saha equation in terms of the photon number density

$$\frac{x_e^2}{1 - x_e} = e^{-B/kT} \frac{\pi^{1/2}}{2^{5/2} \eta_{b\gamma} \zeta(3)} \left( \frac{m_e c^2}{kT} \right)^{3/2}. \tag{3.22}$$

Because $\eta_{b\gamma} \sim 10^{-9}$, the Saha equation implies that the medium only becomes substantially neutral at a temperature of $T \approx 0.3\,\mathrm{eV}$ or at a redshift of $z_* \sim 10^3$. At this point, there are not enough photons even in the Wien tail above the binding energy to ionize hydrogen. We plot the Saha solution in Fig. 3.5.

Near the epoch of recombination, the recombination rates become insufficient to maintain ionization equilibrium. There is also a small contribution from helium recombination that must be added. The current standard for following the non-equilibrium ionization history is the RECFAST code (Seager *et al.*, 2000) which employs the traditional two-level atom calculation of Peebles (1968) but alters the hydrogen case B recombination rate $\alpha_B$ to fit the results of a multilevel atom. More specifically, RECFAST solves a coupled system of equations for the ionization fraction $x_i$ in singly ionized hydrogen and helium ($i = \mathrm{H, He}$)

$$\frac{dx_i}{d\ln a} = \frac{\alpha_B C_i n_{Hp}}{H} \left[ s(x_{\max} - x_i) - x_i x_e \right], \tag{3.23}$$

where $n_{Hp} = (1 - Y_p) n_b$ is the total hydrogen plus proton number density accounting for the helium mass fraction $Y_p$, $x_e \equiv n_e / n_{Hp} = \sum x_i$ is the total ionization fraction, $n_e$ is the free electron density, $x_{\max}$ is the maximum $x_i$ achieved through full ionization,

$$s = \frac{\beta}{n_{Hp}} e^{-B_{1s}/T_b},$$

$$C_i^{-1} = 1 + \frac{\beta \alpha_B e^{-B_{2s}/T_b}}{\Lambda_\alpha + \Lambda_{2s1s}},$$

$$\beta = g_{\mathrm{rat}} \left( \frac{T_b m_e}{2\pi} \right)^{3/2}, \tag{3.24}$$

with $g_{rat}$ the ratio of statistical weights, $T_b$ the baryon temperature, $B_L$ the binding energy of the $L$th level, $\Lambda_\alpha$ the rate of redshifting out of the Lyman-$\alpha$ line corrected for the energy difference between the 2s and 2p states

$$\Lambda_\alpha = \frac{1}{\pi^2} (B_{1s} - B_{2p})^3 \, e^{-(B_{2s}-B_{2p})/T_b} \frac{H}{(x_{max} - x_i)n_{Hp}} \tag{3.25}$$

and $\Lambda_{2s1s}$ as the rate for the two-photon 2s $-$ 1s transition. For reference, for hydrogen $B_{1s} = 13.598\,\mathrm{eV}$, $B_{2s} = B_{2p} = B_{1s}/4$, $\Lambda_{2s1s} = 8.22458\,\mathrm{s}^{-1}$, $g_{rat} = 1$, $x_{max} = 1$. For helium $B_{1s} = 24.583\,\mathrm{eV}$, $B_{2s} = 3.967\,\mathrm{eV}$, $B_{2p} = 3.366\,\mathrm{eV}$, $\Lambda_{2s1s} = 51.3\,\mathrm{s}^{-1}$, $g_{rat} = 4$, $x_{max} = Y_p/[4(1 - Y_p)]$.

Note that if the recombination rate is faster than the expansion rate $\alpha_B C_i n_{Hp}/H \gg 1$, the ionization solutions for $x_i$ reach the Saha equilibrium $\mathrm{d}x_i/\mathrm{d}\ln a = 0$. In this case $s(x_{max} - x_i) = x_i x_e$ or

$$\begin{aligned} x_i &= \frac{1}{2}\left[ \sqrt{(x_{ei} + s)^2 + 4s x_{max}} - (x_{ei} + s) \right] \\ &= x_{max}\left[ 1 - \frac{x_{ei} + x_{max}}{s}\left(1 - \frac{x_{ei} + 2x_{max}}{s}\right) + ... \right], \end{aligned} \tag{3.26}$$

where $x_{ei} = x_e - x_i$ is the ionization fraction excluding the species. The recombination of hydrogenic doubly ionized helium is here handled purely through the Saha equation with a binding energy of $1/4$ the $B_{1s}$ of hydrogen and $x_{max} = Y_p/[4(1 - Y_p)]$. The case B recombination coefficients as a function of $T_b$ are given in Seager *et al.* (2000) as is the strong thermal coupling between $T_b$ and $T_{CMB}$. The multilevel-atom fudge that REC-FAST introduces is to replace the hydrogen $\alpha_B \to 1.14\alpha_B$ independently of cosmology. While this fudge suffices for current observations, which approach the $\sim 1\%$ level, the recombination standard will require improvement if CMB anisotropy predictions are to reach an accuracy of 0.1% (e.g. Rubiño-Martín *et al.*, 2006; Switzer and Hirata, 2008; Wong *et al.*, 2008).

The phenomenology of CMB temperature and polarization anisotropy is primarily governed by the redshift of recombination $z_* = a_*^{-1} - 1$ when most of the contributions originate. This redshift though carries little dependence on standard cosmological parameters. The insensitivity follows from the fact that recombination proceeds rapidly once $B_{1s}/T_b$ has reached a certain threshold as the Saha equation illustrates. Defining the redshift of recombination as the epoch at which the Thomson optical depth during recombination (*i.e.* excluding reionization) reaches unity, $\tau_{rec}(a_*) = 1$, a fit to the recombination calculation gives (Hu, 2005)

$$a_*^{-1} \approx 1089\left(\frac{\Omega_m h^2}{0.14}\right)^{0.0105}\left(\frac{\Omega_b h^2}{0.024}\right)^{-0.028}, \tag{3.27}$$

around a fiducial model of $\Omega_m h^2 = 0.14$ and $\Omega_b h^2 = 0.024$.

The Universe is known to be reionized at low redshifts due to the lack of a Gunn–Peterson trough in quasar absorption spectra. Moreover, large-angle CMB polarization detections (see Fig. 3.19) suggest that this transition back to full ionization occurred around $z \sim 10$ leaving an extended neutral period between recombination and reionization.

## 3.3 Temperature anisotropy from recombination

Spatial variations in the CMB temperature at recombination are seen as temperature anisotropy by the observer today. The temperature anisotropy of the CMB was first detected in 1992 by the COBE Differential Microwave Radiometer (DMR) instrument

*Wayne Hu*

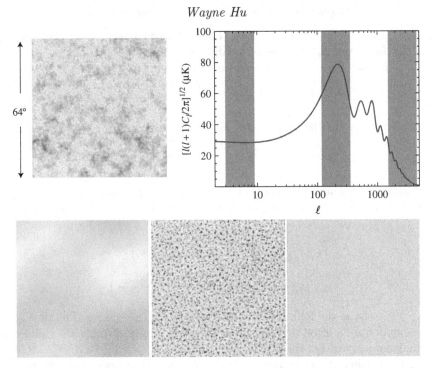

FIGURE 3.6. From temperature maps to power spectrum. The original temperature fluctuation map (top left) corresponding to a simulation of the power spectrum (top right) can be band filtered to illustrate the power spectrum in three characteristic regimes: from left to right, the large-scale gravitational regime of COBE, the first acoustic peak where most of the power lies, and the damping tail where fluctuations are dissipated. Adapted from Hu and White, 2004.

(Smoot *et al.* 1992). These corresponded to variations of order $\Delta T/T \sim 10^{-5}$ across $10°-90°$ on the sky (see Fig. 3.6).

Most of the structure in the temperature anisotropy however is associated with acoustic oscillations of the photon–baryon plasma on $\sim 1°$ scales. Throughout the 1990s constraints on the location of the first peak steadily improved culminating with the determinations of the TOCO (Miller *et al.*, 1999), BOOMERanG, (de Bernardis *et al.*, 2000) and Maxima-1 (Hanany *et al.*, 2000) experiments. Currently from Wilkinson Microwave Anisotropy Probe (WMAP; Spergel *et al.*, 2007) and ground-based experiments, we have precise measurements of the first five acoustic peaks (see Fig. 3.7). Primary fluctuations beyond this scale (below $\sim 10'$) are damped by Silk damping (Silk, 1968) as verified observationally first by the CBI experiment (Padin *et al.*, 2001).

In this section, we deconstruct the basic physics behind these phenomena. We begin with the geometric projection of temperature inhomogeneities at recombination onto the sky of the observer in Section 3.3.1. We continue with the basic equations of fluid mechanics that govern acoustic phenomena in Section 3.3.2. We add gravitational (Section 3.3.3), baryonic (Section 3.3.4), matter–radiation (Section 3.3.5), and dissipational (Section 3.3.6) effects in the sections that follow. Finally we put these pieces back together to discuss the information content of the acoustic peaks in Section 3.3.7.

### 3.3.1 Anisotropy from inhomogeneity

Given that the CMB radiation is blackbody to experimental accuracy (see Fig. 3.4), one can characterize its spatial and angular distribution by its temperature at the position $\mathbf{x}$ of the observer in the direction $\hat{\mathbf{n}}$ on the observer's sky

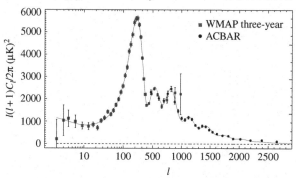

FIGURE 3.7. Temperature power spectrum from recent measurements from WMAP and ACBAR along with the best-fit ΛCDM model. The main features of the temperature power spectrum including the first five acoustic peaks and damping tail have now been measured. Adapted from Reichardt *et al.*, 2009.

$$f(\nu, \hat{\mathbf{n}}, \mathbf{x}) = [\exp(2\pi\nu/T(\hat{\mathbf{n}}; \mathbf{x})) - 1]^{-1}, \qquad (3.28)$$

where $\nu = E/2\pi$ is the observation frequency. The hypothetical observer could be an electron in the intergalactic medium or the true observer on Earth or the second Lagrange point, L2. When the latter is implicitly meant, we will take $\mathbf{x} = \mathbf{0}$. We will occasionally suppress the coordinate $\mathbf{x}$ when this position is to be understood.

For statistically isotropic, Gaussian random temperature fluctuations a harmonic description is more efficient than a real space description. For the angular structure at the position of the observer, the appropriate harmonics are the spherical harmonics. These are the eigenfunctions of the Laplace operator on the sphere and form a complete basis for scalar functions on the sky

$$\Theta(\hat{\mathbf{n}}) = \frac{T(\hat{\mathbf{n}}) - \bar{T}}{\bar{T}} = \sum_{\ell m} \Theta_{\ell m} Y_{\ell m}(\hat{\mathbf{n}}). \qquad (3.29)$$

For statistically isotropic fluctuations, the ensemble average of the temperature fluctuations are described by the power spectrum

$$\langle \Theta_{\ell m}^{*} \Theta_{\ell' m'} \rangle = \delta_{\ell \ell'} \delta_{mm'} C_{\ell}. \qquad (3.30)$$

Moreover, the power spectrum contains all of the statistical information in the field if the fluctuations are Gaussian. Here $C_{\ell}$ is dimensionless but is often shown with units of squared temperature, e.g. $\mu K^2$, by multiplying through by the background temperature today $\bar{T}$. The correspondence between angular size and amplitude of fluctuations and the power spectrum is shown in Fig. 3.6.

Let us begin with the simple approximation that the temperature field at recombination is isotropic but inhomogeneous and the anisotropy viewed at the present is due to the observer seeing different portions of the recombination surface in different directions (see Fig. 3.8). We will develop this picture further in the following sections with gravitational redshifts, dipole or Doppler anisotropic sources, the finite duration of recombination, and polarization.

Under this simple instantaneous recombination approximation, the angular temperature fluctuation distribution is simply a projection of the spatial temperature fluctuation

$$\Theta(\hat{\mathbf{n}}) = \int dD\, \Theta(\mathbf{x}) \delta(D - D_*), \qquad (3.31)$$

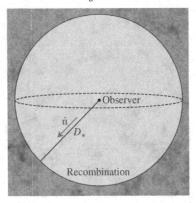

FIGURE 3.8. From inhomogeneity to anisotropy. Temperature inhomogeneities at recombination are viewed at a distance $D_*$ as anisotropy on the observer's sky.

where $D = \int dz/H$ is the comoving distance and $D_*$ denotes the distance a CMB photon travels from recombination. Here $\Theta(\mathbf{x}) = [T(\mathbf{x}) - \bar{T}]/\bar{T}$ is the spatial temperature fluctuation at recombination. Note that the cosmological redshift does not appear in the temperature fluctuation since the background and fluctuation redshift alike.

The spatial power spectrum at recombination can likewise be described by its harmonic modes. In a flat geometry, these are Fourier modes

$$\Theta(\mathbf{x}) = \int \frac{d^3 k}{(2\pi)^3} \Theta(\mathbf{k}) e^{i\mathbf{k}\cdot\mathbf{x}}. \tag{3.32}$$

More generally they are the eigenfunctions of the Laplace operator on the three-dimensional space with constant curvature of a general Friedmann–Robertson–Walker metric (see e.g. Hu, 2003). For a statistically homogeneous spatial distribution, the two-point function is described by the power spectrum

$$\langle \Theta(\mathbf{k})^* \Theta(\mathbf{k}') \rangle = (2\pi)^3 \delta(\mathbf{k} - \mathbf{k}') P(k). \tag{3.33}$$

These relations tell us that the amplitude of the angular and spatial power spectra are related. A useful way of establishing this relation and quantifying the power in general is to describe the contribution per logarithmic interval to the variance of the respective fields:

$$\langle \Theta(\mathbf{x})\Theta(\mathbf{x}) \rangle = \int \frac{d^3 k}{(2\pi)^3} P(k) = \int d\ln k \, \frac{k^3 P(k)}{2\pi^2} \equiv \int d\ln k \, \Delta_T^2(k). \tag{3.34}$$

A scale-invariant spectrum has equal contribution to the variance per e-fold $\Delta_T^2 = k^3 P(k)/2\pi^2 = \text{const}$. To relate this to the amplitude of the angular power spectrum, we expand equation (3.31) in Fourier modes

$$\Theta(\hat{\mathbf{n}}) = \int \frac{d^3 k}{(2\pi)^3} \Theta(\mathbf{k}) e^{i\mathbf{k}\cdot D_*\hat{\mathbf{n}}}. \tag{3.35}$$

The Fourier modes themselves can be expanded in spherical harmonics with the relation

$$e^{ikD_*\cdot\hat{\mathbf{n}}} = 4\pi \sum_{\ell m} i^\ell j_\ell(kD_*) Y_{\ell m}^*(\hat{\mathbf{k}}) Y_{\ell m}(\hat{\mathbf{n}}), \tag{3.36}$$

where $j_\ell$ is the spherical Bessel function. Extracting the multipole moments, we obtain

$$\Theta_{\ell m} = \int \frac{d^3 k}{(2\pi)^3} \Theta(\mathbf{k}) 4\pi i^\ell j_\ell(kD_*) Y_{\ell m}(\mathbf{k}). \tag{3.37}$$

We can then relate the angular and spatial two-point functions (3.30) and (3.33)

$$\langle \Theta_{\ell m}^* \Theta_{\ell' m'} \rangle = \delta_{\ell\ell'} \delta_{mm'} 4\pi \int d\ln k\, j_\ell^2(kD_*) \Delta_T^2(k) = \delta_{\ell\ell'} \delta_{mm'} C_\ell. \tag{3.38}$$

Given a slowly varying, nearly scale-invariant spatial power spectrum we can take $\Delta_T^2$ out of the integral and evaluate it at the peak of the Bessel function $kD_* \approx \ell$. The remaining integral can be evaluated in closed form $\int_0^\infty j_\ell^2(x) d\ln x = 1/[2\ell(\ell+1)]$ yielding the final result

$$C_\ell \approx \frac{2\pi}{\ell(\ell+1)} \Delta_T^2(\ell/D_*). \tag{3.39}$$

Likewise even slowly varying features like the acoustic peaks of Section 3.3.2 also mainly map to multipoles of $\ell \approx kD_*$ due to the delta-function-like behavior of $j_\ell^2$.

It is therefore common to plot the angular power spectrum as

$$\mathcal{C}_\ell \equiv \frac{\ell(\ell+1)}{2\pi} C_\ell \approx \Delta_T^2. \tag{3.40}$$

It is also common to plot $\mathcal{C}_\ell^{1/2} \approx \Delta_T$, the logarithmic contribution to the rms of the field, in units of $\mu$K.

Now let us compare this expression to the variance per log interval in multipole space:

$$\langle T(\hat{n}) T(\hat{n}) \rangle = \sum_{\ell m} \sum_{\ell' m'} \langle T_{\ell m}^* T_{\ell' m'} \rangle Y_{\ell m}^*(\hat{n}) Y_{\ell' m'}(\hat{n})$$

$$= \sum_\ell C_\ell \sum_m Y_{\ell m}^*(\hat{n}) Y_{\ell m}(\hat{n}) = \sum_\ell \frac{2\ell+1}{4\pi} C_\ell. \tag{3.41}$$

For variance contributions from $\ell \gg 1$,

$$\sum_\ell \frac{2\ell+1}{4\pi} C_\ell \approx \int d\ln\ell \frac{\ell(2\ell+1)}{4\pi} C_\ell \approx \int d\ln\ell \frac{\ell(\ell+1)}{2\pi} C_\ell. \tag{3.42}$$

Thus $\mathcal{C}_\ell$ is also approximately the variance per log interval in angular space as well.

### 3.3.2 Acoustic oscillation basics

*Thomson tight coupling.* To understand the angular pattern of temperature fluctuations seen by the observer today, we must understand the spatial temperature pattern at recombination. That in turn requires an understanding of the dominant physical processes in the plasma before recombination.

Thomson scattering of photons off of free electrons is the most important process, given its relatively large cross section (averaged over polarization states)

$$\sigma_T = \frac{8\pi\alpha^2}{3m_e^2} = 6.65 \times 10^{-25} \text{ cm}^2. \tag{3.43}$$

The important quantity to consider is the mean free path of a photon given Thomson scattering and a medium with a free electron density $n_e$. Before recombination when the ionization fraction $x_e \approx 1$ this density is given by

$$n_e = (1 - Y_p) x_e n_b$$
$$= 1.12 \times 10^{-5} (1 - Y_p) x_e \Omega_b h^2 (1+z)^3 \text{cm}^{-3}. \tag{3.44}$$

The comoving mean free path $\lambda_C$ is given by

$$\lambda_C^{-1} \equiv \dot{\tau} \equiv n_e \sigma_T a, \tag{3.45}$$

where the extra factor of $a$ comes from converting physical to comoving coordinates and $Y_p$ is the primordial helium mass fraction. We have also represented this mean free path in terms of an scattering absorption coefficient $\dot{\tau}$ where dots are conformal time $\eta \equiv \int dt/a$ derivatives and $\tau$ is the optical depth.

Near recombination ($z \approx 10^3$, $x_e \approx 1$) and, given $\Omega_b h^2 \approx 0.02$ and $Y_p = 0.24$, the mean free path is

$$\lambda_C \equiv \frac{1}{\dot{\tau}} \sim 2.5\,\mathrm{Mpc}. \qquad (3.46)$$

This scale is almost two orders of magnitude smaller than the horizon at recombination. On scales $\lambda \gg \lambda_C$ photons are tightly coupled to the electrons by Thomson scattering which in turn are tightly coupled to the baryons by Coulomb interactions.

As a consequence, any bulk motion of the photons must be shared by the baryons. In fluid language, the two species have a single bulk velocity $v_\gamma = v_b$ and hence no entropy generation or heat conduction occurs. Furthermore, the shear viscosity of the fluid is negligible. Shear viscosity is related to anisotropy in the radiative pressure or stress and rapid scattering isotropizes the photon distribution. This is also the reason why in equation (3.31) we took the photon distribution to be isotropic but inhomogeneous.

We shall see that the fluid motion corrects this by allowing dipole $\ell = 1$ anisotropy in the distribution but no higher $\ell$ modes. It is only on scales smaller than the diffusion scale that radiative viscosity ($\ell = 2$) and heat conduction becomes sufficiently large to dissipates the bulk motions of the plasma (see Section 3.3.6).

*Zeroth-Order approximation.* To understand the basic physical picture, let us begin our discussion of acoustic oscillations in the tight coupling regime with a simplified system which we will refine as we progress.

First, let us ignore the dynamical impact of the baryons on the fluid motion. Given that the baryon and photon velocity are equal and the momentum density of a relativistic fluid is given by $(\rho+p)v$, where $p$ is the pressure, and $v$ is the fluid velocity, this approximation relates to the quantity

$$R \equiv \frac{(\rho_b + p_b)v_b}{(\rho_\gamma + p_\gamma)v_\gamma} = \frac{\rho_b + p_b}{\rho_\gamma + p_\gamma} = \frac{3\rho_b}{4\rho_\gamma}$$
$$\approx 0.6 \left(\frac{\Omega_b h^2}{0.02}\right)\left(\frac{a}{10^{-3}}\right), \qquad (3.47)$$

where we have used the fact that $\rho_\gamma \propto T^4$ so that its value is fixed by the redshifting background $\bar{T} = 2.725(1+z)\,\mathrm{K}$. Neglect of the baryon inertia and momentum only fails right around recombination.

Next, we shall assume that the background expansion is matter dominated to relate time and scale factor. The validity of this approximation depends on the matter–radiation ratio

$$\frac{\rho_m}{\rho_r} = 3.6 \left(\frac{\Omega_m h^2}{0.15}\right)\left(\frac{a}{10^{-3}}\right), \qquad (3.48)$$

and is approximately valid during recombination and afterwords. One expects from these arguments that order unit differences between the real Universe and our basic description will occur. We will in fact use these differences in the following sections to show how the baryon and matter densities are measured from the acoustic peak morphology.

Finally, we shall consider the effect of pressure forces and neglect gravitational forces. While this is not a valid approximation in and of itself, we shall see that for a photon-dominated system, the error in ignoring gravitational forces exactly cancels with that from ignoring gravitational redshifts that photons experience after recombination (see Section 3.3.3).

*Continuity equation.* Given that Thomson scattering neither creates nor destroys photons, the continuity equation implies that the photon number density only changes due to flows into and out of the volume. In a non-expanding Universe that would require

$$\dot{n}_\gamma + \nabla \cdot (n_\gamma \mathbf{v}_\gamma) = 0. \tag{3.49}$$

Since $n_\gamma$ is the number density of photons per unit physical (not comoving) volume, this equation must be corrected for the expansion. The effect of the expansion can alternately be viewed as that of the Hubble flow diluting the number density everywhere in space. Because number densities scale as $n_\gamma \propto a^{-3}$, the expansion alters the continuity equation as

$$\dot{n}_\gamma + 3n_\gamma \frac{\dot{a}}{a} + \nabla \cdot (n_\gamma \mathbf{v}_\gamma) = 0. \tag{3.50}$$

Since we are interested in small fluctuations around the background, let us linearize the equations $n_\gamma \approx \bar{n}_\gamma + \delta n_\gamma$ and drop terms that are higher than first order in $\delta n_\gamma / n_\gamma$ and $v_\gamma$. Note that $v_\gamma$ is first order in the number density fluctuations since as we shall see in the Euler equation discussion below it is generated from the pressure gradients associated with density fluctuations.

The continuity equation (3.50) for the fluctuations becomes

$$\left( \frac{\delta n_\gamma}{n_\gamma} \right)^{\cdot} = -\nabla \cdot \mathbf{v}_\gamma. \tag{3.51}$$

Since the number density $n_\gamma \propto T^3$, the fractional density fluctuation is related to the temperature fluctuation $\Theta$ as

$$\frac{\delta n_\gamma}{n_\gamma} = 3\frac{\delta T}{T} \equiv 3\Theta. \tag{3.52}$$

Expressing the continuity equation in terms of $\Theta$, we obtain

$$\dot{\Theta} = -\frac{1}{3}\nabla \cdot \mathbf{v}_\gamma. \tag{3.53}$$

Fourier transforming this equation, we get

$$\dot{\Theta} = -\frac{1}{3}i\mathbf{k} \cdot \mathbf{v}_\gamma, \tag{3.54}$$

for the relationship between the Fourier mode amplitudes.

*Euler equation.* Now let us examine the origin of the fluid velocity. In the background, the velocity vanishes due to isotropy. However, Newtonian mechanics dictates that pressure forces will generate particle momentum as $\dot{\mathbf{q}} = \mathbf{F}$. Newton's law must also be modified for the expansion. If we associate the de Broglie wavelength with the inverse momentum, this wavelength also stretches with the expansion. For photons, this accounts for the redshift factor. For non-relativistic matter, this means that bulk velocities decay with the expansion. In either case, we generalize Newton's law to read

$$\dot{\mathbf{q}} + \frac{\dot{a}}{a}\mathbf{q} = \mathbf{F}. \tag{3.55}$$

For a collection of particles, the relevant quantity is the momentum density

$$(\rho_\gamma + p_\gamma)\mathbf{v}_\gamma \equiv \int \frac{d^3q}{(2\pi)^3}\mathbf{q}f, \tag{3.56}$$

and likewise the force becomes a force density. For the photon–baryon fluid, this force density is provided by the pressure gradient. The result is the Euler equation

$$[(\rho_\gamma + p_\gamma)\mathbf{v}_\gamma]\dot{} = -4\frac{\dot{a}}{a}(\rho_\gamma + p_\gamma)\mathbf{v}_\gamma - \nabla p_\gamma, \tag{3.57}$$

where the 4 on the right-hand side comes from combining the redshifting of wavelengths with that of number densities. Since photons have an equation of state $p_\gamma = \rho_\gamma/3$, the Euler equation becomes

$$\frac{4}{3}\rho_\gamma\dot{\mathbf{v}}_\gamma = -\frac{1}{3}\nabla\rho_\gamma,$$
$$\dot{\mathbf{v}}_\gamma = -\nabla\Theta. \tag{3.58}$$

In Fourier space, the Euler equation becomes

$$\dot{\mathbf{v}}_\gamma = -i\mathbf{k}\Theta. \tag{3.59}$$

The factor of $i$ here represents the fact that the temperature maxima and minima are zeros of the velocity in real space due to the gradient relation, i.e. temperature and velocity have a $\pi/2$ phase shift. It is convenient therefore to define the velocity amplitude to absorb this factor

$$\mathbf{v}_\gamma \equiv -iv_\gamma\hat{\mathbf{k}}. \tag{3.60}$$

The direction of the fluid velocity is always parallel to the wavevector $\mathbf{k}$ for linear (scalar) perturbations and hence we can write the Euler equation as

$$\dot{v}_\gamma = k\Theta. \tag{3.61}$$

*Acoustic peaks.* Combining the continuity (3.54) and Euler (3.61) equations to eliminate the fluid velocity, we get the simple harmonic oscillator equation

$$\ddot{\Theta} + c_s^2 k^2 \Theta = 0, \tag{3.62}$$

where the adiabatic sound speed $c_s^2 = 1/3$ for the photon-dominated fluid and more generally is defined as

$$c_s^2 \equiv \frac{\dot{p}_\gamma}{\dot{\rho}_\gamma}. \tag{3.63}$$

The solution to the oscillator equation can be specified given two initial conditions $\Theta(0)$ and $v_\gamma(0)$ or $\dot{\Theta}(0)$,

$$\Theta(\eta) = \Theta(0)\cos(ks) + \frac{\dot{\Theta}(0)}{kc_s}\sin(ks), \tag{3.64}$$

where the sound horizon is defined as

$$s \equiv \int c_s d\eta. \tag{3.65}$$

In real space, these oscillations appear as standing waves for each Fourier mode.

These standing waves continue to oscillate until recombination. At this point the free electron density drops drastically (see Fig. 3.5) and the photons freely stream to the observer. The pattern of acoustic oscillations on the recombination surface seen by the observer becomes the acoustic peaks in the temperature anisotropy.

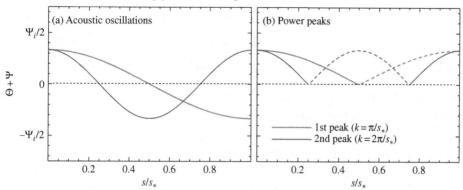

FIGURE 3.9. Acoustic oscillation basics. All modes start from the same initial epoch with time denoted by the sound horizon relative to the sound horizon at recombination $s_*$. (a) Wavenumbers that reach extrema in their effective temperature $\Theta + \Psi$ (accounting for gravitational redshifts Section 3.3.3) at $s_*$ form a harmonic series $k_n = n\pi/s_*$. (b) Amplitude of the fluctuations is the same for the maxima and minima without baryon inertia. Adapted from Hu and Dodelson, 2002.

Let us focus on the adiabatic mode which starts with a finite density or temperature fluctuation and vanishing velocity perturbation. At recombination $\eta_*$, the oscillation reaches (see Fig. 3.9)

$$\Theta(\eta_*) = \Theta(0)\cos(ks_*). \tag{3.66}$$

Considering a spectrum of $k$ modes, the critical feature of these oscillations is that they are temporally coherent. The underlying assumption is that fluctuations of all wavelengths originated at $\eta = 0$ or at least $\eta \ll \eta_*$. Without inflation this would violate causality for long-wavelength fluctuations, i.e. the analog of the horizon problem for perturbations. With inflation, superhorizon modes originate during an inflationary epoch $\eta_i \ll \eta_*$.

Modes caught in the extrema of their oscillation follow a harmonic relation

$$k_n s_* = n\pi, \quad n = 1, 2, 3\ldots \tag{3.67}$$

yielding a fundamental scale or frequency, related to the inverse sound horizon

$$k_A = \pi/s_*. \tag{3.68}$$

Since the power spectrum is proportional to the square of the fluctuation, both maxima and minima contribute peaks in the spectrum. Observational verification of this harmonic series is the primary evidence for inflationary adiabatic initial conditions (Hu and White, 1996).

The fundamental physical scale is translated into a fundamental angular scale by simple projection according to the angular diameter distance $D_A$

$$\theta_A = \lambda_A/D_A,$$
$$\ell_A = k_A D_A, \tag{3.69}$$

[see Eq. (3.39)]. In a flat Universe, the distance is simply $D_A = D \equiv \eta_0 - \eta_* \approx \eta_0$, the horizon distance, and $k_A = \pi/s_* = \sqrt{3}\pi/\eta_*$ so

$$\theta_A \approx \frac{\eta_*}{\eta_0}. \tag{3.70}$$

Furthermore, in a matter-dominated Universe $\eta \propto a^{1/2}$ so $\theta_A \approx 1/30 \approx 2°$ or

$$\ell_A \approx 200. \tag{3.71}$$

We shall see in Section 3.3.7 that radiation and dark energy introduce important corrections to this prediction from their influence on $\eta_*$ and $D_*$, respectively. Nonetheless it is remarkable that this simple argument predicts the basics of acoustic oscillations: their existence, coherence and fundamental scale.

*Acoustic troughs: Doppler effect.* Acoustic oscillations also imply that the plasma is moving relative to the observer. This bulk motion imprints a temperature anisotropy via the Doppler effect

$$\left(\frac{\Delta T}{T}\right)_{\text{dop}} = \hat{\mathbf{n}} \cdot \mathbf{v}_\gamma. \tag{3.72}$$

Averaged over directions

$$\left(\frac{\Delta T}{T}\right)_{\text{rms}} = \frac{v_\gamma}{\sqrt{3}}, \tag{3.73}$$

which given the acoustic solution of Eq. (3.64) implies

$$\frac{v_\gamma}{\sqrt{3}} = -\frac{\sqrt{3}}{k}\dot{\Theta} = \frac{\sqrt{3}}{k}kc_s\,\Theta(0)\sin(ks)$$
$$= \Theta(0)\sin(ks). \tag{3.74}$$

Interestingly, the Doppler effect for the photon-dominated system is of equal amplitude and $\pi/2$ out of phase: extrema of temperature are turning points of velocity (Fig. 3.10).

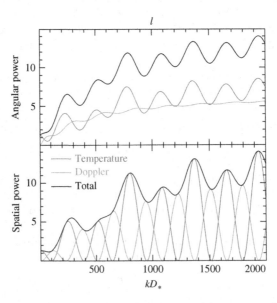

FIGURE 3.10. Doppler effect. The Doppler effect provides fluctuations of comparable strength to the local temperature fluctuations from acoustic oscillations in $k$-space (lower) providing features at the troughs of the latter. In angular space, projection effects smooth the Doppler features leaving an acoustic morphology that reflects the temperature oscillations. The peak height modulation comes from the baryon inertia (Section 3.3.4) and the gradual increase in power with $\ell$ from radiation domination (Section 3.3.5).

If we simply add the $k$-space temperature and Doppler effects in quadrature we would obtain

$$\left(\frac{\Delta T}{T}\right)^2 = \Theta^2(0)[\cos^2(ks) + \sin^2(ks)] = \Theta^2(0). \tag{3.75}$$

In other words there are no preferred $k$-modes at harmonics and a scale-invariant initial temperature spectrum would lead to a scale-invariant spatial power spectrum at recombination. However the Doppler effect carries an angular dependence that changes its projection on the sky $\hat{\mathbf{n}} \cdot \mathbf{v}_\gamma \propto \hat{\mathbf{n}} \cdot \hat{\mathbf{k}}$. In a coordinate system where $\hat{\mathbf{z}} \parallel \hat{\mathbf{k}}$, this angular dependence yields an extra factor of $Y_{10}$ in the analogue of Eq. (3.38). This extra factor can be reabsorbed into the total angular dependence through Clebsch–Gordan recoupling (Hu and White, 1997)

$$Y_{10}Y_{\ell 0} \to Y_{\ell \pm 1 0}. \tag{3.76}$$

The recoupling implied for the radial harmonics changes $j_\ell(x) \to j_\ell'(x)$. The projection kernel $j_\ell'(x)$ lacks a strong feature at $\ell \sim x$ and so Doppler contributions in $k$ are spread out in $\ell$. This is simply a mathematical way of stating that the Doppler effect vanishes when the observer is looking perpendicular to $\mathbf{v} \parallel \mathbf{k}$ whereas it is in that direction that the acoustic peaks in temperature gain most of their contribution. The net effect, including baryonic effects that we discuss below, is that the peak structure is dominated by the local temperature at recombination and not the local fluid motion.

### 3.3.3 Gravito-acoustic oscillations

Thus far we have neglected gravitational forces and redshifts in our discussion of plasma motion. The true system exhibits gravito-acoustic or Jeans oscillations. We were able to employ this swindle to get the basic properties of acoustic oscillations because in a photon-dominated plasma the effect of gravitational forces and gravitational redshifts exactly cancel in constant gravitational potentials. To go beyond the photon-dominated plasma and matter-dominated expansion approximation, we now need to include these effects. Furthermore, we shall see that the gravitational potential perturbations from inflation are also the source of the initial temperature fluctuation.

*Continuity equation and Newtonian curvature.* The photon continuity equation for the number density is altered by gravity since the presence of a gravitational potential alters the coordinate volume. Formally in general relativity this comes from the space–space piece of the metric – a spatial curvature perturbation $\Phi$:

$$ds^2 = a^2[-(1 + 2\Psi)d\eta^2 + (1 + 2\Phi)dx^2] \tag{3.77}$$

for a flat cosmology.

We can think of this curvature perturbation as changing the local scale factor $a \to a(1 + \Phi)$ so that the expansion dilution is generalized to

$$\frac{\dot{a}}{a} \to \frac{\dot{a}}{a} + \dot{\Phi}. \tag{3.78}$$

Hence the full continuity equation is now given by

$$(\delta n_\gamma)^{\cdot} = -3\delta n_\gamma \frac{\dot{a}}{a} - 3n_\gamma \dot{\Phi} - n_\gamma \nabla \cdot \mathbf{v}_\gamma, \tag{3.79}$$

or

$$\dot{\Theta} = -\frac{1}{3}kv_\gamma - \dot{\Phi}. \tag{3.80}$$

*Euler equation and Newtonian forces.* Likewise the gravitational force from gradients in the gravitational potential ($\Psi$, formally the time–time piece of the metric perturbation) modifies the momentum conservation equation. The Newtonian force $\mathbf{F} = -m\nabla\Psi$ generalized to momentum density brings the Euler equation to

$$\dot{v}_\gamma = k(\Theta + \Psi). \tag{3.81}$$

General relativity says that $\Phi$ and $\Psi$ are the relativistic analogs of the Newtonian potential and that $\Phi \approx -\Psi$ in the absence of sources of anisotropic stress or viscosity.

*Photon-dominated oscillator.* We can again combine the continuity equation (3.80) and Euler equation (3.81) to form the forced simple harmonic oscillator system

$$\ddot{\Theta} + c_s^2 k^2 \Theta = -\frac{k^2}{3}\Psi - \ddot{\Phi}. \tag{3.82}$$

Note that the effect of baryon inertia is still absent in this system. To make further progress in understanding the effect of gravity on acoustic oscillations we need to specify the gravitational potential $\Psi \approx -\Phi$ from the Poisson equation and understand its time evolution.

*Poisson equation and constant potentials.* In our matter-dominated approximation, $\Phi$ is generated by matter density fluctuations $\Delta_m$ through the cosmological Poisson equation

$$k^2\Phi = 4\pi G a^2 \rho_m \Delta_m, \tag{3.83}$$

where the difference from the usual Poisson equation comes from the use of comoving coordinates for $k$ ($a^2$ factor), the removal of the background density into the background expansion ($\rho_m \Delta_m$) and finally a coordinate subtlety that enters into the definition of $\Delta_m$. In general, the relativistic Poisson equation would have contributions from the momentum density. Alternatively we can relate

$$\Delta_m = \frac{\delta\rho_m}{\rho_m} + 3\frac{\dot{a}}{a}\frac{v_m}{k} \tag{3.84}$$

to the density fluctuation in a coordinate system that comoves with the matter (formally through a gauge transformation). Beyond the matter-dominated approximation, the Poisson equation would carry contributions from all of the species of energy density and the comoving coordinate system would also reflect the total.

In a matter-dominated epoch (or in fact any epoch when gravitational potential gradients and not stress gradients dominate the momentum equation), the matter Euler equation implies $v_m \sim k\eta\Psi$. The matter continuity equation then implies $\Delta_m \sim -k\eta v_m \sim -(k\eta)^2\Psi$. The Poisson equation then yields $\Phi \sim \Delta_m/(k\eta)^2 \sim -\Psi$. Here we have used the Friedmann equation $H^2 = 8\pi G\rho_m/3$ and $\eta = \int d\ln a/(aH) \sim 1/(aH)$. In other words, the density perturbation $\Delta_m$ grows at exactly the right rate to keep the gravitational potential constant. More formally, if stress perturbations are negligible compared with density perturbations ($\delta p \ll \delta\rho$) then the potential will remain constant in periods where the background equation of state $p/\rho$ is constant. With a varying equation of state, it is the comoving curvature $\zeta$ rather than the Newtonian curvature that is strictly constant (Bardeen, 1980).

*Effective temperature.* If the gravitational potential is constant, we can rewrite the oscillator equation (3.82) as

$$\ddot{\Theta} + \ddot{\Psi} + c_s^2 k^2 (\Theta + \Psi) = 0. \tag{3.85}$$

The solution in Eq. (3.64) for adiabatic initial conditions then generalizes to

$$[\Theta + \Psi](\eta) = [\Theta + \Psi](0)\,\cos(ks). \tag{3.86}$$

Like a mass on spring in a constant gravitational field of the Earth, the solution just represents oscillations around a displaced minimum.

Furthermore, $\Theta + \Psi$ is also the observed temperature fluctuation. Photons lose energy climbing out of gravitational potentials at recombination and so the observer at the present will see

$$\frac{\Delta T}{T} = \Theta + \Psi. \tag{3.87}$$

Therefore, from the perspective of the observer, the acoustic oscillations ignoring both gravitational forces and gravitational redshifts are unchanged: initial perturbations in the effective temperature oscillate around zero with a frequency given by the sound speed. What the consideration of gravity adds is a way of connecting the initial conditions to inflationary curvature fluctuations.

*The Sachs–Wolfe 1/3.* The effective temperature perturbation includes both the local temperature $\Theta$ in Newtonian coordinates and the gravitational redshift factor $\Psi$. For $ks_* \ll 1$, the oscillator is frozen at its initial conditions and the total is called the Sachs–Wolfe effect (Sachs and Wolfe, 1967). The division into two pieces is actually an artifact of the coordinate system. Both pieces are determined by the initial curvature perturbation from inflation.

To see this, let us relate both the Newtonian potential and local temperature fluctuation to a change in time coordinate from comoving coordinates to Newtonian coordinates (formally through a gauge transformation, see White and Hu, 1997 for a more detailed treatment). The Newtonian gravitational potential $\Psi$ is a perturbation to the temporal coordinate [see Eq. (3.77)]

$$\frac{\delta t}{t} = \Psi. \tag{3.88}$$

Given the Friedmann equation, this is equivalent to a perturbation in the scale factor

$$t = \int \frac{da}{aH} \propto \int \frac{da}{a\rho^{1/2}} \propto a^{3(1+w)/2}, \tag{3.89}$$

where $w \equiv p/\rho$. During matter domination $w = 0$ and

$$\frac{\delta a}{a} = \frac{2}{3}\frac{\delta t}{t}. \tag{3.90}$$

Since the CMB temperature is cooling as $T \propto a^{-1}$ a local change in the scale factor changes the local temperature

$$\Theta = -\frac{\delta a}{a} = -\frac{2}{3}\Psi. \tag{3.91}$$

Combining this with $\Psi$ to form the effective temperature gives

$$\Theta + \Psi = \frac{1}{3}\Psi. \tag{3.92}$$

The consequence is that overdense regions where $\Psi$ is negative (potential wells) are cold spots in the effective temperature.

Inflation provides a source for initial curvature fluctuations. Specifically the comoving curvature perturbation $\zeta$ becomes a Newtonian curvature of

$$\Phi = \frac{3}{5}\zeta \qquad (3.93)$$

in the matter-dominated epoch. The initial amplitude of scalar curvature perturbations is usually given as $A_S = \delta_\zeta^2$ which characterizes the variance contribution per e-fold to the curvature near some fiducial wavenumber $k_n$ (see Eq. 3.34)

$$\Delta_\zeta^2 = \frac{k^3 P_\zeta}{2\pi^2} = \delta_\zeta^2 \left(\frac{k}{k_n}\right)^{n-1}, \qquad (3.94)$$

where $n = 1$ for a scale-invariant spectrum. Combining these relations for $n = 1$

$$\frac{\ell(\ell+1)C_\ell}{2\pi} \approx \frac{\delta_\zeta^2}{25}. \qquad (3.95)$$

in the Sachs–Wolfe limit. The $10^{-5}$ fluctuations measured by COBE then correspond to $\delta_\zeta \approx 5 \times 10^{-5}$.

### 3.3.4 Baryonic effects

The next level of detail that we need to add is the inertial effect of baryons in the plasma. Since Eq. (3.47) says that the baryon momentum becomes comparable to the photon momentum near recombination, we can expect order unity corrections on the basic acoustic oscillation picture from baryons. With the precise measurements of the first and second peak, this change in the acoustic morphology has already provided the most sensitive measure of the baryon–photon ratio to date exceeding that of Big Bang nucleosynthesis.

*Baryon loading.* Baryons add extra mass to the photon–baryon plasma or equivalently an enhancement of the momentum density of the plasma given by $R = (p_b + \rho_b)/(p_\gamma + \rho_\gamma)$. Specifically the momentum density of the joint system

$$(\rho_\gamma + p_\gamma)v_\gamma + (\rho_b + p_b)v_b \equiv (1 + R)(\rho_\gamma + p_\gamma)v_{\gamma b} \qquad (3.96)$$

is conserved. For generality, we have introduced the baryon velocity $v_b$ and the momentum-weighted velocity $v_{\gamma b}$ but in the tightly coupled plasma $v_b \approx v_{\gamma b} \approx v_\gamma$ (*cf,* Section 3.3.6).

The Euler equation (3.81) becomes

$$[(1 + R)(\rho_\gamma + p_\gamma)\mathbf{v}_{\gamma b}]^{\cdot} = -4\frac{\dot{a}}{a}(1 + R)(\rho_\gamma + p_\gamma)\mathbf{v}_{\gamma b}$$
$$- \nabla p_\gamma - (1 + R)(\rho_\gamma + p_\gamma)\nabla\Psi. \qquad (3.97)$$

This equation takes the same form as the photon-dominated system except for the $(1+R)$ terms which multiply everything but the pressure gradient terms since the pressure comes predominantly from the photons. We can rewrite the equation more compactly as

$$[(1 + R)v_{\gamma b}]^{\cdot} = k\Theta + (1 + R)k\Psi. \qquad (3.98)$$

*Oscillator with gravity and baryons.* The photon continuity equation (3.80) remains the same so that the oscillator equation becomes

$$[(1 + R)\dot{\Theta}]^{\cdot} + \frac{1}{3}k^2\Theta = -\frac{1}{3}k^2(1 + R)\Psi - [(1 + R)\dot{\Phi}]^{\cdot}. \qquad (3.99)$$

This equation is the final oscillator equation for the tight-coupling or perfect-fluid regime.

We can make several simplifications to illuminate the impact of baryons. First let us continue to use the matter-dominated approximation where $\Psi = -\Phi = \mathrm{const}$. Next let us make the adiabatic approximation where the change in $R$ is slow compared with the frequency of oscillation $\dot{R}/R \ll \omega = kc_s$. In that case, Eq. (3.99) looks like an oscillator equation with a fractional change in the mass given by $R$ and a change in the sound speed

$$c_s^2 = \frac{\dot{p}_\gamma + \dot{p}_b}{\dot{\rho}_\gamma + \dot{\rho}_b} = \frac{1}{3(1+R)}. \tag{3.100}$$

Consequently, the solution in Eq. (3.86) is modified as

$$[\Theta + (1+R)\Psi](\eta) = [\Theta + (1+R)\Psi](0)\cos(ks). \tag{3.101}$$

This solution is reminiscent of that of adding mass to the spring in a constant gravitational field of the Earth.

There are three effects of baryon loading. First the amplitude of oscillations increases by a factor of $1 + 3R$

$$[\Theta + (1+R)\Psi](0) = \frac{1}{3}(1+3R)\Psi(0). \tag{3.102}$$

Next the equilibrium point of the oscillation is now shifted so that relative to zero effective temperature, the even and odd peaks have different amplitudes

$$[\Theta + \Psi]_n = [\pm(1+3R) - 3R]\frac{1}{3}\Psi(0),$$

$$[\Theta + \Psi]_1 - [\Theta + \Psi]_2 = [-6R]\frac{1}{3}\Psi(0). \tag{3.103}$$

In particular, baryon loading increases the heights of the odd peaks over the even peaks (see Fig. 3.11).

Finally, the lowering of the sound speed changes the acoustic scale in Eq. (3.69) as

$$\ell_A \propto \sqrt{1+R}. \tag{3.104}$$

The effects of baryon loading in a full calculation are actually smaller since $R$ is growing in time.

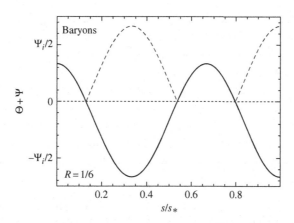

FIGURE 3.11. Acoustic oscillations with baryons. Baryons add inertia to the photon-baryon plasma displacing the zero point of the oscillation and making compressional peaks (minima) larger than rarefaction peaks (maxima). The absolute value of the fluctuation in effective temperature is shown in dotted lines. Adapted from Hu and Dodelson, 2002.

*Baryon–Photon momentum ratio evolution.* One can get a handle on the effect of evolution of $R$ by again equating the system to the analogous physical oscillator. The baryons add inertia or mass to the system and for a slowly varying mass the oscillator equation has an adiabatic invariant

$$\frac{E}{\omega} = \frac{1}{2} m_{\text{eff}} \omega A^2 = \frac{1}{2}(1+R)kc_s A^2 \propto A^2 (1+R)^{1/2} = \text{const.} \qquad (3.105)$$

Amplitude of oscillation $A \propto (1+R)^{-1/4}$ decays adiabatically as the photon–baryon ratio changes. This offsets the gain in the overall amplitude from Eq. (3.102). Coupled with uncertainties in the distance to recombination in interpreting the $\ell_A$ measurement, this leaves the modulation of the peak heights as the effect that provides most of the information about the baryon–photon ratio in the acoustic peaks.

### 3.3.5 Matter–Radiation ratio

Next we want to go beyond the matter-dominated expansion approximation. The Universe only becomes matter dominated in the few e-folds before recombination [see Eq. (3.48)]. Peaks corresponding to wavenumbers that began oscillating earlier carry the effects of the prior epoch of radiation domination. These effects come in through the evolution of the gravitational potential which acts as a forcing function on the oscillator through Eq. (3.99).

*Potential Decay.* The argument given in Section 3.3.3 for the constancy of the gravitational potential depends crucially on gravity being the dominant force affecting the total density. When radiation dominates the total density, radiation stresses become more important than gravity on scales smaller than the sound horizon. The total density fluctuation stops growing and instead oscillates with the acoustic frequency. The Poisson equation in the radiation-dominated epoch

$$k^2 \Phi = 4\pi G a^2 \rho_r \Delta_r \qquad (3.106)$$

then implies that $\Phi$ oscillates and decays with an amplitude $\propto a^{-2}$ (see Fig. 3.12). As an aside, the relativistic stresses of dark energy make the gravitational potential decay again during the acceleration epoch and lead to the so called integrated Sachs–Wolfe effect.

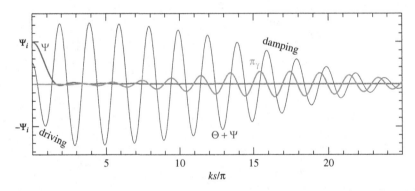

FIGURE 3.12. Acoustic oscillations with gravitational forcing and dissipational damping. For a mode that enters the sound horizon during radiation domination, the gravitational potential decays at horizon crossing and drives the acoustic amplitude higher. As the photon diffusion length increases and becomes comparable to the wavelength, radiative viscosity $\pi_\gamma$ is generated from quadrupole anisotropy leading to dissipation and polarization. Adapted from Hu and Dodelson, 2002.

*Radiation driving.* An examination of Fig. 3.12 shows that the time evolution of the gravitational potential is in phase with the acoustic oscillations themselves and act as a driving force on the acoustic oscillations. We can estimate the effect on the amplitude of oscillations in the limit that the force is fully coherent. In that case we can take the continuity equation (3.80) and simply integrate it

$$[\Theta + \Psi](\eta) = [\Theta + \Psi](0) + \Delta\Psi - \Delta\Phi$$
$$= \frac{1}{3}\Psi(0) - 2\Psi(0) = \frac{5}{3}\Psi(0). \tag{3.107}$$

This estimate gives an acoustic amplitude that is five times that of the Sachs–Wolfe effect. This enhancement only occurs for modes that begin oscillating during the radiation-dominated epoch, i.e. the higher peaks. The net effect is a gradual ramp up of the acoustic oscillation amplitude across the horizon wavenumber at matter radiation equality (see Fig. 3.10).

In fact the coherent approximation is exact for a photon–baryon fluid but must be corrected for the neutrino contribution to the radiation density.

*External potential approach.* For pedagogical purposes it is sometimes useful to go beyond the coherent approximation. Neutrino corrections are one example; isocurvature initial conditions are another.

We have seen that the solutions to the homogeneous equation for the oscillator equation (3.82) are

$$(1 + R)^{-1/4}\cos(ks), \qquad (1 + R)^{-1/4}\sin(ks) \tag{3.108}$$

in the adiabatic or high-frequency limit. Considering the potentials as external, we can solve for the temperature perturbation as (Hu and Sugiyama, 1995)

$$(1 + R)^{1/4}\Theta(\eta) = \Theta(0)\cos(ks) + \frac{\sqrt{3}}{k}\left[\dot{\Theta}(0) + \frac{1}{4}\dot{R}(0)\Theta(0)\right]\sin(ks)$$
$$+ \frac{\sqrt{3}}{k}\int_0^\eta d\eta'(1 + R')^{3/4}\sin[ks - ks']F(\eta'), \tag{3.109}$$

where

$$F = -\ddot{\Phi} - \frac{\dot{R}}{1 + R}\dot{\Phi} - \frac{k^2}{3}\Psi. \tag{3.110}$$

By including the neutrino effects in the gravitational potential, we can show from this approach that radiation driving actually creates an acoustic amplitude that is close to four times the Sachs–Wolfe effect.

### 3.3.6 Damping

The final piece in the acoustic oscillation puzzle is the damping of power beyond $\ell \sim 10^3$ shown in Fig. 3.6. Up until this point, we have considered the oscillations in the tight coupling approximation where the photons and baryons respond to pressure and gravity as a single perfect fluid. Fluid imperfections are associated with the Compton mean free path in Eq. (3.46). Dissipation becomes strong at the diffusion scale, the distance a photon can random walk in a given time $\eta$ (Silk, 1968),

$$\lambda_D = \sqrt{N}\lambda_C = \sqrt{\eta/\lambda_C}\,\lambda_C = \sqrt{\eta\lambda_C}. \tag{3.111}$$

This scale is the geometric mean between the horizon and mean free path. Given that $\lambda_D/\eta_* \sim$ few percent, we expect that the $n \geq 3$ peaks to be affected by dissipation. To improve on this estimate, we develop the microphysical description of dissipation next.

*Continuity equations.* To treat the photons and baryons as separate systems, we now need to supplement the photon continuity equation with the baryon continuity equation

$$\dot{\Theta} = -\frac{k}{3}v_\gamma - \dot{\Phi}, \quad \dot{\delta}_{\mathrm{b}} = -kv_{\mathrm{b}} - 3\dot{\Phi}. \tag{3.112}$$

The baryon equation follows from number conservation with $\rho_{\mathrm{b}} = m_{\mathrm{b}}n_{\mathrm{b}}$ and $\delta_{\mathrm{b}} \equiv \delta\rho_{\mathrm{b}}/\rho_{\mathrm{b}}$.

*Euler and Navier–Stokes equations.* The momentum conservation equations must also be separated into photon and baryon pieces

$$\dot{v}_\gamma = k(\Theta + \Psi) - \frac{k}{6}\pi_\gamma - \dot{\tau}(v_\gamma - v_{\mathrm{b}}),$$

$$\dot{v}_{\mathrm{b}} = -\frac{\dot{a}}{a}v_{\mathrm{b}} + k\Psi + \dot{\tau}(v_\gamma - v_{\mathrm{b}})/R,$$

where the photons gain an anisotropic stress term $\pi_\gamma$ from radiation viscosity. The baryon equation follows from the same derivation as in Section 3.3.2 where the redshift of the momentum is carried by the bulk velocity instead of the redshifting temperature. Finally there is a momentum exchange term from Compton scattering. Note that the total momentum in the system is conserved and hence the scattering terms come with opposite sign.

*Viscosity.* Radiative shear viscosity is equivalent to quadrupole moments in the temperature field. These quadrupole moments are generated by radiation streaming from hot to cold regions much like how temperature inhomogeneity is converted to anisotropy in Fig. 3.8.

In the tight coupling limit where $\dot{\tau}/k$, the optical depth through a wavelength of the fluctuation is high; one therefore expects

$$\pi_\gamma \sim v_\gamma \frac{k}{\dot{\tau}}, \tag{3.113}$$

since it must be generated by streaming and suppressed by scattering. A more detailed calculation from the Boltzmann or radiative transfer equation says (Kaiser, 1983)

$$\pi_\gamma \approx 2A_v v_\gamma \frac{k}{\dot{\tau}}, \tag{3.114}$$

where $A_v = 16/15$ once polarization effects are incorporated

$$\dot{v}_\gamma = k(\Theta + \Psi) - \frac{k}{3}A_v \frac{k}{\dot{\tau}}v_\gamma. \tag{3.115}$$

The oscillator equation with viscosity becomes

$$c_s^2 \frac{\mathrm{d}}{\mathrm{d}\eta}(c_s^{-2}\dot{\Theta}) + \frac{k^2 c_s^2}{\dot{\tau}}A_v\dot{\Theta} + k^2 c_s^2\Theta = -\frac{k^2}{3}\Psi - c_s^2\frac{\mathrm{d}}{\mathrm{d}\eta}(c_s^{-2}\dot{\Phi}),$$

As in a mechanical oscillator, a term that depends on $\dot{\Theta}$ provides a dissipational term to the solutions.

*Heat conduction.* Relative motion between the photons and baryons also damps oscillations. By expanding the continuity and momentum conservation equations in the small number $k/\dot{\tau}$ one obtains for the full oscillator equation

$$c_s^2 \frac{\mathrm{d}}{\mathrm{d}\eta}(c_s^{-2}\dot{\Theta}) + \frac{k^2 c_s^2}{\dot{\tau}}[A_v + A_h]\dot{\Theta} + k^2 c_s^2 \Theta = -\frac{k^2}{3}\Psi - c_s^2 \frac{\mathrm{d}}{\mathrm{d}\eta}(c_s^{-2}\dot{\Phi})$$

where

$$A_h = \frac{R^2}{1+R}. \tag{3.116}$$

*Dispersion relation.* We can solve the damped oscillator equation in the adiabatic approximation by taking a trial solution $\Theta \propto \exp(\mathrm{i}\int \omega \mathrm{d}\eta)$ to obtain the dispersion relation

$$\omega = \pm k c_s \left[1 \pm \frac{\mathrm{i}}{2}\frac{k c_s}{\dot{\tau}}(A_v + A_h)\right]. \tag{3.117}$$

The imaginary term in the dispersion relation gives an exponential damping of the oscillation amplitude

$$\exp(\mathrm{i}\int \omega \mathrm{d}\eta) = \mathrm{e}^{\pm \mathrm{i}ks}\exp\left[-k^2 \int \mathrm{d}\eta \frac{1}{2}\frac{c_s^2}{\dot{\tau}}(A_v + A_h)\right]$$

$$= \mathrm{e}^{\pm \mathrm{i}ks}\exp\left[-(k/k_\mathrm{D})^2\right], \tag{3.118}$$

where the diffusion wavenumber is given by

$$k_\mathrm{D}^{-2} = \int \mathrm{d}\eta \frac{1}{\dot{\tau}}\frac{1}{6(1+R)}\left(\frac{16}{15} + \frac{R^2}{(1+R)}\right). \tag{3.119}$$

Note that in both the high and low $R$ limits

$$\lim_{R\to 0} k_\mathrm{D}^{-2} = \frac{1}{6}\frac{16}{15}\int \mathrm{d}\eta \frac{1}{\dot{\tau}},$$

$$\lim_{R\to\infty} k_\mathrm{D}^{-2} = \frac{1}{6}\int \mathrm{d}\eta \frac{1}{\dot{\tau}}. \tag{3.120}$$

Hence the dissipation scale is

$$\lambda_\mathrm{D} = \frac{2\pi}{k_\mathrm{D}} \sim \frac{2\pi}{\sqrt{6}}(\eta\dot{\tau}^{-1})^{\frac{1}{2}} \tag{3.121}$$

and comparable to the geometric mean between horizon and mean free path as expected from the random walk argument. For a baryon density of $\Omega_\mathrm{b}h^2 \approx 0.02$, radiation viscosity is responsible for most of the dissipation and we show the correspondence between viscosity generation and dissipation in Fig. 3.12.

Since the diffusion length changes rapidly through recombination and the medium changes from optically thick to optically thin, the damping estimates above are only qualitative. A full Boltzmann (radiative transfer solution) shows a more gradual, but still exponential, damping of roughly

$$\mathcal{D}_\ell \approx \exp[-(\ell/\ell_\mathrm{D})^{1.25}], \tag{3.122}$$

with a damping scale of $\ell_\mathrm{D} = 2\pi D_*/\lambda_\mathrm{D}$ and (Hu, 2005)

$$\frac{\lambda_\mathrm{D}}{\mathrm{Mpc}} \approx 64.5 \left(\frac{\Omega_\mathrm{m}h^2}{0.14}\right)^{-0.278}\left(\frac{\Omega_\mathrm{b}h^2}{0.024}\right)^{-0.18}, \tag{3.123}$$

for small changes around the central values of $\Omega_\mathrm{m}h^2$ and $\Omega_\mathrm{b}h^2$.

This envelope also accounts for enhanced damping due to the finite duration of recombination. Instead of a delta function in the projection equation (3.31) we have the visibility function $\dot{\tau}\mathrm{e}^{-\tau}$ that acts as a smearing out of any contributions that have wavelengths shorter than the thickness of the recombination surface and survive dissipation.

### 3.3.7 Information from the peaks

In the preceding sections we have examined the physical processes involved in the formation of the acoustic peaks and explained their sensitivity to the energy content and expansion rate of the Universe. Converting the measurements into parameter constraints of course requires a more accurate numerical description. Numerical codes that solve the Einstein–Boltzmann radiative transfer equations for the CMB and matter (Peebles and Yu, 1970; Bond and Efstathiou, 1984; Vittorio and Silk, 1984) are now accurate at the $\sim 1\%$ level for the acoustic peaks for publically available codes (Seljak and Zaldarriaga, 1996; Lewis *et al.*, 2000). Their numerical precision on the other hand is substantially better and approaches the 0.1% level required for cosmic variance limited measurements out to $\ell \sim 10^3$. The accuracy is now limited by the input physics, mainly recombination (see Section 3.2.3). In this section, we relate the qualitative discussion of the previous sections to the quantitative information content of the peaks.

*First peak: curvature and dark energy.* The comparison between the predicted acoustic peak scale $\lambda_A$ and its angular extent provides a measurement of the angular diameter distance to recombination. The angular diameter distance in turn depends on the spatial curvature and expansion history of the Universe.

Sensitivity to the expansion history during the acceleration epoch comes about through the radial distance a photon travels along the line of sight

$$D_* = \int_0^{z_*} \frac{dz}{H(z)} = \eta_0 - \eta_*. \tag{3.124}$$

With the matter and radiation energy densities measured, the remaining contributor to the expansion rate $H(z)$ is the dark energy.

However in a curved Universe, the apparent or angular diameter distance $D_A$ in Eq. (3.69) is no longer the distance a photon travels radially along the line of sight. The radius of curvature of the space is given in terms of the total density $\Omega_{\rm tot}$ in units of the critical density as

$$R^{-2} = H_0^2(\Omega_{\rm tot} - 1). \tag{3.125}$$

A positive curvature space has $\Omega_{\rm tot} > 1$ and a real radius of curvature. A negatively curved space has $\Omega_{\rm tot} < 1$ and an imaginary radius of curvature. The positively curved space is shown in Fig. 3.13. The curvature makes a transverse distance $L$ related to its angular extent $d\alpha$ as $L = d\alpha D_A$ with

$$D_A = R \sin(D/R). \tag{3.126}$$

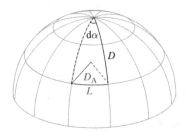

FIGURE 3.13. Angular diameter distance and curvature. In a non-flat (here closed) Universe, the apparent or angular diameter distance $D_A = L d\alpha$ does not equal the radial distance traveled by the photon. Objects in a closed Universe are further than they appear whereas in an open Universe they are closer than they appear.

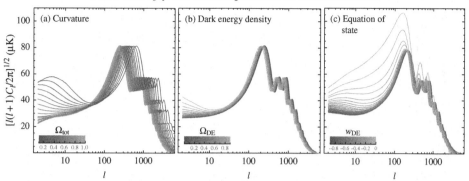

FIGURE 3.14. Curvature and dark energy. Given a fixed physical scale for the acoustic peaks (fixed $\Omega_b h^2$ and $\Omega_m h^2$) the observed angular position of the peaks provides a measure of the angular diameter distance and the parameters it depends on: curvature, dark energy density and dark energy equation of state. Changes at low $\ell$ multipoles are due to the decay of the gravitational potential after matter domination from the integrated Sachs–Wolfe effect.

The same formula applies for negatively curved spaces but is more conveniently expressed with the relation

$$R\sin(D/R) = |R|\sinh(D/|R|) \qquad (3.127)$$

for imaginary $R$. In a positively curved geometry $D_A < D$ and objects are further than they appear. In a negatively curved Universe $R$ is imaginary and $R\sin(D/R) = i|R|\sin(D/i|R|) = |R|\sinh(D/|R|)$, and $D_A > D$ objects are closer than they appear. Since the detection of the first acoustic peak it has been clear that the Universe is close to being spatially flat (Miller *et al.*, 1999; de Bernardis *et al.*, 2000; Hanany *et al.*, 2000). How close and how well-measured $D_*$ is for dark energy studies depends on the calibration of the physical scale $\lambda_A = 2s_*$, i.e. the sound horizon at recombination.

The sound horizon,

$$s_* = \frac{2\sqrt{3}}{3}\sqrt{\frac{a_*}{R_*\Omega_m H_0^2}}\ln\frac{\sqrt{1+R_*}+\sqrt{R_*+r_*R_*}}{1+\sqrt{r_*R_*}}, \qquad (3.128)$$

in turn depends on two things; the baryon–photon momentum density ratio,

$$R_* \equiv \frac{3}{4}\frac{\rho_b}{\rho_\gamma}\bigg|_{a_*} = 0.729\left(\frac{\Omega_b h^2}{0.024}\right)\left(\frac{a_*}{10^{-3}}\right), \qquad (3.129)$$

and the expansion rate prior to recombination, which is determined by the matter–radiation ratio

$$r_* \equiv \frac{\rho_r}{\rho_m}\bigg|_{a_*} = 0.297\left(\frac{\Omega_m h^2}{0.14}\right)^{-1}\left(\frac{a_*}{10^{-3}}\right)^{-1}. \qquad (3.130)$$

The calibration of these two quantities involves the higher acoustic peaks. The bottom line is that the limiting factor in the calibration is the precision with which the matter density is known

$$\frac{\delta D_{A*}}{D_{A*}} \approx \frac{1}{4}\frac{\delta(\Omega_m h^2)}{\Omega_m h^2}. \qquad (3.131)$$

In principle the Planck satellite can achieve a 1% measurement of the matter density or a $\sim 0.25\%$ measure of distance. Current errors on the distance are $\sim 1$–2%.

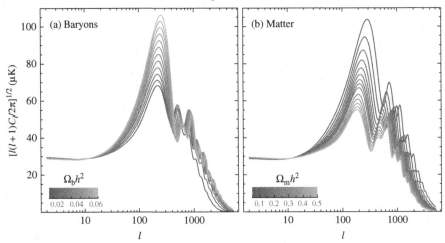

FIGURE 3.15. Baryons and matter. Baryons change the relative heights of the even and odd peaks through their inertia in the plasma. The matter–radiation ratio also changes the overall amplitude of the oscillations from driving effects. Adapted from Hu and Dodelson, 2002.

*Second peak: baryons.* The baryon–photon ratio controls the even–odd modulation of peak heights through the baryon loading effect (see Section 3.3.4). The second peak represents rarefaction of the acoustic wave in a gravitational potential and hence is suppressed in amplitude by the baryon inertia. The dependence of the spectrum on the baryon density $\Omega_b h^2$ is shown in Fig. 3.15. Since the first tentative detection of the second peak (de Bernardis *et al.*, 2000) the CMB has placed limits on the baryon density. Currently its measurement, largely from WMAP (Spergel *et al.*, 2007), gives $\Omega_b h^2 = 0.0227 \pm 0.0006$ (Reichardt *et al.*, 2009). This is well enough constrained that associated errors on the sound horizon and shape of the matter power spectrum (see below) are small.

*Third peak: dark matter.* The third peak begins to show the effects of the matter–radiation ratio on the overall amplitude of the acoustic peaks. Furthermore, decay in the gravitational potential during radiation domination would reduce the baryon loading effect and change the peak height ratios of the second and third peaks (*e.g.* Hu *et al.*, 2001). The dependence of the spectrum on the baryon density $\Omega_m h^2$ is shown in Fig. 3.15. Constraints on the third peak from the DASI experiment (Pryke *et al.*, 2002) represented the first direct evidence for dark matter at the epoch of recombination. Current constraints from a combination of WMAP and higher-resolution ground- and balloon-based data yield $\Omega_m h^2 = 0.135 \pm 0.007$ (Reichardt *et al.*, 2009). Since this parameter controls the error on the distance to recombination through Eq. (3.131) and the matter power spectrum (see below), it is important to improve the precision of its measurement with the third and higher peaks.

*Damping tail: consistency.* Under the standard thermal history of Section 3.2 and matter content, the parameters that control the first three peaks also determine the structure of the damping tail at $\ell > 10^3$: namely, the angular diameter distance to recombination $D_*$, the baryon density $\Omega_b h^2$ and the matter density $\Omega_m h^2$. When the damping tail was first discovered by the CBI experiment (Padin *et al.*, 2001), it supplied compelling support for the standard theoretical modeling of the physics at recombination outlined here. Currently the best constraints on the damping tail are from the ACBAR experiment (Reichardt *et al.*, 2009, see Fig. 3.7). Consistency between the low-order peaks and the damping tail can be used to make precision tests of recombination and any physics beyond the standard model at that epoch. For example, damping tail measurements can be used to constrain the evolution of the fine structure constant.

*Matter power spectrum: shape and amplitude.* The acoustic peaks also determine the shape and amplitude of the matter power spectrum. First, acoustic oscillations are shared by the baryons. In particular, the plasma motion kinematically produces enhancements of density near recombination (see Euler and Navier–Stokes equations in Section 3.3.6)

$$\delta_b \approx -k\eta_* v_b(\eta_*) \approx -k\eta_* v_\gamma(\eta_*). \tag{3.132}$$

This enhancement then imprints into the matter power spectrum at an amplitude reduced by $\rho_b/\rho_m$ due to the small baryon fraction (Hu and Sugiyama, 1996). Second, the gravitational potentials that the cold dark matter perturbations fall in are evolving through the plasma epoch due to the processes described in Section 3.3.5. On scales above the horizon, relativistic stresses are never important and the gravitational potential remains constant. The Poisson equation then implies that

$$\Delta \sim (k\eta)^2 \Phi(0), \tag{3.133}$$

and in particular, the density perturbation at horizon crossing where $k\eta \sim 1$ is $\Delta = \Delta_H \approx \Phi(0)$. For a fluctuation that crosses the horizon during radiation domination, the total density perturbation is Jeans stabilized until matter-radiation equality

$$\eta_{eq} \approx 114 \left( \frac{\Omega_m h^2}{0.14} \right)^{-1} \text{Mpc}. \tag{3.134}$$

Thereafter, relativistic stresses again become irrelevant and the potential remains unchanged until matter ceases to dominate the expansion

$$\Phi \approx (k\eta_{eq})^{-2}\Delta_H \sim (k\eta_{eq})^{-2}\Phi(0). \tag{3.135}$$

The transfer in shape from the initial conditions due to baryon oscillations and matter–radiation equality is usually encapsulated into a transfer function $T(k)$ – see Fig. 3.16. Given an initial power spectrum of the form (3.94), the evolution through to matter domination transforms the potential power spectrum to $k^3 P_\Phi/2\pi^2 \propto k^{n-1}T^2(k)$ where

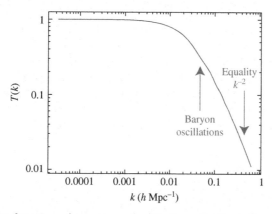

FIGURE 3.16. Transfer function. Acoustic and radiation physics is imprinted on the matter power spectrum as quantified by the transfer function $T(k)$. The former is responsible for baryon oscillations in the spectrum and the latter a suppression of growth due to Jeans stability for scales smaller than the horizon at matter-radiation equality.

$T(k) \propto k^{-2}$ beyond the wavenumber at matter–radiation equality. This scaling is slightly modified due to the logarithmic growth of dark matter fluctuations during the radiation epoch when the radiation density is Jeans stable. The matter power spectrum and potential power spectrum are related by the Poisson equation and so carry the same shape. Specifically

$$\frac{k^3 P_m(k, a)}{2\pi^2} = \frac{4}{25}\delta_\zeta^2 \left(\frac{G(a)a}{\Omega_m}\right)^2 \left(\frac{k}{H_0}\right)^4 \left(\frac{k}{k_{norm}}\right)^{n-1} T^2(k), \qquad (3.136)$$

where we have included a factor $G(a)$ to account for the decay in the potential during the acceleration epoch when relativistic stresses are again important (see e.g. Hu, 2005). This factor only depends on time and not scale as long as the scale in question is within the Jeans scale of the accelerating component. In this limit, $G(a)$ is determined by the solution to

$$\frac{d^2 G}{d\ln a^2} + \left(4 + \frac{d\ln H}{d\ln a}\right)\frac{dG}{d\ln a} + \left[3 + \frac{d\ln H}{d\ln a} - \frac{3}{2}\Omega_m(a)\right]G = 0, \qquad (3.137)$$

with an initial conditions of $G(\ln a_{md}) = 1$ and $G'(\ln a_{md}) = 0$ at an epoch $a_{md}$ when the Universe is fully matter dominated.

The transfer function $T(k)$, with $k$ in Mpc$^{-1}$, depends only on the baryon density $\Omega_b h^2$ and the matter density $\Omega_m h^2$ which are well determined by the CMB acoustic peaks. Features in the matter power spectrum, especially the baryon oscillations, then serve as standard rulers for distance measurements (Eisenstein *et al.*, 1998). For example, its measurement in a local redshift survey where the distance is calibrated in $h$ Mpc$^{-1}$ would give the Hubble constant $h$. Detection of the features requires a Gpc$^3$ of volume and so precise, purely local, measurements are not feasible. Nonetheless, the first detection of these oscillations by the SDSS LRG redshift survey out to $z \sim 0.4$ provide remarkably tight constraints on the acceleration of the expansion (Eisenstein *et al.*, 2005).

The CMB also determines the initial normalization $\delta_\zeta$ and so provides a means by which to test the effect of the acceleration on the growth function $G(a)$. The precision of this determination is largely set by reionization. The opacity provided by electrons after reionization suppress the observed amplitude of the peaks relative to the initial amplitude and hence

$$\delta_\zeta \approx 4.6 e^{-(0.1-\tau)} \times 10^{-5}, \qquad (3.138)$$

where $\tau$ is the optical depth to recombination. Note that this is the normalization at $k = 0.05$ Mpc$^{-1}$ and even with uncertainties in the optical depth of $\delta\tau \sim 0.03$ it exceeds the precision of the COBE normalization [cf. Eq. (3.95)]. Finally, with the matter and baryon transfer effects determined, the acoustic spectrum also constrains the tilt. The WMAP team provided the first hints of a small deviation from scale-invariance (Spergel *et al.*, 2007) and the current constraints are $n \approx 0.965 \pm 0.015$.

Combining these factors into the conventional measure of the amplitude of matter fluctuations today

$$\sigma_8^2 \equiv \int \frac{dk}{k}\frac{k^3 P(k, a = 1)}{2\pi^2} W_\sigma^2(kr)$$

$$\sigma_8 \approx \frac{\delta_\zeta}{5.59 \times 10^{-5}} \left(\frac{\Omega_b h^2}{0.024}\right)^{-1/3} \left(\frac{\Omega_m h^2}{0.14}\right)^{0.563}$$

$$\times (3.123h)^{(n-1)/2} \left(\frac{h}{0.72}\right)^{0.693} \frac{G_0}{0.76}, \qquad (3.139)$$

where $W_\sigma(x) = 3x^{-3}(\sin x - x\cos x)$ is the Fourier transform of a top-hat window of radius $r = 8h^{-1}$ Mpc.

## 3.4 Polarization anisotropy from recombination

Thomson scattering of quadrupolarly anisotropic but unpolarized radiation generates linear polarization. As we have seen in Section 3.3.1, $\ell \geq 2$ anisotropy develops only in optically thin conditions. Given that polarization also requires scattering to be generated, the polarization anisotropy is generically much smaller than the temperature anisotropy. The main source of polarization from recombination is associated with the acoustic peaks in temperature. This source was first detected by the DASI experiment (Kovac *et al.*, 2002) and recent years have seen increasingly precise measurements (see Fig. 3.17).

Since acoustic polarization arises from linear scalar perturbations, they possess a symmetry that relates the direction of polarization to the wavevector or change in the polarization amplitude. As we shall see in the next section, this symmetry is manifest in the the absence of $B$-modes (Kamionkowski *et al.*, 1997; Zaldarriaga and Seljak, 1997). $B$-modes at recombination can be generated from the quadrupole moment of a gravitational wave. This yet-to-be-detected signal would be invaluable for early Universe studies involving the inflationary origin of perturbations.

We begin by reviewing the Stokes parameter description of polarization (Section 3.4.1) and its relation to $E$ and $B$ harmonic representation (Section 3.4.2). We continue with a discussion of polarized Thomson scattering in Section 3.4.3. In Sections 3.4.4 and 3.4.5 we discuss the polarization signatures of acoustic oscillations and gravitational waves.

### *3.4.1 Statistical description*

The polarization field can be analyzed in a way very similar to the temperature field, save for one complication. In addition to its strength, polarization also has an orientation, depending on relative strength of two linear polarization states.

The polarization field is defined locally in terms of Stokes parameters. In general, the polarization state of radiation in direction $\hat{n}$ is described by the intensity matrix

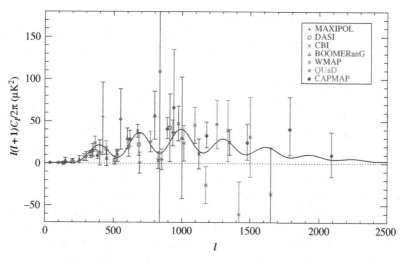

FIGURE 3.17. $E$-mode polarization power spectrum measurements. Adapted from Bischoff *et al.*, 2008. Reproduced by permission of the AAS.

$\langle E_i(\hat{\mathbf{n}})E_j^*(\hat{\mathbf{n}})\rangle$, where $\mathbf{E}$ is the electric field vector in the transverse plane and the brackets denote time averaging. As a $2 \times 2$ hermitian matrix, it can be decomposed into the Pauli basis

$$\mathbf{P} = C\,\langle \mathbf{E}(\hat{\mathbf{n}})\,\mathbf{E}^\dagger(\hat{\mathbf{n}})\rangle$$
$$= \Theta(\hat{\mathbf{n}})\boldsymbol{\sigma}_0 + Q(\hat{\mathbf{n}})\,\boldsymbol{\sigma}_3 + U(\hat{\mathbf{n}})\,\boldsymbol{\sigma}_1 + V(\hat{\mathbf{n}})\,\boldsymbol{\sigma}_2, \qquad (3.140)$$

where

$$\boldsymbol{\sigma}_0 = \begin{pmatrix} 1 & 0 \\ 0 & 1 \end{pmatrix}, \quad \boldsymbol{\sigma}_2 = \begin{pmatrix} 0 & -1 \\ 1 & 0 \end{pmatrix} i,$$
$$\boldsymbol{\sigma}_1 = \begin{pmatrix} 0 & 1 \\ 1 & 0 \end{pmatrix}, \quad \boldsymbol{\sigma}_3 = \begin{pmatrix} 1 & 0 \\ 0 & -1 \end{pmatrix}. \qquad (3.141)$$

Orthogonality of the Pauli matrices says that the Stokes parameters are recovered from the polarization matrix as $\mathrm{Tr}(\boldsymbol{\sigma}_i \mathbf{P})/2$. The Stokes $Q$ and $U$ parameters define the linear polarization state whereas $V$ defines the circular polarization state. We have chosen the proportionality constant so that all the Stokes parameters are in temperature fluctuation units.

From this description, we see that $Q$ represents polarization aligned with one of the principal axes of the transverse coordinate system whereas $U$ represents polarization at $45°$ to these axes.

### 3.4.2 EB Harmonic description

The disadvantage of the Stokes $Q$ and $U$ representation is that the distinction between the two depends on the coordinate system for the two transverse directions on the sky. Under a rotation of this basis by $\theta$

$$Q' \pm iU' = e^{\mp 2i\theta}[Q \pm iU]. \qquad (3.142)$$

For a harmonic decomposition, it is more useful to choose this basis to be given by the wavevector itself. For small sections of the sky, the harmonic decomposition becomes a Fourier transform and we can define the $E$ and $B$ harmonics as

$$E(\mathbf{l}) \pm iB(\mathbf{l}) = \int d\hat{\mathbf{n}}[Q'(\hat{\mathbf{n}}) \pm iU'(\hat{\mathbf{n}})]e^{-i\mathbf{l}\cdot\hat{\mathbf{n}}} \qquad (3.143)$$
$$= e^{\mp 2i\phi_{\mathbf{l}}} \int d\hat{\mathbf{n}}[Q(\hat{\mathbf{n}}) \pm iU(\hat{\mathbf{n}})]e^{-i\mathbf{l}\cdot\hat{\mathbf{n}}},$$

where in the second line we have rotated $Q$ and $U$ back to a fixed coordinate system with the angle $\phi_{\mathbf{l}}$ that the Fourier vector makes with the $\mathbf{x}$ axis.

  • For linear scalar fluctuations, we have seen that the only direction for a given harmonic mode is set by the direction of the wavevector itself: velocity fields point in this direction or its opposite, quadrupole moments are symmetric about this axis, etc. This means that symmetry requires that such sources only generate $Q'$ or $E$ for each mode. This symmetry also holds once all the modes are superimposed back to the full polarization field: linear scalar perturbations generate only $E$-modes where the polarization direction is related to the direction in which the polarization amplitude changes.

To generalize this decomposition to the full curved sky, we need to replace plane waves, the tensor eigenfunctions of the Laplace operator in a flat space, to the correct tensor eigenfunctions for the curved space. These are called the spin-2 spherical harmonics:

$$\nabla^2_{\pm 2}Y_{\ell m}[\boldsymbol{\sigma}_3 \mp i\boldsymbol{\sigma}_1] = -[l(l+1)-4]_{\pm 2}Y_{\ell m}[\boldsymbol{\sigma}_3 \mp i\boldsymbol{\sigma}_1]. \qquad (3.144)$$

They obey the usual orthogonality and completeness relations

$$\int \mathrm{d}\hat{\mathbf{n}}\, {}_sY^*_{\ell m}(\hat{\mathbf{n}})\, {}_sY_{\ell m}(\hat{\mathbf{n}}) = \delta_{\ell\ell'}\delta_{mm'},$$

$$\sum_{\ell m} {}_sY^*_{\ell m}(\hat{\mathbf{n}})\, {}_sY_{\ell m}(\hat{\mathbf{n}}') = \delta(\phi - \phi')\delta(\cos\theta - \cos\theta'). \tag{3.145}$$

We can therefore decompose the linear polarization field just like the temperature field

$$[Q(\hat{\mathbf{n}}) \pm iU(\hat{\mathbf{n}})] = -\sum_{\ell m}[E_{\ell m} \pm iB_{\ell m}]_{\pm 2}Y_{\ell m}(\hat{\mathbf{n}}). \tag{3.146}$$

Likewise the power spectra are given by

$$\langle E^*_{\ell m}E_{\ell m}\rangle = \delta_{\ell\ell'}\delta_{mm'}C^{EE}_\ell, \tag{3.147}$$

$$\langle B^*_{\ell m}B_{\ell m}\rangle = \delta_{\ell\ell'}\delta_{mm'}C^{BB}_\ell, \tag{3.148}$$

and the cross correlation of $E$-polarization with temperature by

$$\langle \Theta^*_{\ell m}E_{\ell m}\rangle = \delta_{\ell\ell'}\delta_{mm'}C^{\Theta E}_\ell. \tag{3.149}$$

Other cross correlations vanish if parity is conserved.

### 3.4.3 Thomson scattering

The differential cross section for Thomson scattering

$$\frac{\mathrm{d}\sigma}{\mathrm{d}\Omega} = \frac{3}{8\pi}|\hat{\mathbf{E}}' \cdot \hat{\mathbf{E}}|^2\sigma_T, \tag{3.150}$$

is polarization dependent and hence scattering generates polarization. Here $\hat{\mathbf{E}}'$ and $\hat{\mathbf{E}}$ denote the incoming and outgoing directions of the electric field or polarization vector. Consider incoming radiation in the $\hat{\mathbf{x}}$ direction scattered at right angles into the $-\hat{\mathbf{y}}$ direction (see Fig. 3.18). Heuristically, incoming radiation shakes an electron in the direction of its electric field vector or polarization $\hat{\mathbf{e}}'$ causing it to radiate with an outgoing polarization parallel to that direction. However since the outgoing polarization $\hat{\mathbf{e}}$ must be orthogonal to the outgoing direction, incoming radiation that is polarized parallel to the outgoing direction cannot scatter leaving only one polarization state.

The incoming radiation however comes from all angles. If it were completely isotropic in intensity, radiation coming along the $\hat{\mathbf{z}}$ would provide the polarization state that is missing from that coming along $\hat{\mathbf{x}}$ leaving the net outgoing radiation unpolarized. Only

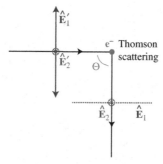

FIGURE 3.18. Thomson scattering geometry. A quadrupole anisotropy in the incoming radiation leads to linear polarization. For scattering at $\Theta = \pi/2$, only one component of the initially unpolarized radiation ($\hat{\mathbf{E}}'_2$) is scattered leaving one outgoing state ($\hat{\mathbf{E}}_1$) unpopulated.

a quadrupole temperature anisotropy in the radiation generates a net linear polarization from Thomson scattering. As we have seen, a quadrupole can only be efficiently generated if the Universe is optically thin to Thomson scattering at a given perturbation.

### 3.4.4 Acoustic polarization

Acoustic oscillations in the dissipation regime provide the conditions necessary for polarization. Recall that radiative viscosity in the plasma is equivalent to quadrupole anisotropy in the photons. Since the quadrupole is of order (see Fig. 3.12)

$$\pi_\gamma \sim \frac{k v_\gamma}{\dot\tau} \sim \left(\frac{k}{k_D}\right)\frac{v_\gamma}{k_D \eta_*}, \tag{3.151}$$

the polarization spectrum rises as $l/l_D$ to peak at the damping scale with an amplitude of about 10% of the temperature fluctuations before falling due to the elimination of the acoustic source itself due to damping. Since $v_\gamma$ is out of phase with the temperature,

$$\Theta + \Psi \propto \cos(ks); \quad v_\gamma \propto \sin(ks), \tag{3.152}$$

the polarization peaks are also out of phase with the temperature peaks. Furthermore, the phase relation also tells us that the polarization is correlated with the temperature perturbations. The correlation power $C_l^{\Theta E}$ being the product of the two, exhibits oscillations at twice the acoustic frequency

$$(\Theta + \Psi)(v_\gamma) \propto \cos(ks)\sin(ks) \propto \sin(2ks). \tag{3.153}$$

As in the case of the damping, predicting the precise value requires numerical codes (Bond and Efstathiou, 1987) since $\dot\tau$ changes so rapidly near recombination. Nonetheless the detailed predictions bear these qualitative features.

Like the damping scale, the acoustic polarization spectrum is uniquely predicted from the temperature spectrum once $\Omega_b h^2$, $\Omega_m h^2$ and the initial conditions are specified. Polarization thus represents a sharp test on the assumptions of the recombination physics and power-law curvature fluctuations in the initial conditions used in interpreting the temperature peaks. For example, features in the initial power spectrum appear more distinct in the polarization than the temperature due to projection effects (see e.g. Hu and Okamoto, 2004).

### 3.4.5 Gravitational waves

During the break down of tight coupling that occurs at last scattering, any gravitational waves present will also imprint a local quadrupole anisotropy to the photons and hence a linear polarization to the CMB (Polnarev, 1985). These contribute to the $BB$ power and their detection would provide invaluable information on the origin of the fluctuations. Specifically, in simple inflationary models their amplitude gives the energy scale of inflation. The gravitational wave amplitude $h$ oscillates and decays once inside the horizon, so the associated polarization source scales as $\dot h/\dot\tau$ and so peaks at the $l \approx 100$ horizon scale and not the damping scale at recombination (see Fig. 3.19). This provides a useful scale separation of the various polarization effects.

If the energy scale of inflation is near the $10^{16}$ GeV scale then the signal is potentially detectable by the next generation of polarization experiments.

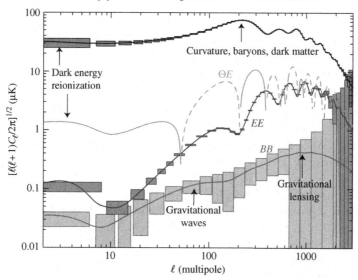

FIGURE 3.19. Polarized landscape. While the $E$-spectrum and $\Theta E$ cross correlation are increasingly well measured, the $B$-spectrum from inflationary gravitational waves (shown here near the maximal value allowed by the temperature spectrum) and gravitational lensing remains undetected. Shown here are projected error bars associated with Planck sample variance and detector noise. Adapted from Hu and Dodelson, 2002.

## 3.5 Future perspectives

The one and a half decades since the discovery of CMB anisotropy by the Cosmic Background Explorer Differential Wave Spectrometer (COBE DMR) has seen remarkable progress that ushered in the current epoch of precision cosmology. From the preliminary detections of degree scale power to the current measurements of five acoustic peaks, the damping tail and acoustic polarization, the milestones in the observational verification of our theoretical understanding of the Universe at recombination have steadily been overtaken. Correspondingly, the measurements of the energy density contents of the Universe at recombination and the distance to and hence expansion rate since recombination have improved from order unity constraints to several-percent-level determinations.

The next generation of experiments, including the Planck satellite and ground-based polarization measurements, will push these determinations to the 1% level and beyond and perhaps detect the one outstanding prediction of the recombination epoch: the $B$-mode polarization of gravitational waves from inflation. Likewise the enhanced range of precision measurements of the power spectrum will bring measurements of the spectrum of scalar perturbations to the percent level and further test the physics of inflation.

Beyond the recombination epoch reviewed here, these experiments will also test secondary temperature and polarization anisotropy from reionization, lensing and galaxy clusters as well as the Gaussianity of the initial conditions. Combined, these measurements will test the standard cosmological model to unprecedented precision.

*Acknowledgments*

I thank the organizers R. Rebolo and J. A. Rubiño-Martín and the students of the XIX Canary Island Winter School of Astrophysics.

## REFERENCES

ALPHER, R. A., and HERMAN, R. C. (1948). *Phys. Rev.* **74**, 1737.

BARDEEN, J. M. (1980). *Phys. Rev.* **D22**, 1882.

BISCHOFF, C., HYATT, L., MCMAHON, J. J. *et al.* (2008). *Astrophys. J.* **684**, 771.

BOND, J. R. and EFSTATHIOU, G. (1984). *Astrophys. J. Lett.* **285**, L45.

BOND, J. R. and EFSTATHIOU, G. (1987). *Mon. Not. R. Astron. Soc.* **226**, 655.

DANESE, L. and DE ZOTTI, G. (1982). *Astron. Astrophys.* **107**, 39.

DE BERNARDIS, P., ADE, P. A. R., BOCK, J. J. *et al.* (2000). *Nature* **404**, 955.

DICKE, R. H., PEEBLES, P. J. E., ROLL, P. G. and WILKINSON, D. T. (1965). *Astrophys. J.* **142**, 414.

EISENSTEIN, D. J., ZEHAVI, I., HOGG, D. W. *et al.* (2005). *Astrophys. J.* **633**, 560.

EISENSTEIN, D., HU, W. and TEGMARK, M. (1998). *Astrophys. J. Lett.* **504**, L57.

FIXSEN, D. J., CHENG, E. S., GALES, J. M. *et al.* (1996). *Astrophys. J.* **473**, 576.

GAMOW, G. (1948). *Phys. Rev.* **74**, 505.

HANANY, S., ADE, P. A. R., BALBI, A. *et al.* (2000). *Astrophys. J. Lett.* **545**, L5.

HU, W. (1995). Ph.D Thesis.

HU, W. (2003). *ICTP* **14**, 149.

HU, W. (2005). *ASPC* **339**, 215.

HU, W. and DODELSON, S. (2002). *Ann. Rev. Astron. Astrophys.* **40**, 171.

HU, W. and OKAMOTO, T. (2004). *Phys. Rev.* **D69**, 043004.

HU, W. and SILK, J. (1993). *Phys. Rev.* **D48**, 485.

HU, W. and SUGIYAMA, N. (1995). *Astrophys. J.* **444**, 489.

HU, W. and SUGIYAMA, N. (1996). *Astrophys. J.* **471**, 542.

HU, W. and WHITE, M. (1996). *Phys. Rev. Lett.* **77**, 1687.

HU, W. and WHITE, M. (1997). *Phys. Rev. D* **56**, 596.

HU, W. and WHITE, M. J. (2004). *Sci. Am.* **290N2**, 32.

HU, W., FUKUGITA, M., ZALDARRIAGA, M. and TEGMARK, M. (2001). *Astrophys. J.* **549**, 669.

ILLARIONOV, A. F. and SUNYAEV, R. A. (1975). *Soviet Astron.* **18**, 691.

KAISER, N. (1983). *Mon. Not. R. Astron. Soc.* **202**, 1169.

KAMIONKOWSKI, M., KOSOWSKY, A. and STEBBINS, A. (1997). *Phys. Rev.* **D55**, 7368.

KOVAC, J., LEITCH, E. M., CARLSTROM, J. E. *et al.* (2002). *Nature* **420**, 772.

LEWIS, A., CHALLINOR, A. and LASENBY, A. (2000). *Astrophys. J.* **538**, 473.

MATHER, J. C., FIXSEN, D. J., SHAFER, R. A., MOSIER, C. and WILKINSON, D. T. (1999). *Astrophys. J.* **512**, 511.

MILLER, A. D., CALDWELL, R., DEVLIN, M. J. *et al.* (1999). *Astrophys. J. Lett.* **524**, L1.

PADIN, S., CARTWRIGHT, J. K., MASON, B. S. *et al.* (2001). *Astrophys. J. Lett.* **549**, L1.

PEEBLES, P. J. E. (1968). *Astrophys. J.* **153**, 1.

PEEBLES, P. J. E. and YU, J. T. (1970). *Astrophys. J.* **162**, 815.

POLNAREV, A. G. (1985). *Soviet Astronomy* **29**, 607.

PRYKE, C., HALVERSON, N. W., LEITCH, E. M. *et al.* (2002). *Astrophys. J.* **568**, 46.

REICHARDT, C. L., ADE, P. A. R., BOCK, J. J. *et al.* (2009). *Astrophys. J.* **694**, 1200.

RUBIÑO-MARTÍN, J. A., CHLUBA, J. and SUNYAEV, R. A. (2006). *Mon. Not. R. Astron. Soc.* **371**, 1939.

SACHS, R. K. and WOLFE, A. M. (1967). *Astrophys. J.* **147**, 73.

SEAGER, S., SASSELOV, D. D. and SCOTT, D. (2000). *Astrophys. J. Suppl.* **128**, 407.

SELJAK, U. and ZALDARRIAGA, M. (1996). *Astrophys. J.* **469**, 437.

SILK, J. (1968). *Astrophys. J.* **151**, 459.

SMOOT, G. F., BENNETT, C. L., KOGUT, A. *et al.* (1992). *Astrophys. J. Lett.* **396**, L1.

SPERGEL, D. N., BEAN, R., DORÉ, O. *et al.* (2007). *Astrophys. J. Suppl.* **170**, 377.

SUNYAEV, R. A. and ZELDOVICH, Y. B. (1970). *Astrophys. Space Sci.* **7**, 20.

SWITZER, E. R. and HIRATA, C. M. (2008). *Phys. Rev. D* **77**, 083008.

TYTLER, D., O'MEARA, J. M., SUZUKI, N. and LUBIN, D. (2000). *Physica Scripta Volume T* **85**, 12.

VITTORIO, N. and SILK, J. (1984). *Astrophys. J. Lett.* **285**, L39.

WHITE, M. and HU, W. (1997). *Astron. Astrophys.* **321**, 8.

WONG, W. Y., MOSS, A. and SCOTT, D. (2008). *Mon. Not. R. Astron. Soc.* **386**, 1023.

ZALDARRIAGA, M. and SELJAK, U. (1997). *Phys. Rev. D* **55**, 1830.

ZELDOVICH, Y. B. and SUNYAEV, R. A. (1969). *Astrophys. Space Sci.* **4**, 301.

# 4. CMB fluctuations in the post-recombination Universe

## MATTHIAS BARTELMANN

## Abstract

These lectures describe three different effects affecting the propagation of CMB photons from the shell of last scattering to the observer: the (thermal) Sunyaev–Zeldovich effect, i.e. the inverse Compton scattering of CMB photons off hot electrons; gravitational lensing by large-scale cosmic structures; and the integrated Sachs–Wolfe effect. Emphasis is given on a physical explanation of these effects and their derivation.

## 4.1 The thermal Sunyaev–Zeldovich effect

### 4.1.1 Physical ingredients

Let us begin with the Sunyaev–Zeldovich (SZ) effect (Sunyaev and Zeldovich, 1972). The physical process behind it is simply explained (cf. the left panel of Fig. 4.1); CMB photons passing through a hot plasma suffer inverse Compton scattering off hot electrons. Their energies are on average increased, but their number remains unchanged. The number of photons at low energies is reduced and equally increased at high energies. Galaxy clusters, among the most prominent cosmic reservoirs of hot electrons, therefore cast shadows on the CMB at low frequencies and shine at high frequencies (Fig. 4.1).

Cosmic microwave background photons were released when their average energy was near $0.3\,\text{eV}$, and they have by now cooled to $\sim 0.3\,\text{meV}$. Typical electron energies in clusters are a few keV, or seven orders of magnitude larger. This implies that the process can be studied in the limit of Thomson scattering. The polarisation-averaged, differential Thomson cross section is

$$\frac{\mathrm{d}\sigma}{\mathrm{d}\Omega} = \frac{r_\mathrm{e}^2}{2}\left(1 + \cos^2\theta\right), \tag{4.1}$$

where $r_\mathrm{e}$ is the classical electron radius,

$$r_\mathrm{e} = \frac{e^2}{mc^2} \approx 2.8 \times 10^{-13}\,\text{cm}, \tag{4.2}$$

and $\theta$ is the scattering angle.

Momentum conservation in the scattering process demands

$$\frac{h\nu}{c}\boldsymbol{e} + \boldsymbol{p} = \frac{h\nu'}{c}\boldsymbol{e}' + \boldsymbol{p}', \tag{4.3}$$

where primed quantities are taken after the scattering and unprimed quantities before. The photon propagation direction is given by the unit vector $\boldsymbol{e}$, $\boldsymbol{p}$ is the electron momentum and $\nu$ the photon energy. Safely ignoring terms quadratic in $\nu$, (4.3) implies

$$c^2\boldsymbol{p}'^2 = c^2\boldsymbol{p}^2 + 2h\nu c(\boldsymbol{e} - \boldsymbol{e}')\boldsymbol{p} - 2h\delta\nu c\boldsymbol{e}'\boldsymbol{p}, \tag{4.4}$$

where the frequency change $\delta\nu = \nu' - \nu$ was introduced.

Energy conservation, $E + h\nu = E' + h\nu'$, and the relativistic energy-momentum relation,

$$E^2 = c^2\boldsymbol{p}^2 + m^2c^4, \tag{4.5}$$

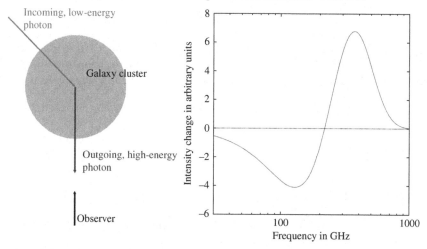

FIGURE 4.1. *Left*: The thermal Sunyaev–Zeldovich effect: low-energy photons are inversely Compton-scattered to much higher energies by hot electrons in galaxy clusters. *Right*: The spectrum of the thermal Sunyaev–Zeldovich effect.

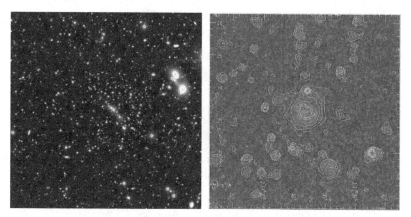

FIGURE 4.2. Cluster Cl 0016 in the optical (*left*, from Clowe *et al.*, 2000) and its thermal SZ effect (*right*, adapted from Hughes and Birkinshaw, 1998, contours) superposed on its X-ray emission. Reproduced by permission of the AAS.

allows

$$c^2 \boldsymbol{p'}^2 = c^2 \boldsymbol{p}^2 - 2Eh\delta\nu, \tag{4.6}$$

again suppressing terms quadratic in $\nu$. Equating (4.4) and (4.6) gives

$$h\delta\nu = -\frac{h\nu c \boldsymbol{p}(\boldsymbol{e} - \boldsymbol{e}')}{E - c\boldsymbol{p}\boldsymbol{e}'} \approx -\frac{h\nu}{mc}\boldsymbol{p}(\boldsymbol{e} - \boldsymbol{e}') \tag{4.7}$$

for the energy change of the photon due to the collision. Again, terms quadratic in the photon energy have been neglected, and the final approximation is valid for non-relativistic electrons, $cp \ll E$.

### 4.1.2 The Fokker–Planck approach

In the absence of collisions, Liouville's theorem asserts that the phase-space density $f(\boldsymbol{w})$ remains unchanged, where $\boldsymbol{w}$ abbreviates all phase-space coordinates. With collisions, the phase-space density changes according to

*Matthias Bartelmann*

$$\frac{\mathrm{d}f}{\mathrm{d}t} = \text{gain} - \text{loss}, \tag{4.8}$$

where gain and loss terms are due to scattering into and out of the phase-space element $\mathrm{d}\boldsymbol{w}$ under consideration. Let $\psi(\boldsymbol{w}, \Delta\boldsymbol{w})\mathrm{d}\boldsymbol{w}\mathrm{d}t$ be the transition probability due to scattering by $\Delta\boldsymbol{w}$ from $\boldsymbol{w} \rightarrow \boldsymbol{w} + \Delta\boldsymbol{w}$ within the time interval $\mathrm{d}t$. Then, the gain term is

$$\int \mathrm{d}\Delta\boldsymbol{w}\,\psi(\boldsymbol{w} - \Delta\boldsymbol{w}, \Delta\boldsymbol{w})f(\boldsymbol{w} - \Delta\boldsymbol{w}), \tag{4.9}$$

while the loss term is

$$\int \mathrm{d}\Delta\boldsymbol{w}\,\psi(\boldsymbol{w}, \Delta\boldsymbol{w})f(\boldsymbol{w}). \tag{4.10}$$

Inserting (4.9) and (4.10) into (4.8) yields the master equation

$$\frac{\mathrm{d}f}{\mathrm{d}t} = \int \mathrm{d}\Delta\boldsymbol{w}\,[\psi(\boldsymbol{w} - \Delta\boldsymbol{w}, \Delta\boldsymbol{w})f(\boldsymbol{w} - \Delta\boldsymbol{w}) - \psi(\boldsymbol{w}, \Delta\boldsymbol{w})f(\boldsymbol{w})] \tag{4.11}$$

describing the change of the phase-space density. Equation (4.7) shows that the energy change is very small if the electrons are non-relativistic, $cp \ll mc^2$. It is then useful to expand the integrand in (4.11) into a Taylor series to second order,

$$\begin{aligned}
\psi(\boldsymbol{w} - \Delta\boldsymbol{w}, \Delta\boldsymbol{w})f(\boldsymbol{w} - \Delta\boldsymbol{w}) &= \psi(\boldsymbol{w}, \Delta\boldsymbol{w})f(\boldsymbol{w}) \\
&\quad - \frac{\partial}{\partial w_i}\,[\psi(\boldsymbol{w}, \Delta\boldsymbol{w})f(\boldsymbol{w})]\,\Delta w_i \\
&\quad + \frac{1}{2}\frac{\partial^2}{\partial w_i \partial w_j}\,[\psi(\boldsymbol{w}, \Delta\boldsymbol{w})f(\boldsymbol{w})]\,\Delta w_i \Delta w_j. \tag{4.12}
\end{aligned}$$

Inserting this into the master equation (4.11) gives the Fokker–Planck equation

$$\frac{\mathrm{d}f}{\mathrm{d}t} = -\frac{\partial}{\partial w_j}\,\left[f(\boldsymbol{w})D_1^i(\boldsymbol{w})\right] + \frac{1}{2}\frac{\partial^2}{\partial w_i \partial w_j}\,\left[f(\boldsymbol{w})D_2^{ij}(\boldsymbol{w})\right], \tag{4.13}$$

which approximates scattering as a diffusion process in phase space. The diffusion coefficients are

$$D_1^i(\boldsymbol{w}) = \int \mathrm{d}\Delta\boldsymbol{w}\,\psi(\boldsymbol{w}, \Delta\boldsymbol{w})\Delta w_i,$$

$$D_2^{ij}(\boldsymbol{w}) = \int \mathrm{d}\Delta\boldsymbol{w}\,\psi(\boldsymbol{w}, \Delta\boldsymbol{w})\Delta w_i \Delta w_j. \tag{4.14}$$

### 4.1.3 Derivation of the spectrum

These coefficients can now be worked out keeping only the frequency change $\delta\nu$ of the photons as the single relevant coordinate of the change $\Delta\boldsymbol{w}$ in the phase-space coordinates and integrating over all others. We first introduce the angle $\vartheta$ between $(\boldsymbol{e} - \boldsymbol{e}')$ and the electron momentum $\boldsymbol{p}$ as the polar angle of a spherical coordinate system for integration over all electron momenta. Since $\boldsymbol{e}$ and $\boldsymbol{e}'$ are both unit vectors enclosing the angle $\theta$,

$$|\boldsymbol{e} - \boldsymbol{e}'| = \sqrt{2 - 2\cos\theta}, \tag{4.15}$$

and

$$(\boldsymbol{e} - \boldsymbol{e}') \cdot \boldsymbol{p} = p\cos\vartheta\sqrt{2(1 - \cos\theta)}. \tag{4.16}$$

Adopting a thermal, non-relativistic electron momentum distribution,

$$N(p) = \frac{1}{(2\pi mkT_e)^{3/2}}\exp\left(-\frac{p^2}{2mkT_e}\right), \tag{4.17}$$

the mean energy change of a scattered photon averaged over the electron distribution can be found combining (4.7), (4.16) and (4.17). However, its leading-order term vanishes, which signals that we have to evaluate it expanding the photon energy change (4.7) at least to second order in the electron momentum $\boldsymbol{p}$. However, the leading-order term of the mean-squared energy change is finite,

$$\langle (h\delta\nu)^2 \rangle = 2\pi \left( \frac{h\nu}{mc} \right)^2 \int \sin\vartheta \, d\vartheta \int p^2 dp \, N(p) \, p^2 \cos^2\vartheta \, 2(1 - \cos\theta)$$

$$= 4mkT_{\rm e} \left( \frac{h\nu}{mc} \right)^2 (1 - \cos\theta), \tag{4.18}$$

for photons scattered by an angle $\theta$. The calculation of the diffusion coefficient $D_1$ is thus substantially more complicated than of $D_2$, hence we omit the detailed calculation of $D_1$ here for brevity.

Evaluating the second diffusion coefficient, we need to take the scattering probability into account, given by the product of the electron number density $n_{\rm e}$ and the differential Thomson cross section (4.1). Substituting $\langle (h\delta\nu)^2 \rangle$ from (4.18) for $\Delta w_i \Delta w_j$ in (4.14), we find

$$D_2^{ij} = 8\pi n_{\rm e} \frac{r_{\rm e}^2}{2} mkT_{\rm e} \left( \frac{h\nu}{mc} \right)^2 \int \sin\theta d\theta \, (1 - \cos\theta)(1 + \cos^2\theta)$$

$$= 2n_{\rm e}\sigma_{\rm T} \frac{kT_{\rm e}}{mc^2} (h\nu)^2, \tag{4.19}$$

where the total Thomson cross section $\sigma_{\rm T} = 8\pi r_{\rm e}^2/3$ was inserted.

Neglecting $D_1$, the Fokker–Planck equation (4.13) then reduces to

$$\frac{df}{dt} = \frac{1}{2} \frac{\partial^2}{\partial (h\nu)^2} \left[ f(h\nu) D_2(h\nu) \right], \tag{4.20}$$

where $f(h\nu)$ is the phase-space distribution of the photons. We substitute the time variable by the Compton-$y$ parameter,

$$dy = n_{\rm e}\sigma_{\rm T} \frac{kT_{\rm e}}{mc^2} c dt, \tag{4.21}$$

and the photon energy by the dimensionless energy parameter $x = h\nu/(kT)$, where $T$ is the photon temperature. Then, the Fokker–Planck equation can be integrated over $dy$ to yield the change

$$\Delta f = y \left( x^2 \frac{d^2 f}{dx^2} + 4x \frac{df}{dx} + 2f \right) \tag{4.22}$$

of the photon phase-space distribution, $f = [\exp(x) - 1]^{-1}$. Taking the precise expression for $D_1$ into account changes this equation to

$$\Delta f = y \left( x^2 \frac{d^2 f}{dx^2} + 4x \frac{df}{dx} \right). \tag{4.23}$$

The resulting intensity change

$$\Delta I(x) \propto x^3 \Delta f \tag{4.24}$$

reveals the shape characteristic for the thermal SZ effect shown in the right panel of Fig. 4.1: it is negative below, and positive above, $x = 3.83$, corresponding to $\nu_0 = 217\,{\rm GHz}$. Since $\nu_0$ is determined by the ratio of the photon frequency to the photon temperature, it is higher by $1 + z$ in the rest frame of clusters at redshifts $z > 0$, but subsequently redshifted as the photons then propagate from the cluster to us. Thus, the observed zero transition of the thermal SZ effect occurs at $217\,\mathrm{GHz}$ irrespective of the cluster redshift. This distinctive spectral property is likely to become very important for future cluster searches in the sub-mm regime, e.g. with the Planck satellite, which has a dedicated frequency band centred on $217\,\mathrm{GHz}$ specifically to identify the zero transition of the thermal SZ signal.

## 4.2 Cluster detection

### 4.2.1 Concept of linear filtering

The previous derivation has shown that the intensity change at a frequency $x$ due to the thermal SZ effect can be factorised into a spectral dependence $g(x)$ and the Compton-$y$ parameter integrated over the line-of-sight,

$$\Delta I_{\mathrm{SZ}}(x) = g(x) \int n_e \sigma_\mathrm{T} \frac{kT_e}{mc^2} \mathrm{d}l = g(x)\, y(\boldsymbol{\theta}), \tag{4.25}$$

where the integrated Compton-$y$ parameter was written as a function of position on the sky $\boldsymbol{\theta}$ in the last step. Given a template for the internal electron and temperature distribution in a cluster, $y(\boldsymbol{\theta})$ can be seen as a template for the Comptonisation projected on the sky. Assuming typical radial profile shapes for $n_e$ and $T_e$, we can factorise $y(\boldsymbol{\theta})$ into an amplitude and a shape template $\tau(\boldsymbol{\theta})$,

$$y(\boldsymbol{\theta}) = A\tau(\boldsymbol{\theta}). \tag{4.26}$$

Given an experiment with a frequency response function $F_i(x)$ in the frequency band $i$ and a beam shape $B(\boldsymbol{\theta})$, the measured Compton-$y$ signal is

$$S(\boldsymbol{\theta}) = A\,(F_i * g)(x)\,(B * \tau)(\boldsymbol{\theta}), \tag{4.27}$$

i.e. the thermal-SZ spectrum $g(x)$ is convolved with the frequency response and the template $\tau$ with the beam shape. We write

$$(B * \tau)(\boldsymbol{\theta}) = \int B(\boldsymbol{\theta} - \boldsymbol{\theta}')\tau(\boldsymbol{\theta}')\mathrm{d}^2\theta' = \tau_\mathrm{B}(\boldsymbol{\theta}) \tag{4.28}$$

and $A_i = A(F_i * g)$ for the amplitude in the frequency band $i$, and thus find

$$S(\boldsymbol{\theta}) = A_i\tau_\mathrm{B}(\boldsymbol{\theta}). \tag{4.29}$$

The data stream will be the sum of signal and noise,

$$D = A\tau + N, \tag{4.30}$$

where the noise is characterised by its power spectrum,

$$\left\langle \hat{N}^*(\boldsymbol{k}')\hat{N}(\boldsymbol{k}) \right\rangle = (2\pi)^2\delta(\boldsymbol{k} - \boldsymbol{k}')P_N(k), \tag{4.31}$$

where the hat denotes the Fourier transform and the asterisk the complex conjugate, and $\boldsymbol{k}$ is a two-dimensional wavevector on (tangential planes to) the sky. Our goal is now to extract the signal amplitude $A$ by filtering the data linearly with a filter $\psi$ such that (i) the result is unbiased and (ii) the signal-to-noise ratio is optimised (Haehnelt and Tegmark, 1996). Thus, we require that the estimated amplitude

$$A_{\mathrm{est}} = \int D(\boldsymbol{\theta})\psi(\boldsymbol{\theta})\mathrm{d}^2\theta \tag{4.32}$$

faithfully reproduce the true amplitude $A$ when averaged over a (hypothetical) sample of identical clusters whose data streams only differ by the noise realisation,

$$b = \langle A_{\text{est}} - A \rangle = 0, \tag{4.33}$$

where $b$ abbreviates the bias. We further require that the variance

$$\sigma^2 = \left\langle (A_{\text{est}} - A)^2 \right\rangle = A^2 - 2A\langle A_{\text{est}} \rangle + \langle A_{\text{est}}^2 \rangle \tag{4.34}$$

be minimised.

### 4.2.2 Derivation of an optimal filter shape

If the filter is indeed unbiased, $\langle A_{\text{est}} \rangle = A$, which implies the normalisation constraint

$$\int \tau(\theta)\psi(\theta)\mathrm{d}^2\theta = 1. \tag{4.35}$$

Moreover, the mean-squared estimated amplitude is

$$\langle A_{\text{est}}^2 \rangle = \left\langle \int \psi(A\tau + N)\mathrm{d}^2\theta' \int \psi(A\tau + N)\mathrm{d}^2\theta \right\rangle$$
$$= A^2 + \left\langle \left( \int \psi N \mathrm{d}^2\theta \right)^2 \right\rangle \tag{4.36}$$

assuming that the noise vanishes on average, thus

$$\sigma^2 = \left\langle \left( \int \psi N \mathrm{d}^2\theta \right)^2 \right\rangle. \tag{4.37}$$

Using now the Fourier convolution theorem, we find

$$\left\langle \left( \int \psi N \mathrm{d}^2\theta \right)^2 \right\rangle = \left\langle \int \frac{\mathrm{d}^2 k}{(2\pi)^2} \int \frac{\mathrm{d}^2 k'}{(2\pi)^2} \left( \hat{N}^* \hat{\psi} \right) \left( \hat{N} \hat{\psi}^* \right) \right\rangle$$
$$= \int \frac{\mathrm{d}^2 k}{(2\pi)^2} P_N(k)|\hat{\psi}|^2. \tag{4.38}$$

Now we need to ensure that $\sigma^2$ is minimised under the constraint $b = 0$. We thus introduce the functional

$$L[\psi] = \sigma^2 + \lambda b \tag{4.39}$$

with a Lagrange multiplier $\lambda$ and require that the variation of $L$ with respect to $\psi$ vanish. By definition of $b$,

$$\delta b = A \int \tau \delta\psi \mathrm{d}^2\theta = A \int \frac{\mathrm{d}^2 k}{(2\pi)^2} \hat{\tau}^* \delta\hat{\psi}, \tag{4.40}$$

where Parseval's theorem was used in the last step. Similarly, we find from (4.38)

$$\delta\sigma^2 = 2 \int \frac{\mathrm{d}^2 k}{(2\pi)^2} P_N(k)\hat{\psi}^* \delta\hat{\psi}. \tag{4.41}$$

Requiring that $\delta L = 0$ for any variation, $\delta\hat{\psi}$ then requires

$$\hat{\psi} = -\lambda \frac{A\hat{\tau}}{P_N} \tag{4.42}$$

for the filter shape. The Lagrange multiplier is determined by the normalisation constraint as (4.35),

$$\lambda = - \left[ A \int \frac{\mathrm{d}^2 k}{(2\pi)^2} \frac{|\hat{\tau}|^2}{P_N} \right]^{-1}, \tag{4.43}$$

showing that the unbiased, optimised filter $\psi$ is uniquely determined by

$$\hat{\psi} = \frac{\hat{\tau}}{P_N} \left[ \int \frac{\mathrm{d}^2 k}{(2\pi)^2} \frac{|\hat{\tau}|^2}{P_N} \right]^{-1} \tag{4.44}$$

in Fourier space. Its shape is very intuitive: for a flat noise power spectrum, it would mimic the shape of the template function $\hat{\tau}$. Any features in the power spectrum will lower the filter where the noise is high and enhance it where the noise is low. For a recent extension of the linear filtering concept to the full sky and multiple frequency bands, see Schäfer *et al.* (2006) and references therein.

## 4.3 Gravitational lensing

### *4.3.1 Light deflection by an isolated lens*

An isolated gravitational lens with Newtonian gravitational potential $\Phi$ embedded in a locally flat space-time deforms the Minkowskian metric to

$$\mathrm{d}s^2 = - \left( 1 + \frac{2\Phi}{c^2} \right) c^2 \mathrm{d}t^2 + \left( 1 - \frac{2\Phi}{c^2} \right) c^2 \mathrm{d}\boldsymbol{x}^2, \tag{4.45}$$

provided the lens is weak, $\Phi \ll c^2$, and moves with velocities much smaller than the speed of light (see Narayan and Bartelmann, 1999 for an introduction). The propagation condition for light, $\mathrm{d}s = 0$, yields the speed of light in the gravitational field,

$$c' = \frac{|\mathrm{d}\boldsymbol{x}|}{\mathrm{d}t} = c \left( 1 + \frac{2\Phi}{c^2} \right), \tag{4.46}$$

to first order in $\Phi/c^2$. Since $\Phi \to 0$ at infinity, $\Phi < 0$ within the lens, and $c' < c$. A weak gravitational field thus has the index of refraction

$$n = \frac{c}{c'} = 1 - \frac{2\Phi}{c^2} > 1. \tag{4.47}$$

Fermat's principle asserts that the light travel time $\tau$ along the actual light path between two fixed points is extremal,

$$\delta\tau = \delta \int \frac{n[\boldsymbol{x}(\lambda)]}{c} \frac{\mathrm{d}|\boldsymbol{x}|}{\mathrm{d}\lambda} \mathrm{d}\lambda = 0, \tag{4.48}$$

where $\lambda$ is an affine curve parameter along the light path. We thus introduce the Lagrange function

$$L = n(\boldsymbol{x}) \left( \dot{\boldsymbol{x}}^2 \right)^{1/2}, \tag{4.49}$$

where the overdot denotes the derivative with respect to $\lambda$, and use Euler's equation to find the equation of motion

$$\frac{\mathrm{d}}{\mathrm{d}\lambda} \left[ n(\boldsymbol{x}) \frac{\dot{\boldsymbol{x}}}{|\dot{\boldsymbol{x}}|} \right] - \boldsymbol{\nabla} n(\boldsymbol{x}) = 0. \tag{4.50}$$

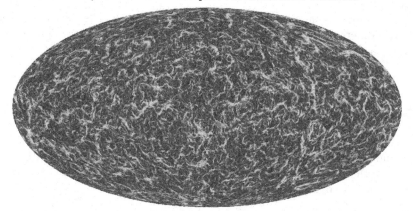

FIGURE 4.3. Simulated, full-sky map of the deflection angle caused by matter inhomogeneities between the observer and the CMB (Carbone *et al.*, 2008).

If we rescale the otherwise arbitrary curve parameter $\lambda$ such that $\dot{\boldsymbol{x}} = \boldsymbol{e}$ is the unit tangent vector to the light ray, (4.50) yields

$$n\dot{\boldsymbol{e}} + \boldsymbol{e}(\boldsymbol{\nabla} n \cdot \boldsymbol{e}) - \boldsymbol{\nabla} n = n\dot{\boldsymbol{e}} - \boldsymbol{\nabla}_{\perp} n = 0, \tag{4.51}$$

where $\boldsymbol{\nabla}_{\perp} n = \boldsymbol{\nabla} n - \boldsymbol{e}(\boldsymbol{\nabla} n \cdot \boldsymbol{e})$ is the gradient of $n$ perpendicular to the light ray. This equation shows that the tangential vector $\boldsymbol{e}$ changes by

$$\dot{\boldsymbol{e}} = \boldsymbol{\nabla}_{\perp} \ln n = -\frac{2}{c^2} \boldsymbol{\nabla}_{\perp} \Phi. \tag{4.52}$$

The total deflection angle would be the line-of-sight integral over $\dot{\boldsymbol{e}}$.

### 4.3.2 Cosmic lensing

In a homogeneous and isotropic space-time, the equation of geodesic deviation, or Jacobi equation, of general relativity can be simplified to read

$$\ddot{\boldsymbol{x}} - K\boldsymbol{x} = 0, \tag{4.53}$$

where $\boldsymbol{x}$ is the comoving transverse separation between two neighboring light rays, and

$$K = \left(\frac{H_0}{c}\right)^2 (\Omega_{\text{m}0} + \Omega_{\Lambda 0} - 1) \tag{4.54}$$

is the spatial curvature of a model universe with Hubble constant $H_0$, matter density parameter $\Omega_{\text{m}0}$ and cosmological constant $\Omega_{\Lambda 0}$. As a simple oscillator equation, (4.53) has trigonometric or hyperbolic solutions. For $K = 0$, as we shall assume henceforth, the transverse separation $\boldsymbol{x}$ grows linearly along the light rays, as it must in Euclidean geometry. In the presence of perturbations, (4.52) and (4.53) can be combined as

$$\ddot{\boldsymbol{x}} - K\boldsymbol{x} = -\frac{2}{c^2} \boldsymbol{\nabla}_{\perp} \Phi. \tag{4.55}$$

If the two light rays start from the same point enclosing the angle $\boldsymbol{\theta}$, the solution of (4.55) for $K = 0$ is

$$\boldsymbol{x}(\boldsymbol{\theta}, w) = w\boldsymbol{\theta} - \frac{2}{c^2} \int_0^w \mathrm{d}w'(w - w') \boldsymbol{\nabla}_{\perp} \Phi, \tag{4.56}$$

where $w$ is the comoving angular distance along the light path. The total deflection angle quantifies the difference between the mutual separations of the two light rays in the unperturbed and in the perturbed cases,

$$\boldsymbol{\alpha}(\boldsymbol{\theta}, w) = \frac{w\boldsymbol{\theta} - \boldsymbol{x}(\boldsymbol{\theta}, w)}{w} = \frac{2}{c^2} \int_0^w dw' \frac{w - w'}{w} \boldsymbol{\nabla}_\perp \Phi. \tag{4.57}$$

The integral should be carried out along the actual light path, but can typically be approximated by an integral along a fiducial, unperturbed light path (Born's approximation; see Bartelmann and Schneider, 2001, for a review on weak cosmological lensing, and Lewis and Challinor, 2006, for a thorough review on lensing of the CMB).

If we replace the physical coordinates perpendicular to the light path by the angular coordinates $\boldsymbol{\theta}$, the perpendicular gradient transforms to

$$\boldsymbol{\nabla}_\perp \rightarrow \frac{1}{w'} \boldsymbol{\nabla}_\theta. \tag{4.58}$$

Thus,

$$\boldsymbol{\alpha} = \boldsymbol{\nabla}_\theta \psi(\boldsymbol{\theta}), \quad \psi(\boldsymbol{\theta}) = \frac{2}{c^2} \int_0^w dw' \frac{w - w'}{ww'} \Phi, \tag{4.59}$$

defining the effective lensing potential $\psi$.

### 4.3.3 Statistics of cosmological light deflection

It is important to note that deflections themselves cannot be observed. We cannot distinguish between the sky as we know it and the sky isotropically deflected by some fixed angle because that would just shift the origin of the angular coordinates on the sky by a fixed amount. Only differences between deflection angles of neighboring light rays have a physical meaning, which is quantified by the two-point or higher-order correlation properties of the deflection angle.

Introducing the power spectrum of the Newtonian potential,

$$\left\langle \hat{\Phi}^*(\boldsymbol{k}') \hat{\Phi}(\boldsymbol{k}) \right\rangle = (2\pi)^3 P_\Phi(k) \delta(\boldsymbol{k} - \boldsymbol{k}'), \tag{4.60}$$

we find the correlation function of the effective lensing potential by expanding the potential into a Fourier series and using (4.60),

$$\langle \psi(\boldsymbol{\theta}) \psi(\boldsymbol{\theta}') \rangle = \frac{4}{c^4} \int dw' \int dw'' \left( \frac{w - w'}{ww'} \right) \left( \frac{w - w''}{ww''} \right)$$
$$\times \int \frac{d^3 k}{(2\pi)^3} P_\Phi(k) e^{i\boldsymbol{k}(\boldsymbol{x}' - \boldsymbol{x}'')}. \tag{4.61}$$

The vectors $\boldsymbol{x}'$ and $\boldsymbol{x}''$ have the components $\boldsymbol{x}' = (w', w'\boldsymbol{\theta})$ and $\boldsymbol{x}'' = (w'', w''\boldsymbol{\theta})$ along and transverse to the line of sight. We now expand the phase factor into spherical harmonics, using

$$e^{i\boldsymbol{k}\cdot\boldsymbol{x}} = 4\pi \sum_{lm} i^l j_l(kw) Y_{lm}^*(\boldsymbol{\theta}_x) Y_{lm}(\boldsymbol{\theta}_k), \tag{4.62}$$

where $j_l$ is the spherical Bessel function of order $l$, and $\boldsymbol{\theta}_x$ and $\boldsymbol{\theta}_k$ are the direction angles of $\boldsymbol{x}$ and $\boldsymbol{k}$. This gives

$$\int \frac{d^3 k}{(2\pi)^3} P_\Phi(k) e^{i\boldsymbol{k}(\boldsymbol{x}' - \boldsymbol{x}'')} = 16\pi^2 \int_0^\infty \frac{k^2 dk}{(2\pi)^3} P_\Phi(k)$$
$$\times \sum_{lm} j_l(kw') j_l(kw'') Y_{lm}^*(\boldsymbol{\theta}') Y_{lm}(\boldsymbol{\theta}'') \tag{4.63}$$

in (4.61). Expanding the lensing potential into spherical harmonics,

$$\psi(\boldsymbol{\theta}) = \sum_{lm} \psi_{lm} Y_{lm}(\boldsymbol{\theta}), \tag{4.64}$$

we see that the angular power spectrum of the lensing potential is

$$
\begin{aligned}
C_l^\psi &= \langle \psi_{lm}^* \psi_{lm} \rangle \\
&= \frac{8}{\pi c^4} \int_0^w dw' \int_0^w dw'' \left( \frac{w - w'}{ww'} \right) \left( \frac{w - w''}{ww''} \right) \\
&\quad \times \int_0^\infty k^2 dk P_\Phi(k) j_l(kw') j_l(kw'').
\end{aligned}
\tag{4.65}
$$

If typical structures responsible for the lensing effect are small compared to cosmological scales, $k^{-1} \ll w$ or $kw \gg 1$, we can approximate

$$\int_0^\infty k^2 dk P_\Phi(k) j_l(kw') j_l(kw'') \approx \frac{\pi}{2w'^2} P_\Phi\left( \frac{l}{w'} \right) \delta(w' - w''). \tag{4.66}$$

This is called Limber's approximation, which leaves the angular power spectrum in the form

$$C_l^\psi = \frac{4}{c^4} \int_0^w \frac{dw'}{w'^2} \left( \frac{w - w'}{ww'} \right)^2 P_\Phi\left( \frac{l}{w'} \right). \tag{4.67}$$

The Poisson equation in physical coordinates $r$,

$$\nabla_r^2 \Phi = 4\pi G \rho, \tag{4.68}$$

transforms to

$$\nabla^2 \Phi = \frac{3}{2} H_0^2 \Omega_{m0} \frac{\delta}{a} \tag{4.69}$$

in comoving coordinates, when the density is replaced by the relative density perturbation $\delta$ with respect to the mean cosmic density. In Fourier space, the Laplacian $\nabla^2$ is replaced by a factor $-k^2$, so the power spectra of the density and its potential are related by

$$P_\Phi(k) = \frac{9}{4} H_0^4 \Omega_{m0}^2 \frac{P_\delta(k)}{a^2 k^4}. \tag{4.70}$$

Inserting this result into the angular power spectrum of the lensing potential yields

$$C_l^\psi = \frac{9\Omega_{m0}^2}{l^4} \frac{H_0^4}{c^4} \int_0^w \frac{dw'}{a^2} \left( \frac{w - w'}{w} \right)^2 P_\Phi\left( \frac{l}{w'} \right). \tag{4.71}$$

### 4.3.4 Effects of lensing on the CMB temperature

Gravitational lensing of the CMB has the effect that the relative temperature fluctuation *observed* in direction $\boldsymbol{\theta}$,

$$\tau_{\mathrm{obs}}(\boldsymbol{\theta}) = \frac{T(\boldsymbol{\theta}) - \bar{T}}{\bar{T}}, \tag{4.72}$$

is the *intrinsic* intensity in the deflected direction $\boldsymbol{\theta} - \boldsymbol{\alpha}$,

$$\tau_{\mathrm{obs}}(\boldsymbol{\theta}) = \tau[\boldsymbol{\theta} - \boldsymbol{\alpha}(\boldsymbol{\theta})]. \tag{4.73}$$

Since typical scales in $\tau$ are substantially larger than those in $\boldsymbol{\alpha}$, we can Taylor-expand (4.73) and write

$$\tau_{\text{obs}}(\boldsymbol{\theta}) = \tau(\boldsymbol{\theta}) - \boldsymbol{\alpha} \cdot \boldsymbol{\nabla}\tau(\boldsymbol{\theta}) + \frac{1}{2}\frac{\partial^2 \tau}{\partial\theta_i \partial\theta_j}\alpha_i\alpha_j. \tag{4.74}$$

These terms are more easily expressed in Fourier space. Since the deflection angle is the gradient of the lensing potential (4.59),

$$\boldsymbol{\alpha} = \int \frac{\mathrm{d}^2 l}{(2\pi)^2}(\mathrm{i}\boldsymbol{l}\hat{\psi})\mathrm{e}^{\mathrm{i}\boldsymbol{l}\cdot\boldsymbol{\theta}} \quad \text{and}$$

$$\boldsymbol{\nabla}\tau(\boldsymbol{\theta}) = \int \frac{\mathrm{d}^2 l}{(2\pi)^2}(\mathrm{i}\boldsymbol{l}\hat{\tau})\mathrm{e}^{\mathrm{i}\boldsymbol{l}\cdot\boldsymbol{\theta}} \tag{4.75}$$

in terms of the Fourier transforms $\hat{\psi}$ and $\hat{\tau}$. Consequently, the product $\boldsymbol{\alpha} \cdot \boldsymbol{\nabla}\tau$ can be written as

$$\boldsymbol{\alpha} \cdot \boldsymbol{\nabla}\tau = -\int \frac{\mathrm{d}^2 l_1}{(2\pi)^2} \int \frac{\mathrm{d}^2 l_2}{(2\pi)^2}\boldsymbol{l}_1 \cdot \boldsymbol{l}_2\hat{\psi}(\boldsymbol{l}_1)\hat{\tau}(\boldsymbol{l}_2)\mathrm{e}^{\mathrm{i}(\boldsymbol{l}_1+\boldsymbol{l}_2)\cdot\boldsymbol{\theta}}, \tag{4.76}$$

which has the Fourier transform

$$-\int \frac{\mathrm{d}^2 l_1}{(2\pi)^2}\boldsymbol{l}_1 \cdot (\boldsymbol{l} - \boldsymbol{l}_1)\hat{\psi}(\boldsymbol{l}_1)\hat{\tau}(\boldsymbol{l} - \boldsymbol{l}_1). \tag{4.77}$$

Likewise, we can write

$$\frac{1}{2}\frac{\partial^2 \tau}{\partial\theta_i \partial\theta_j}\alpha_i\alpha_j = \frac{1}{2}\int \frac{\mathrm{d}^2 l_1}{(2\pi)^2}l_{1i}l_{1j}\hat{\tau}(\boldsymbol{l}_1) \tag{4.78}$$

$$\times \int \frac{\mathrm{d}^2 l_2}{(2\pi)^2}l_{2i}\hat{\psi}(\boldsymbol{l}_2)\int \frac{\mathrm{d}^2 l_3}{(2\pi)^2}l_{3j}\hat{\psi}(\boldsymbol{l}_3)\mathrm{e}^{\mathrm{i}(\boldsymbol{l}_1+\boldsymbol{l}_2+\boldsymbol{l}_3)\cdot\boldsymbol{\theta}},$$

whose Fourier transform is

$$\frac{1}{2}\int \frac{\mathrm{d}^2 l_1}{(2\pi)^2} \int \frac{\mathrm{d}^2 l_2}{(2\pi)^2}(\boldsymbol{l}_1 \cdot \boldsymbol{l}_2)[\boldsymbol{l}_1 \cdot (\boldsymbol{l} - \boldsymbol{l}_1 - \boldsymbol{l}_2)]\hat{\tau}(\boldsymbol{l}_1)\hat{\psi}(\boldsymbol{l}_2)\hat{\psi}(\boldsymbol{l} - \boldsymbol{l}_1 - \boldsymbol{l}_2). \tag{4.79}$$

Imagine now we Fourier transform (4.74) using the expressions (4.77) and (4.79), square the result, keep terms up to second order in the lensing potential, and average them to obtain the power spectrum of the observed CMB temperature fluctuations. The first term squared yields the intrinsic CMB power spectrum. The second term squared, and the product of the first and the third terms, lead to a convolution of the power spectra for $\tau$ and $\psi$. The product of the first and the second terms vanishes because the mean deflection angle does, and other terms are of higher than second order in $\psi$. Somewhat lengthy, but straightforward calculation yields

$$C^\tau_{l,\text{obs}} = C^\tau_l\left(1 - l^2 R_\psi\right) + \int \frac{\mathrm{d}^2 l_1}{(2\pi)^2}\left[\boldsymbol{l}_1 \cdot (\boldsymbol{l} - \boldsymbol{l}_1)\right]^2 C^\psi_{l_1}C^\tau_{|\boldsymbol{l}-\boldsymbol{l}_1|}, \tag{4.80}$$

where

$$R_\psi = \frac{1}{3\pi}\int_0^\infty l^3\mathrm{d}l\, C^\psi_l \approx 3 \times 10^{-7} \tag{4.81}$$

quantifies the total power in the deflection angle.

We thus see two effects of lensing on the CMB temperature power spectrum (Seljak, 1996). First, its amplitude is lowered by a factor increasing quadratically with the multipole order. Second, it is convolved with the power spectrum of the lensing potential,

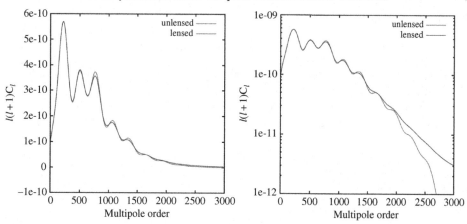

FIGURE 4.4. Cosmic lensing smoothes the peaks in the CMB power spectrum (*left*) and creates CMB structures well in the Silk damping tail (*right*; curves produced with CMBfast (Seljak and Zaldarriaga, 1996).

which gives rise to a slight smoothing. For $l \gtrsim 2000$, the factor $(1 - l^2 R_\psi)$ becomes negative, which signals the break down of the Taylor approximation. There, however, the amplitude of $C_l^\tau$ is already so low that this is practically irrelevant. For large $l$, we can thus ignore the first term in (4.80) and approximate $(\boldsymbol{l} - \boldsymbol{l}_1) \approx \boldsymbol{l}$ in the second term. This gives

$$C_{l,\mathrm{obs}}^\tau \approx l^2 C_l^\psi R_\tau \quad \text{with} \quad R_\tau = \frac{1}{3\pi} \int_0^\infty l^3 \mathrm{d}l\, C_l^\tau \approx 10^{-9}\, \mu\mathrm{K}^2. \tag{4.82}$$

Lensing thus creates power in the Silk damping tail of the CMB (Metcalf and Silk, 1997).

### 4.3.5 Spin-weighted fields on the sphere

Before we can speak about lensing of the CMB polarisation, we need to clarify the notion of spin-weighted fields on the sphere, how they can be described without reference to a specific coordinate frame tangential to the sphere, and how their statistics can be worked out (Zaldarriaga and Seljak, 1997; Castro *et al.*, 2005; Heavens *et al.*, 2006). A function $_s f(\boldsymbol{\theta})$ is said to have spin-weight $s$ if

$$_s f(\boldsymbol{\theta}) \to \mathrm{e}^{-\mathrm{i}s\psi}\,_s f(\boldsymbol{\theta}) \tag{4.83}$$

under right-handed rotations of the local coordinate frame by an angle $\psi$. Suppose the local coordinate frame has basis vectors $\{\boldsymbol{e}_1, \boldsymbol{e}_2\}$, then the two vectors

$$\boldsymbol{m} = \frac{1}{\sqrt{2}}(\boldsymbol{e}_1 + \mathrm{i}\boldsymbol{e}_2), \quad \boldsymbol{m}^\dagger = \frac{1}{\sqrt{2}}(\boldsymbol{e}_1 - \mathrm{i}\boldsymbol{e}_2) \tag{4.84}$$

have spin-weight $\pm 1$, respectively, as is easily verified by applying a rotation matrix to $\boldsymbol{m}$ and $\boldsymbol{m}^\dagger$. If $F$ is a rank-$s$ tensor on the sphere, it defines a spin-$s$ field when applied to the vectors $\boldsymbol{m}$, and a spin-$(-s)$ field when applied to the vectors $\boldsymbol{m}^\dagger$,

$$_s f = F(\boldsymbol{m}, \dots \boldsymbol{m}), \quad _{-s} f = F(\boldsymbol{m}^\dagger, \dots, \boldsymbol{m}^\dagger). \tag{4.85}$$

Taking the covariant derivative of a rank-$s$ tensor $F$ returns a rank-$(s+1)$ tensor $\nabla F$ which defines a spin-$(s+1)$ field

$$\eth_s f = -\sqrt{2}(\nabla F)(\boldsymbol{m}, \dots, \boldsymbol{m}, \boldsymbol{m}) \tag{4.86}$$

and a spin-$(s-1)$ field

$$\eth^\dagger{}_s f = -\sqrt{2}(\nabla F)(\boldsymbol{m}^\dagger, \ldots, \boldsymbol{m}^\dagger, \boldsymbol{m}^\dagger). \tag{4.87}$$

The operator $\eth$ is called "edth" (Newman and Penrose, 1966). It raises the spin by 1, while its adjoint $\eth^\dagger$ lowers the spin by 1.

It is straightforward to express the edth operator in Cartesian coordinates, where the covariant reduces to the partial derivative. If $F$ has rank-1 (i.e. $F$ is a vector), for example,

$$(\nabla F)(\boldsymbol{m}, \boldsymbol{m}) = \frac{1}{2}(\partial_1 + i\partial_2)(F_1 + iF_2), \tag{4.88}$$

where $F_i = F(\boldsymbol{e}_i)$ as usual. Since

$$_1 f = F(\boldsymbol{m}) = \frac{1}{\sqrt{2}}(F_1 + iF_2) \tag{4.89}$$

in this case, we find

$$\eth_1 f = -\frac{1}{\sqrt{2}}(\partial_1 + i\partial_2)_1 f. \tag{4.90}$$

On the sphere, the metric is described by

$$g = \begin{pmatrix} 1 & 0 \\ 0 & \sin^2\theta \end{pmatrix} \tag{4.91}$$

in the usual basis $\{\boldsymbol{e}_\theta, \boldsymbol{e}_\phi\}$, and the covariant derivative is given in terms of the two non-vanishing Christoffel symbols

$$\Gamma^1_{22} = -\sin\theta\cos\theta, \quad \Gamma^2_{12} = \cot\theta. \tag{4.92}$$

The edth operator on the sphere and its anjoint then become

$$\eth_s f = -\sqrt{2}\sin^s\theta \left(\partial_1 + \frac{i}{\sin\theta}\partial_2\right)\sin^{-s}\theta_s f$$
$$\eth^\dagger{}_s f = -\sqrt{2}\sin^{-s}\theta \left(\partial_1 - \frac{i}{\sin\theta}\partial_2\right)\sin^s\theta_s f. \tag{4.93}$$

We can now use the edth operator to define the spin-weighted spherical harmonics,

$$_sY_{lm} = \sqrt{\frac{(l-s)!}{(l+s)!}}\,\eth^s Y_{lm} \quad (0 \le s \le l)$$
$$_sY_{lm} = \sqrt{\frac{(l+s)!}{(l-s)!}}\,(-1)^s\,\eth^{\dagger-s} Y_{lm} \quad (-l \le s \le 0). \tag{4.94}$$

As their name suggests, the $_sY_{lm}$ have spin $s$. They satisfy the usual orthonormality and completeness relations and the additional convenient rules

$$_sY^*_{lm} = (-1)^s{}_{-s}Y_{lm},$$
$$\eth_s Y_{lm} = \sqrt{(l-s)(l+s+1)}\,{}_{s+1}Y_{lm},$$
$$\eth^\dagger{}_s Y_{lm} = \sqrt{(l+s)(l-s+1)}\,{}_{s-1}Y_{lm},$$
$$\eth^\dagger\eth_s Y_{lm} = -(l-s)(l+s+1)_s Y_{lm}. \tag{4.95}$$

### 4.3.6 Polarisation, E- and B-modes

The polarisation is usually described by the Stokes parameters $I$, $U$, $Q$, and $V$. The intensity $I$ corresponds to the CMB temperature here, for which we have already studied gravitational lensing effects. The circular polarisation $V$ is ignored here because Thomson scattering does not produce any. The remaining parameters $Q$ and $U$ are the components of the trace-free, rank-2 tensor

$$P = \begin{pmatrix} Q & U \\ U & -Q \end{pmatrix}. \tag{4.96}$$

As discussed above, this tensor defines the spin-2 field

$$P(\boldsymbol{m}, \boldsymbol{m}) = \frac{1}{2}(P_{11} - P_{22}) + \frac{i}{2}(P_{12} + P_{21}) = Q + iU \tag{4.97}$$

and the spin-$(-2)$ field

$$P(\boldsymbol{m}^\dagger, \boldsymbol{m}^\dagger) = \frac{1}{2}(P_{11} - P_{22}) - \frac{i}{2}(P_{12} + P_{21}) = Q - iU. \tag{4.98}$$

They are thus conveniently decomposed into spin-$(\pm 2)$ spherical harmonics,

$$(Q + iU) = \sum_{lm} a_{2lm}\, {}_2Y_{lm}, \quad (Q - iU) = \sum_{lm} a_{-2lm}\, {}_{-2}Y_{lm}. \tag{4.99}$$

Applying $\eth^2$ to $(Q - iU)$ will raise the spin from $-2$ to zero,

$$\eth^2 (Q - iU) = \sum_{lm} a_{-2lm} \sqrt{\frac{(l+2)!}{(l-2)!}}\, Y_{lm} \tag{4.100}$$

according to (4.95), while applying $\eth^{\dagger 2}$ to $(Q + iU)$ will lower the spin from 2 to zero,

$$\eth^{\dagger 2}(Q + iU) = \sum_{lm} a_{2lm} \sqrt{\frac{(l+2)!}{(l-2)!}}\, Y_{lm}. \tag{4.101}$$

The expansion coefficients are

$$a_{-2lm} = \int d\Omega\, (Q - iU)_{-2}Y_{lm}^*$$

$$= \sqrt{\frac{(l-2)!}{(l+2)!}} \int d\Omega\, \eth^2 (Q - iU)\, Y_{lm}^* \tag{4.102}$$

and

$$a_{2lm} = \int d\Omega\, (Q + iU)_2 Y_{lm}^*$$

$$= \sqrt{\frac{(l-2)!}{(l+2)!}} \int d\Omega\, \eth^{\dagger 2}(Q + iU)\, Y_{lm}^*. \tag{4.103}$$

Imagine observing the celestial sphere once inside-out and once outside-in. North will remain north, but east and west will be interchanged by this change of viewpoint. That is, the basis vectors $\{\boldsymbol{e}_1, \boldsymbol{e}_2\}$ will transform into $\{\boldsymbol{e}_1, -\boldsymbol{e}_2\}$. Since

$$Q = P(\boldsymbol{e}_1, \boldsymbol{e}_1), \quad U = P(\boldsymbol{e}_1, \boldsymbol{e}_2), \tag{4.104}$$

with the polarisation tensor (4.96),

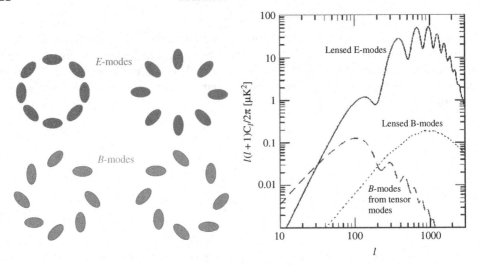

FIGURE 4.5. Left: illustration of $E$- and $B$-modes. Right: lensing moves part of the $E$-mode polarisation power into $B$-mode polarisation (Zaldarriaga and Seljak, 1998).

$$Q \to Q, \quad U \to -U \qquad (4.105)$$

under this parity transformation.

Since $Q$ and $U$ can be combined from the spin-$\pm 2$ fields $(Q \pm \mathrm{i}U)$ as

$$Q = \frac{1}{2}\left[(Q + \mathrm{i}U) + (Q - \mathrm{i}U)\right], \quad U = \frac{\mathrm{i}}{2}\left[(Q + \mathrm{i}U) - (Q - \mathrm{i}U)\right], \qquad (4.106)$$

the linear combinations of spherical-harmonic coefficients

$$a_{Elm} = -\frac{1}{2}\left(a_{2lm} + a_{-2lm}\right), \quad a_{Blm} = -\frac{\mathrm{i}}{2}\left(a_{2lm} - a_{-2lm}\right) \qquad (4.107)$$

define a parity-conserving $E$-mode and a parity-changing $B$-mode. The terminology reminds as of the electric and magnetic fields because the electric field, as a gradient field, is invariant under parity changes, while the magnetic field, as a curl field, is not. A simple example for a $B$-mode is a cyclone on Earth: a satellite looking at it from above will see it with opposite parity as a sailor looking at it from below. The distinction between $E$- and $B$-modes is important because scalar density perturbations cannot create $B$-modes, so the primordial CMB polarisation is expected to be free of $B$-modes.

In real (as opposed to harmonic) space, we define

$$\tilde{E} = \sum_{lm} \sqrt{\frac{(l+2)!}{(l-2)!}}\, a_{Elm} Y_{lm}, \qquad (4.108)$$

and likewise for $\tilde{B}$. Substituting $a_{Elm}$ from (4.107) and $a_{\pm 2lm}$ from (4.102) and (4.103) yields

$$\tilde{E} = -\frac{1}{2}\left[\eth^{\dagger 2}(Q + \mathrm{i}U) + \eth^{2}(Q - \mathrm{i}U)\right] \quad \text{and}$$

$$\tilde{B} = \frac{\mathrm{i}}{2}\left[\eth^{\dagger 2}(Q + \mathrm{i}U) - \eth^{2}(Q - \mathrm{i}U)\right]. \qquad (4.109)$$

By construction, both these fields have spin-0, leaving them invariant under rotations of the local coordinate frame on the sphere. Thus, the separation into $E$- and $B$-modes according to (4.109) separates parity-conserving and non-conserving modes *and* expresses them in a way independent of the local coordinate orientation.

### 4.3.7 Effects of lensing on the CMB polarisation

We can now return to the small-scale limit in which the sky is locally approximated by a plane. The spherical-harmonic transform can then be replaced by the ordinary Fourier transform, i.e.

$$\sum_{lm} a_{lm} Y_{lm} \rightarrow \int d^2 l \hat{f}(l) e^{i l \cdot \theta} \tag{4.110}$$

for a function $f$ whose spherical-harmonic coefficients are $a_{lm}$. The edth operator and its square simplify to their expressions for Cartesian coordinates,

$$\eth = \partial_1 + i\partial_2, \quad \eth^\dagger = \partial_1 - i\partial_2,$$
$$\eth^2 = \partial_1^2 - \partial_2^2 + 2i\partial_1\partial_2, \quad \eth^{\dagger 2} = \partial_1^2 - \partial_2^2 - 2i\partial_1\partial_2. \tag{4.111}$$

From (4.108), we see that the spherical-harmonic coefficients of $\tilde{E}$ and $\tilde{B}$ are

$$a_{\tilde{E}lm} = \sqrt{\frac{(l+2)!}{(l-2)!}} a_{Elm}, \quad a_{\tilde{B}lm} = \sqrt{\frac{(l+2)!}{(l-2)!}} a_{Blm}, \tag{4.112}$$

which implies

$$\hat{E} = \frac{\hat{\tilde{E}}}{l^2}, \quad \hat{B} = \frac{\hat{\tilde{B}}}{l^2} \tag{4.113}$$

in Fourier space. From $\tilde{E}$ and $\tilde{B}$ as given in (4.109), we thus find, in the small-scale limit,

$$\hat{E} = \hat{Q} \cos 2\phi + \hat{U} \sin 2\phi, \quad \hat{B} = \hat{U} \cos 2\phi - \hat{Q} \sin 2\phi, \tag{4.114}$$

where $\phi$ is the angle between the wave vector $l$ and an arbitrary $x_1$ axis. The sine and cosine terms appear because of the identities

$$\cos 2\phi = \frac{l_1^2 - l_2^2}{l^2}, \quad \sin 2\phi = \frac{2l_1 l_2}{l^2}, \tag{4.115}$$

which arise from applying the operators $\eth^{\dagger 2}$ and $\eth^2$ to the Fourier transforms $\hat{Q}$ and $\hat{U}$. Note that (4.114) imply

$$\hat{Q} = \hat{E} \cos 2\phi - \hat{B} \sin 2\phi, \quad \hat{U} = \hat{B} \cos 2\phi + \hat{E} \sin 2\phi. \tag{4.116}$$

The linear combinations $\hat{E} \pm i\hat{B}$ are thus related to $\hat{Q} + i\hat{U}$ by

$$\hat{E} \pm i\hat{B} = e^{\mp 2i\phi}(\hat{Q} + i\hat{U}), \tag{4.117}$$

while

$$\hat{Q} \pm i\hat{U} = e^{\pm 2i\phi}(\hat{E} + i\hat{B}). \tag{4.118}$$

Ignoring $\hat{B}$ because it cannot be produced when the CMB is released, we can assume

$$\hat{Q} \pm i\hat{U} = e^{\pm 2i\phi}\hat{E}. \tag{4.119}$$

The following steps are conceptually simple, but require lengthy calculations. First, we replace the temperature contrast $\tau$ by the combinations $Q \pm iU$ of Stokes parameters in (4.74). According to (4.119), this implies replacing $\hat{\tau}(l_1)$ by $e^{\pm 2i\phi_1}\hat{E}(l_1)$ in the expressions (4.77) and (4.79). The results need to be multiplied by $e^{\mp 2i\phi}$ to find the observed, i.e. lensed, combinations $(\hat{E} \pm i\hat{B})_{\mathrm{obs}}$. Having obtained those, we can form the power spectra

$$\left\langle (\hat{E} + i\hat{B})^*(\hat{E} + i\hat{B}) \right\rangle = C_l^E + C_l^B \qquad (4.120)$$

and

$$\left\langle (\hat{E} - i\hat{B})^*(\hat{E} + i\hat{B}) \right\rangle = C_l^E - C_l^B. \qquad (4.121)$$

Their sum and difference yield the separate power spectra $C_l^E$ and $C_l^B$,

$$
\begin{aligned}
C_{l,\mathrm{obs}}^E &= (1 - l^2 R_\psi) C_l^E \\
&\quad + \int \frac{\mathrm{d}^2 l_1}{(2\pi)^2} \left[ l_1 (l - l_1) \right]^2 \cos^2 2(\phi_1 - \phi) C_{|l-l_1|}^\psi C_{l_1}^E, \\
C_{l,\mathrm{obs}}^B &= \int \frac{\mathrm{d}^2 l_1}{(2\pi)^2} \left[ l_1 (l - l_1) \right]^2 \sin^2 2(\phi_1 - \phi) C_{|l-l_1|}^\psi C_{l_1}^E.
\end{aligned}
\qquad (4.122)
$$

Thus, the observable $E$-mode power spectrum differs from the temperature power spectrum (4.80) merely by the phase factor $\cos^2 2(\phi_1 - \phi)$ in the integral, where $\phi$ and $\phi_1$ are the angles enclosed by $l$ and $l_1$ with the $x_1$ axis. This phase factor differs from unity when the Fourier modes of the lensing potential and the intrinsic $E$-mode polarisation are not aligned. It is very important that any such misalignment will create $B$-mode from $E$-mode polarisation, as (4.122) demonstrates: when $\phi_1 \neq \phi$, the phase factor $\sin^2 2(\phi_1 - \phi)$ differs from zero, and the observed $B$-mode power spectrum is different from zero although any intrinsic $B$-modes were explicitly ignored. Thus, we have arrived at a third effect of gravitational lensing on the CMB: lensing will create $B$- from $E$-mode polarisation (Zaldarriaga and Seljak, 1998).

Structures in the CMB are typically much larger than structures in the gravitationally lensing density field. Thus, we can take the limit $l \ll l_1$ and approximate

$$
\begin{aligned}
C_{l,\mathrm{obs}}^B &\approx \int \frac{\mathrm{d}^2 l_1}{(2\pi)^2} l_1^2 \sin^2 2(\phi_1 - \phi) C_{l_1}^\psi C_{l_1}^E \\
&= \frac{8}{15\pi} \int l_1^5 \mathrm{d}l_1 C_{l_1}^\psi C_{l_1}^E,
\end{aligned}
\qquad (4.123)
$$

which becomes independent of $l$.

### 4.3.8 Recovery of the lensing potential

We do not know the appearance of the unlensed CMB, but lensing produces characteristic patterns on the CMB which may reveal its presence. In particular, lensing mixes Fourier modes and thus correlates the formerly unrelated modes

$$\hat{\tau}_{\mathrm{obs}}(l) \quad \text{and} \quad \hat{\tau}_{\mathrm{obs}}(l + L). \qquad (4.124)$$

To lowest order in the lensing potential $\psi$,

$$\hat{\tau}_{\mathrm{obs}}(l) = \hat{\tau}(l) - \int \frac{\mathrm{d}^2 l_1}{(2\pi)^2} l_1 (l - l_1) \hat{\psi}(l_1) \hat{\tau}(l - l_1), \qquad (4.125)$$

so the expected correlation of different modes turns out to be

$$
\begin{aligned}
\langle \hat{\tau}_{\mathrm{obs}}(l) \hat{\tau}_{\mathrm{obs}}^*(l - L) \rangle &= (2\pi)^2 C_l^\tau \delta(L) \\
&\quad + \left[ L(L - l) C_{|l-L|}^\tau + L \cdot l C_l^\tau \right] \hat{\psi}(L).
\end{aligned}
\qquad (4.126)
$$

To lowest order in $\hat{\psi}$, lensing creates off-diagonal terms in the CMB temperature power spectrum, i.e. it couples different fluctuation modes which would otherwise be independent.

This motivates the construction of a filter $w(l, L)$, of which we require that the convolution

$$N(L) \int \frac{\mathrm{d}^2 l}{(2\pi)^2} \left[\hat{\tau}_{\mathrm{obs}}(l)\hat{\tau}_{\mathrm{obs}}^*(l - L)\right] w(l, L) = \hat{\psi}_{\mathrm{est}}(L) \qquad (4.127)$$

returns an unbiased estimate $\hat{\psi}_{\mathrm{est}}(L)$ of the lensing potential. The requirement that the filter be unbiased,

$$\left\langle \hat{\psi}_{\mathrm{est}} \right\rangle = \hat{\psi}, \qquad (4.128)$$

puts a normalisation constraint on the prefactor $N(L)$,

$$N^{-1}(L) = \int \frac{\mathrm{d}^2 l}{(2\pi)^2} \left[L(L - l)C_{|L-l|}^\tau + L \cdot l C_l^\tau\right] w(l, L). \qquad (4.129)$$

In a way very similar to the construction of a linear filter for the detection of Sunyaev–Zeldovich clusters, the shape of the filter $w(l, L)$ follows from the requirement that the significance of its detections be optimised, i.e. that the variance

$$\left\langle \left(\hat{\psi}_{\mathrm{est}} - \hat{\psi}\right)^2 \right\rangle = \left\langle |\hat{\psi}_{\mathrm{est}}|^2 \right\rangle + \left\langle |\hat{\psi}|^2 \right\rangle \approx \left\langle |\hat{\psi}_{\mathrm{est}}|^2 \right\rangle \qquad (4.130)$$

be minimised. Inserting the expression (4.127) yields

$$\left\langle \hat{\psi}_{\mathrm{est}}^*(L)\hat{\psi}_{\mathrm{est}}(L') \right\rangle = N^2(L)\delta(L - L') \int \frac{\mathrm{d}^2 l_1}{(2\pi)^2} C_{l_1}^\tau C_{|l_1 - L|}^\tau w^2(l_1, L). \qquad (4.131)$$

Varying this with respect to the filter $w(l, L)$ under the bias constraint (4.128), taking into account that the normalisation factor $N(L)$ also depends on $w(l, L)$, and setting the variation to zero returns the filter

$$w(l, L) = \frac{L(L - l)C_{|l_1 - L|}^\tau + L \cdot l C_l^\tau}{C_{|l_1 - L|}^\tau C_l^\tau}, \qquad (4.132)$$

whose application to the squared temperature field will return an unbiased, optimised estimate of the lensing potential (Hu, 2001). This demonstrates one example for the recovery of the lensing effects on the CMB (see also Hirata and Seljak, 2003; Okamoto and Hu, 2003).

## 4.4 The integrated Sachs–Wolfe effect

An immediate consequence of the equivalence principle is that light emitted in a gravitational field suffers the redshift

$$z = \frac{\Delta\Phi}{c^2} \qquad (4.133)$$

when it is received at a gravitational potential differing by $\Delta\Phi$ from the gravitational potential at the point of emission. Since the redshift is defined as the relative wavelength change, it can be identified with the relative temperature change of a blackbody spectrum. Thus,

$$\frac{\mathrm{d}\Phi}{c^2} = \frac{\mathrm{d}T}{T}, \qquad (4.134)$$

and the total temperature change accrued as the photons pass through the peculiar potential landscape is

$$\frac{\Delta T}{T} = \tau = \int \frac{\mathrm{d}\Phi}{c^2} = \int \frac{\dot{\Phi}}{c^2} \mathrm{d}t. \tag{4.135}$$

This temperature change is intuitive to understand. Photons falling into a gravitational potential gain energy which they lose again as they climb out of the potential. If the potential is unchanged as the photon travels through it, the net effect vanishes, but it does not if the potential changes while the photon passes through it. Equation (4.135) thus quantifies the total temperature change accumulated as the photon propagates.

We have seen already in (4.69) that the Poisson equation relating the peculiar gravitational potential to the density contrast in comoving coordinates is

$$\nabla^2 \Phi = \frac{3}{2} H_0^2 \Omega_{\mathrm{m}0} \frac{\delta}{a}. \tag{4.136}$$

In linear theory, which is quite sufficient as we shall confirm later, the density contrast grows as

$$\delta(a) = \delta_0 D_+(a), \tag{4.137}$$

where the linear growth factor $D_+(a)$ is normalised such that it reaches unity today, $D_+(1) = 1$, and $\delta_0$ is the present density contrast at $a = 1$. Inserting this into Poisson's equation (4.136) and Fourier transforming it, we find

$$-k^2 \hat{\Phi} = \frac{3}{2} H_0^2 \Omega_{\mathrm{m}0} \hat{\delta}_0 \frac{D_+(a)}{a}. \tag{4.138}$$

The time evolution of the potential is thus given by

$$\dot{\hat{\Phi}} = -\frac{3}{2} H_0^2 \Omega_{\mathrm{m}0} \frac{\hat{\delta}_0}{k^2} \frac{\mathrm{d}}{\mathrm{d}t} \left[ \frac{D_+(a)}{a} \right]. \tag{4.139}$$

This integrated Sachs–Wolfe effect (Sachs and Wolfe, 1967) thus creates the secondary CMB temperature fluctuation

$$\hat{\tau} = -\frac{3\Omega_{\mathrm{m}0}}{2} \left( \frac{H_0}{c} \right)^2 \frac{\hat{\delta}_0}{k^2} \int_{a_{\mathrm{CMB}}}^{1} \frac{\mathrm{d}}{\mathrm{d}t} \left[ \frac{D_+(a)}{a} \right] \mathrm{d}t, \tag{4.140}$$

where $a_{\mathrm{CMB}} \approx 10^{-3}$ is the scale factor at the release of the CMB. The integral is easily carried out and yields

$$\hat{\tau} = -\frac{3\Omega_{\mathrm{m}0}}{2} \left( \frac{H_0}{c} \right)^2 \frac{\hat{\delta}_0}{k^2} \left[ 1 - \frac{D_+(a_{\mathrm{CMB}})}{a_{\mathrm{CMB}}} \right]. \tag{4.141}$$

In an Einstein–de Sitter Universe, $D_+(a) = a$, and the integrated Sachs–Wolfe effect vanishes completely.

The power spectrum of the temperature fluctuations caused by the integrated Sachs–Wolfe effect is then

$$C_{l,\mathrm{ISW}}^{\tau} = \frac{9\Omega_{\mathrm{m}0}}{4} \left( \frac{H_0}{c} \right)^4 \frac{P_{\delta 0}}{k^4} \left[ 1 - \frac{D_+(a_{\mathrm{CMB}})}{a_{\mathrm{CMB}}} \right]^2. \tag{4.142}$$

Obviously, the factor $k^{-4}$ strongly suppresses the small-scale modes, and with them any non-linear contribution. The integrated Sachs–Wolfe effect is therefore only relevant at large scales.

This concludes our tour through the most important extragalactic CMB foreground effects. We have seen that, and how, a hot electron gas between the CMB and the observer gives rise to a characteristic spectral distortion of the CMB, which can be exploited for constructing linear filters for galaxy clusters. We have derived how gravitational lensing by large-scale structures deflects light rays from distant sources and how this affects the CMB temperature and polarisation power spectra, and we have finally discussed how a changing gravitational potential gives rise to large-scale, secondary CMB fluctuations. Emphasis was laid on deriving these effects in detail, sacrificing completeness. Nonetheless, I hope that this overview will be useful for graduate students and researchers wishing to start from the physical foundation of the secondary effects, and to get acquainted with the necessary formalism. I apologise for the eclectic references. Again, they should not at all be seen as an exhaustive, but as a selected list of fundamental, pioneering or review papers suggested for a next step of further reading.

## Acknowledgements

Finally, I am grateful to the organisers of the XIXth Canary Island Winter School, in particular Rafael Rebolo and Alberto Rubiño-Martín, for providing such a wonderful, rewarding and pleasant occasion both for students and lecturers.

## REFERENCES

BARTELMANN, M. and SCHNEIDER P. (2001). *Phys. Rep.* **340**, 291.

CASTRO, P. G., HEAVENS, A. F. and KITCHING, T. D. (2005). *Phys. Rev. D* **72**, 023516.

CARBONE, C., SPRINGEL, V., BACCIGALUPI, C., BARTELMANN, M. and MATARRESE, S. (2008). *Mon. N. R. Astron. Soc.* **388**, 1618.

CLOWE, D., LUPPINO, G. A., KAISER, N. and GIOIA, I. M. (2000). *Astrophys. J.* **539**, 540.

HAEHNELT, M. G. and TEGMARK, M. (1996). *Mon. Not. R. Astron. Soc.* **279**, 545.

HEAVENS, A. F., KITCHING, T. D. and TAYLOR, A. N. (2006). *Mon. Not. R. Astron. Soc.* **373**, 105.

HIRATA, C. M. and SELJAK, U. (2003). *Phys. Rev. D* **68**, 083002.

HUGHES, J. P. and BIRKINSHAW, M. (1998). *Astrophys. J.* **501**, 1.

HU, W. (2001). *Astrophys. J. Lett.* **557**, L79.

LEWIS, A. and CHALLINOR, A. (2006). *Phys. Rep.* **429**, 1.

METCALF, R. B. and SILK, J. (1997). *Astrophys. J.* **489**, 1.

NARAYAN, R. and BARTELMANN, M. (1999), in *Formation of Structure in the Universe*, ed. A. Dekel and J. P. Ostriker (Cambridge University Press, Cambridge).

NEWMAN, E. T. and PENROSE, R. (1966). *J. Math. Phys.* **7**, 863.

OKAMOTO, T. and HU, W. (2003). *Phys. Rev. D* **67**, 083002.

SACHS, R. K. and WOLFE, A. M. (1967). *Astrophys. J.* **147**, 73.

SCHÄFER, B. M., PFROMMER, C., HELL, R. M. and BARTELMANN, M. (2006). *Mon. Not. R. Astron. Soc.* **370**, 1713.

SELJAK, U. (1996). *Astrophys. J.* **463**, 1.

SELJAK, U. and ZALDARRIAGA, M. (1996). *Astrophys. J.* **469**, 437.

SUNYAEV, R. A. and ZELDOVICH, Y. B. (1972). *Comm. Astrophys. Space Phys.* **4**, 173.

ZALDARRIAGA, M. and SELJAK, U. (1997). *Phys. Rev. D* **55**, 1830.

ZALDARRIAGA, M. and SELJAK, U. (1998). *Phys. Rev. D* **58**, 023003.

# 5. Statistical techniques for data analysis in cosmology

## LICIA VERDE

## Abstract

In these lectures, some statistical tools are introduced, which any cosmologist should know about in order to be able to understand recently published results from the analysis of cosmological data sets. This is not a complete and rigorous introduction to statistics, but rather a "bag of tricks" useful for anybody new to cosmology. Many useful applications are left as exercises.

## 5.1 Introduction

Statistics is everywhere in cosmology, today more than ever: cosmological data sets are getting ever larger, data from different experiments can be compared and combined; as the statistical error bars shrink, the effect of systematics need to be described, quantified and accounted for. As the data sets improve, the parameter space we want to explore also grows. In the last five years there were more than 370 papers with "statistic-" in the title!

There are many excellent books on statistics: however when I started using statistics in cosmology I could not find all the information I needed in the same place. So here I have tried to put together a "starter kit" of statistical tools useful for cosmology. This will not be a rigorous introduction with theorems, proofs etc. The goal of these lectures is to (a) be a practical manual: to give you enough knowledge to be able to understand cosmological data analysis and/or to find out more by yourself and (b) give you a "bag of tricks" (hopefully) useful for your future work.

Many useful applications are left as exercises (often with hints) and are thus proposed in the main text.

I will start by introducing probability and statistics from a "cosmologist point of view." Then I will continue with the description of random fields (ubiquitous in cosmology), followed by an introduction to Monte Carlo methods including Monte Carlo error estimates and Monte Carlo Markov chains. I will conclude with the Fisher matrix technique, useful for quickly forecasting the performance of future experiments.

## 5.2 Probabilities

### 5.2.1 What's probability: Bayesian vs. Frequentist

Probability can be interpreted as a *frequency*,

$$\mathcal{P} = \frac{n}{N}, \tag{5.1}$$

where $n$ stands for the successes and $N$ for the total number of trials.

Or it can be interpreted as a lack of information: if I knew everything, I know that an event is surely going to happen, then $\mathcal{P} = 1$, if I know it is not going to happen then $\mathcal{P} = 0$ but in other cases I can use my judgment and/or information from frequencies to estimate $\mathcal{P}$. The world is divided into "frequentists" and "Bayesians". In general, cosmologists are "Bayesians" and high energy physicists are "frequentists".

For "frequentists", events are just frequencies of occurrence: probabilities are only defined as the quantities obtained in the limit when the number of independent trials tends to infinity.

Bayesians interpret probabilities as the degree of belief in a hypothesis: they use judgment, prior information, probability theory, etc...

As we do cosmology we will be Bayesian.

### 5.2.2 Dealing with probabilities

In probability theory, probability distributions are fundamental concepts. They are used to calculate confidence intervals, for modeling purposes etc. We first need to introduce the concept of the random variable in statistics (and in cosmology). Depending on the problem at hand, the random variable may be the face of a dice, the number of galaxies in a volume $\delta V$ of the Universe, the CMB temperature in a given pixel of a CMB map, the measured value of the power spectrum $P(k)$, etc. The probability that $x$ (your random variable) can take a specific value is $\mathcal{P}(x)$, where $\mathcal{P}$ denotes the probability distribution. The properties of $\mathcal{P}$ are:

(i) $\mathcal{P}(x)$ is a non-negative, real number for all real values of $x$.
(ii) $\mathcal{P}(x)$ is normalized so that $\int \mathrm{d}x \mathcal{P}(x) = 1$.[1]
(iii) For mutually exclusive events $x_1$ and $x_2$, $\mathcal{P}(x_1 + x_2) = \mathcal{P}(x_1) + \mathcal{P}(x_2)$, i.e. the probability of $x_1$ or $x_2$ to happen is the sum of the individual probabilities. $\mathcal{P}(x_1 + x_2)$ is also written as $\mathcal{P}(x_1 \cup x_2)$ or $\mathcal{P}(x_1.OR.x_2)$.
(iv) In general:

$$\mathcal{P}(a,b) = \mathcal{P}(a)\mathcal{P}(b|a); \qquad \mathcal{P}(b,a) = \mathcal{P}(b)\mathcal{P}(a|b). \tag{5.2}$$

The probability of $a$ and $b$ to happen is the probability of $a$ times the conditional probability of $b$ given $a$. Here we can also make the (apparently tautological) identification $\mathcal{P}(a,b) = \mathcal{P}(b,a)$. For independent events then $\mathcal{P}(a,b) = \mathcal{P}(a)\mathcal{P}(b)$.

---
Exercises
---

(1) "Will it be sunny tomorrow?" Answer in the frequentist way and in the Bayesian way.[2]
(2) Produce some examples for rule (iv) above.

---

While frequentists only consider distributions of events, Bayesians consider hypotheses as "events," giving us Bayes' theorem:

$$\mathcal{P}(H|D) = \frac{\mathcal{P}(H)\mathcal{P}(D|H)}{\mathcal{P}(D)}, \tag{5.3}$$

where $H$ stands for hypothesis (generally the set of parameters specifying your model, although many cosmologists now also consider models themselves) and $D$ stands for data. The parameter $\mathcal{P}(H|D)$ is called the *posterior* distribution; $\mathcal{P}(H)$ is called the *prior* and $\mathcal{P}(D|H)$ is called *likelihood*.

Note that this is nothing but Eq. (5.2) with the apparently tautological identity $\mathcal{P}(a,b) = \mathcal{P}(b,a)$ and with substitutions: $b \longrightarrow H$ and $a \longrightarrow D$.

---

[1] For discrete distribution $\int \longrightarrow \sum$
[2] These lectures were given in the Canary Islands, in other locations the answer may differ...

Despite its simplicity, Eq. (5.3) is a really important equation!!! The usual points of heated discussion follow: "How do you choose $\mathcal{P}(H)$?" "Does the choice affects your final results?" (Yes, in general it will). "Isn't this then a bit subjective?"

─────────────────────────── Exercises ───────────────────────────

(3) Consider a positive definite quantity (like for example the tensor-to-scalar ratio $r$ or the optical depth to the last scattering surface $\tau$). What prior should one use? A flat prior in the variable? Or a logarithmic prior (i.e. flat prior in the log of the quantity)? For example, CMB analysis may use a flat prior in $\ln r$, and in $Z = \exp(-2\tau)$. How is this related to using a flat prior in $r$ or in $\tau$? It will be useful to consider the following: effectively we are comparing $\mathcal{P}(x)$ with $\mathcal{P}(f(x))$, where $f$ denotes a function of $x$. For example $x$ is $\tau$ and $f(x)$ is $\exp(-2\tau)$. Recall that: $\mathcal{P}(f) = \mathcal{P}(x(f)) \left|\frac{\mathrm{d}f}{\mathrm{d}x}\right|^{-1}$. The Jacobian of the transformation appears here to conserve probabilities.

(4) Compare Fig. 21 of (Spergel *et al.*, 2007) with Fig. 13 of (Spergel *et al.*, 2003). Consider the WMAP-only contours. Clearly, the 2007 paper uses more data than the 2003 paper, so why is it that the constraints look worst? If you suspect the prior you are correct! Find out which prior has changed, and why it makes such a difference.

(5) Under which conditions does the choice of prior not matter? (Hint: compare the WMAP papers of 2003 and 2007 for the flat $\Lambda$CDM case).

───────────────────────────────────────────────────────────────

### 5.2.3 Moments and cumulants

Moments and cumulants are used to characterize the probability distribution. In the language of probability distribution *averages* are defined as follows:

$$\langle f(x) \rangle = \int \mathrm{d}x\, f(x) \mathcal{P}(x). \tag{5.4}$$

These can then be related to "expectation values" (see later). For now let us just introduce the moments: $\hat{\mu}_m = \langle x^m \rangle$ and, of special interest, the central moments: $\mu_m = \langle (x - \langle x \rangle)^m \rangle$.

─────────────────────────── Exercise ───────────────────────────

(7) Show that $\hat{\mu}_0 = 1$ and that the average $\langle x \rangle = \hat{\mu}_1$. Also show that $\mu_2 = \langle x^2 \rangle - \langle x \rangle^2$.

───────────────────────────────────────────────────────────────

Here, $\mu_2$ is the variance, $\mu_3$ is called the skewness, $\mu_4$ is related to the kurtosis. If you deal with the statistical nature of initial conditions (i.e. primordial non-Gaussianity) or non-linear evolution of Gaussian initial conditions, you will encounter these quantities again (and again...).

The parameters $\mu_2$ and $\mu_3$ are moments but also cumulants. Up to the skewness, central moments and cumulants coincide. For higher-order terms things become more complicated. Moments and cumulants are not identical any more. To keep things as simple as possible let's just consider the Gaussian distribution (see below) as a reference. While moments of order higher than three are non-zero for both Gaussian and non-Gaussian distribution, the *cumulants* are introduced so that cumulants of order higher than two are zero for a Gaussian distribution. Moments for a non-Gaussian distribution differ from the moments of a Gaussian one because they have extra contributions from cumulants of order greater than two. In fact, for a Gaussian distribution all moments of order higher than two are specified by $\mu_1$ and $\mu_2$. Or, in other words, the mean and the variance completely specify a Gaussian distribution. This is not the case for a non-Gaussian distribution. For a non-Gaussian distribution, the relation between central moments and cumulants $\kappa$ for the first six orders is as follows.

$$\mu_1 = 0, \tag{5.5}$$

$$\mu_2 = \kappa_2, \tag{5.6}$$

$$\mu_3 = \kappa_3, \tag{5.7}$$

$$\mu_4 = \kappa_4 + 3(\kappa_2)^2, \tag{5.8}$$

$$\mu_5 = \kappa_5 + 10\kappa_3\kappa_2, \tag{5.9}$$

$$\mu_6 = \kappa_6 + 15\kappa_4\kappa_2 + 10(\kappa_3)^2 + 15(\kappa_2)^3. \tag{5.10}$$

### 5.2.4 Useful trick: the generating function

The generating function allows one, among other things, to compute quickly moments and cumulants of a distribution. Define the generating function as

$$Z(k) = \langle \exp(ikx) \rangle = \int dx \, \exp(ikx)\mathcal{P}(x), \tag{5.11}$$

which may sound familiar as it is a sort of Fourier transform. Note that this can be written as an infinite series (by expanding the exponential) giving (exercise)

$$Z(k) = \sum_{n=0}^{\infty} \frac{(ik)^n}{n!} \hat{\mu}_n. \tag{5.12}$$

So far nothing special, but now the neat trick is that the *moments* are obtained as:

$$\hat{\mu}_n = (-i^n)\frac{d^n}{dk^n}Z(k)|_{k=0}, \tag{5.13}$$

and the *cumulants* are obtained by doing the same operation on $\ln Z$. While this seems just a neat trick for now it will be very useful shortly.

### 5.2.5 Two useful distributions

Two distributions are widely used in cosmology: these are the Poisson distribution and the Gaussian distribution.

### 5.2.5.1 The Poisson distribution

The Poisson distribution describes an independent point process: photon noise, radioactive decay, galaxy distribution for very few galaxies, point sources, etc. It is an example of a discrete probability distribution. For cosmological applications it is useful to think of a Poisson process as follows. Consider a random process (for example a random distribution of galaxies in space) of average density $\rho$. Divide the space into infinitesimal cells, of volume $\delta V$ so small that their occupation can only be 0 or 1 and the probability of having more than one object per cell is 0. Then the probability of having one object in a given cell is $\mathcal{P}_1 = \rho\delta V$ and the probability of getting no object in the cell is therefore $\mathcal{P}_0 = 1 - \rho\delta V$. Thus for one cell the generating function is $Z(k) = \sum_n \mathcal{P}_n \exp(ikn) = 1 + \rho\delta V(\exp(ik) - 1)$ and for a volume $V$ with $V/\delta V$ cells, we have $Z(k) = [1 + \rho\delta V(\exp(ik) - 1)]^{V/\delta V} \sim \exp[\rho V(\exp(ik) - 1)]$. With the substitution $\rho V \longrightarrow \lambda$, we obtain

$$Z(k) = \exp[\lambda(\exp(ik) - 1)] = \sum_{n=0}^{\infty} \frac{\lambda^n}{n!} \exp(-\lambda) \exp(ikn).$$

Thus the Poisson probability distribution we recover is:

$$\mathcal{P}_n = \frac{\lambda^n}{n!} \exp[-\lambda]. \tag{5.14}$$

─────────────────────────── Exercise ───────────────────────────

(8) Show that for the Poisson distribution we have $\langle n \rangle = \lambda$ and that $\sigma^2 = \lambda$.

### 5.2.5.2 *The Gaussian distribution*

The Gaussian distribution is extremely useful because of the "central limit theorem." The central limit theorem states that the sum of many independent and identically distributed random variables will be approximately Gaussianly distributed. The conditions for this to happen are quite mild: the variance of the distribution one starts off with has to be finite. The proof is remarkably simple.

Let's take $n$ events with probability distributions $\mathcal{P}(x_i)$ and $< x_i >= 0$ for simplicity, and let Y be their sum. What is $\mathcal{P}(Y)$? The generating function for $Y$ is the product of the generating functions for the $x_i$:

$$Z_Y(k) = \sum_{m=0}^{m=\infty} \left[ \frac{(ik)^m}{m!} \mu^m \right]^n \simeq \left( 1 - \frac{1}{2} \frac{k^2 < x^2 >}{n} + \cdots \right)^n \tag{5.15}$$

for $n \longrightarrow \infty$ then $Z_Y(k) \longrightarrow \exp[-1/2k^2 < x^2 >]$. By recalling the definition of generating function, Eq. (5.11), we can see that the probability distribution which generated this $Z$ is

$$\mathcal{P}(Y) = \frac{1}{\sqrt{2\pi < x^2 >}} \exp \left[ -\frac{1}{2} \frac{Y^2}{< x^2 >} \right] ; \tag{5.16}$$

that is a Gaussian!

─────────────────────────── Exercises ───────────────────────────

(9) Verify that higher-order cumulants are zero for the Gaussian distribution.
(10) Show that the "central limit theorem" holds for the Poisson distribution.

Beyond the central limit theorem, the Gaussian distribution is very important in cosmology as we believe that the initial conditions, the primordial perturbations generated from inflation, had a distribution very close to Gaussian (although it is crucial to test this experimentally).

We should also remember that thanks to the central limit theorem, when we estimate parameters in cosmology in many cases we approximate our data as having a Gaussian distribution, even if we know that each data point is *NOT* drawn from a Gaussian distribution. The central limit theorem simplifies our lives every day.

There are exceptions though. Let us for example consider $N$ independent data points drawn from a Cauchy distribution:

$$\mathcal{P}(x) = \left[ \pi\sigma(1 + \frac{(x - \bar{x})^2}{\sigma^2}) \right]^{-1}.$$

This is a proper probability distribution as it integrates to unity, but moments diverge. One can show that the numerical mean of a finite number $N$ of observations is finite but the "population mean" [the one defined through the integral of Eq. (5.4) with $f(x) = x$] is not. Note also that the scatter in the average of $N$ data points drawn from this distribution is the same as the scatter in 1-point: the scatter never diminishes regardless of the sample size.

## 5.3 Modeling of data and statistical inference

To illustrate this let us follow the example from Wall and Jenkins (2003). If you have an urn with $N$ red balls and $M$ blue balls and you draw from the urn, probability theory can tell you what the chances are of you picking a red ball given that you have so far drawn $m$ blue and $n$ red ones. However in practice what you want to do is to use probability to tell you what the distribution of the balls in the urn is, having drawn a few from it!

In other words, if you knew everything about the Universe, probability theory could tell you what the probabilities are of getting a given outcome for an observation. However, especially in cosmology, you want to make few observations and draw conclusions about the Universe. With the added complication that experiments in cosmology are not quite like experiments in the laboratory: you can't poke the Universe and see how it reacts, and in many cases you can't repeat the observation, and you can only see a small part of the Universe! Keeping this caveat in mind let's push ahead.

Given a set of observations, often you want to fit a model to the data, where the model is described by a set of parameters $\boldsymbol{\alpha}$. Sometimes the model is physically motivated (say CMB angular power spectra, etc.) or a convenient function (e.g. initial studies of large-scale structure were fitting galaxy correlation functions with power laws). Then you want to define a merit function, that measures the agreement between the data and the model: by adjusting the parameters to maximize the agreement one obtains the *best fit parameters*. Of course, because of measurement errors, there will be errors associated with the parameter determination. To be useful, a fitting procedure should provide (a) best-fit parameters (b) error estimates on the parameters and (c) possibly a statistical measure of the goodness of fit. When (c) suggests that the model is a bad description of the data, then (a) and (b) make no sense.

Remember at this point Bayes' theorem: while you may want to ask: "What is the probability that a particular set of parameters is correct?", what you can ask to a *"figure of merit"* is "Given a set of parameters, what is the probability that this data set could have occurred?". This is the likelihood. You may want to estimate parameters by maximizing the likelihood and somehow identify the likelihood (probability of the data given the parameters) with the likelihood of the model parameters.

### 5.3.1 Chi-square, goodness of fit and confidence regions

Following, Press *et al.* (1992, ch. 15), it is easier to introduce model fitting and parameter estimation using the least-squares example. Let's say that $D_i$ are our data points and $y(\boldsymbol{x}_i|\boldsymbol{\alpha})$ a model with parameters $\boldsymbol{\alpha}$ (Fig. 5.1). For example if the model is a straight line then $\boldsymbol{\alpha}$ denotes the slope and intercept of the line.

The least squares is given by:

$$\chi^2 = \sum_i w_i [D_i - y(\boldsymbol{x}_i|\boldsymbol{\alpha})]^2 \tag{5.17}$$

and you can show that the minimum variance weights are $w_i = 1/\sigma_i^2$.

———————————————— Exercise ————————————————

(11) If the points are correlated, how does this equation change?

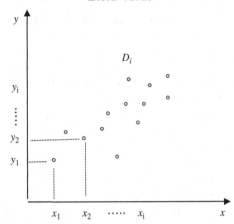

FIGURE 5.1. Example of linear fit to the data points $D_i$ in two dimensions.

Best-fit-value parameters are the parameters that minimize the $\chi^2$. Note that a numerical exploration can be avoided; by solving $\partial \chi^2 / \partial \alpha_i \equiv 0$ you can find the best fit parameters.

### 5.3.1.1 Goodness of fit

In particular, if the measurement errors are Gaussianly distributed, and (as in this example) the model is a linear function of the parameters, then the probability distribution of the $\chi^2$ at the minimum follows a $\chi^2$ distribution for $\nu \equiv n - m$ degrees of freedom (where $m$ is the number of parameters and $n$ is the number of data points). The probability that the observed $\chi^2$, even for a correct model, is less than a value $\hat{\chi}^2$ is $\mathcal{P}(\chi^2 < \hat{\chi}^2, \nu) = \mathcal{P}(\nu/2, \hat{\chi}^2/2) = \Gamma(\nu/2, \hat{\chi}^2/2)$, where $\Gamma$ stands for the incomplete Gamma function. Its complement, $Q = 1 - \mathcal{P}(\nu/2, \hat{\chi}^2/2)$ is the probability that the observed $\chi^2$ exceed by chance $\hat{\chi}^2$ even for a correct model (see Press *et al.*, 1992, ch. 6.3 and 15.2 for more details). It is common that the chi-square distribution holds even for models that are non-linear in the parameters and even in more general cases (see an example later).

The computed probability $Q$ gives a quantitative measure of the goodness of fit when evaluated at the best-fit parameters (i.e. at $\chi^2_{\min}$). If $Q$ is a very small probability then
(a) the model is wrong and can be rejected; or
(b) the errors are really larger than stated; or
(c) the measurement errors were not Gaussianly distributed.
If you know the actual error distribution you may want to *Monte Carlo simulate* synthetic data sets, subject them to your actual fitting procedure, and determine both the probability distribution of your $\chi^2$ statistic and the accuracy with which model parameters are recovered by the fit (see Section 5 on Monte Carlo methods).

On the other hand $Q$ may be too large; if it is too near 1 then something's also up:
(a) errors may have been overestimated; or
(b) the data are correlated and correlations were ignored in the fit.
(c) In principle it may be that the distribution you are dealing with is more compact than a Gaussian distribution, but this is almost never the case. So make sure you exclude cases (a) and (b) before you invest a lot of time in exploring option (c).

*Postscript*: the "Chi-by eye" rule is that the minimum $\chi^2$ should be roughly equal to the number of data minus the number of parameters (giving rise to the widespread use of the so-called "reduced chi-square"). Can you – possibly rigorously – justify this statement?

### 5.3.1.2 Confidence region

Rather than presenting the full probability distribution of errors it is useful to present confidence limits or confidence regions: a region in the $m$-dimensional parameter space ($m$ being the number of parameters) that contain a certain percentage of the total probability distribution. Obviously you want a suitably compact region around the best-fit value. It is customary to choose 68.3%, 95.4%, 99.7%. Ellipsoidal regions have connections with the normal (Gaussian) distribution but in general things may be very different. A natural choice for the shape of confidence intervals is given by constant $\chi^2$ boundaries. For the observed data set the value of parameters $\alpha_0$ minimize the $\chi^2$, denoted by $\chi^2_{min}$. If we perturb $\alpha$ away from $\alpha_0$ the $\chi^2$ will increase. From the properties of the $\chi^2$ distribution it is possible to show that there is a well-defined relation between confidence intervals, formal standard errors and $\Delta\chi^2$. We report here the $\Delta\chi^2$ for the conventionals 1, 2 and 3-$\sigma$ as a function of the number of parameters for the joint confidence levels.

| $\sigma$ | $p(\%)$ | 1 | 2 | 3 |
|---|---|---|---|---|
| 1-$\sigma$ | 68.3 | 1.00 | 2.30 | 3.53 |
| | 90 | 2.71 | 4.61 | 6.25 |
| 2-$\sigma$ | 95.4 | 4.00 | 6.17 | 8.02 |
| 3-$\sigma$ | 99.73 | 9.00 | 11.8 | 14.2 |

In general, let's spell out the following prescription. If $\mu$ is the number of fitted parameters for which you want to plot the joint confidence region and $p$ is the confidence limit desired, you need to find the $\Delta\chi^2$ such that the probability of a chi-square variable with $\mu$ degrees of freedom being less than $\Delta\chi^2$ is $p$. For general values of $p$ this is given by $Q$ described above (for the standard 1, 2, 3-$\sigma$ see table above).

*P.S.* Frequentists use $\chi^2$ a lot.

### 5.3.2 Likelihoods

One can be more sophisticated than $\chi^2$, if $\mathcal{P}(D)$ ($D$ represents data) is known. Remember from the Bayes' theorem (Eq. 5.3), the probability of the data given the model (hypothesis) is the likelihood. If we set $\mathcal{P}(D) = 1$ (after all, you got the data) and ignore the prior, by maximizing the likelihood we find the most likely hypothesis or, often, the most likely parameters of a given model.

Note that we have ignored $\mathcal{P}(D)$ and the prior so in general this technique does not give you a goodness of fit, nor an absolute probability of the model, only relative probabilities. Frequentists rely on $\chi^2$ analyses where a goodness of fit can be established.

In many cases (thanks to the central limit theorem) the likelihood can be well approximated by a multivariate Gaussian:

$$\mathcal{L} = \frac{1}{(2\pi)^{n/2} |\det C|^{1/2}} \exp\left[ -\frac{1}{2} \sum_{ij} (D-y)_i C_{ij}^{-1} (D-y)_j \right], \qquad (5.18)$$

where $C_{ij} = \langle (D_i - y_i)(D_j - y_j) \rangle$ is the covariance matrix.

───────────────── Exercise ─────────────────

(12) When are likelihood analyses and $\chi^2$ analyses the same?

*5.3.2.1 Confidence levels for likelihood*

For Bayesian statistics, confidence regions are found as regions $R$ in *model space* such that $\int_R \mathcal{P}(\boldsymbol{\alpha}|D)\mathrm{d}\boldsymbol{\alpha}$ is, say, 0.68 for 68% confidence level and 0.95 for 95% confidence. Note that this encloses the prior information. To report results independently of the prior the likelihood ratio is used. In this case compare the likelihood at a particular point in model space $\mathcal{L}(\boldsymbol{\alpha})$ with the value of the maximum likelihood $\mathcal{L}_{\mathrm{max}}$. Then a model is said to be acceptable if

$$-2\ln\left[\frac{\mathcal{L}(\boldsymbol{\alpha})}{\mathcal{L}_{\mathrm{max}}}\right] \leq \text{threshold.} \tag{5.19}$$

Then the threshold should be calibrated by calculating the distribution of the likelihood ratio in the case where a particular model is the true model. There are some cases, however, when the value of the threshold is the corresponding confidence limit for a $\chi^2$ with $m$ degrees of freedom, for $m$ number of parameters.

——————————————— Exercise ———————————————

(13) In what cases?[3]

_____

*5.3.3 Marginalization, combining different experiments*

Of all the model parameters $\alpha_i$ some of them may be uninteresting. Typical examples of nuisance parameters are calibration factors, the galaxy bias parameter, etc., but also it may be that we are interested on constraints on only one cosmological parameter at a time rather than on the *joint* constraints on two or more parameters simultaneously. One then marginalizes over the uninteresting parameters by integrating the posterior distribution:

$$P(\alpha_1\ldots\alpha_j|D) = \int \mathrm{d}\alpha_{j+1}\ldots\mathrm{d}\alpha_m P(\boldsymbol{\alpha}|D) \tag{5.20}$$

if there are in total $m$ parameters and we are interested in $j$ of them ($j < m$). Note that if you have two independent experiments, the combined likelihood of the two experiments is just the product of the two likelihoods. (of course if the two experiments are non independent then one would have to include their covariance). In many cases one of the two experiments can be used as a prior. A word of caution is in order here. We can always combine independent experiments by multiplying their likelihoods, and if the experiments are good and sound and the model used is a good and complete description of the data, all is well. However it is always important to: (a) think about the priors one is using and to quantify their effects; (b) make sure that results from independent experiments are consistent. By multiplying likelihood from inconsistent experiments you can always get some sort of results but it does not mean that the result actually makes sense.

Sometimes you may be interested in placing a prior on the uninteresting parameters before marginalization. The prior may come from a previous measurement or from your "belief."

Typical examples of this are: marginalization over calibration uncertainty, over point-source amplitude or over beam errors for CMB studies. For example for marginalization

_____

[3] Solution: the data must have Gaussian errors, the model must depend linearly on the parameters, the gradients of the model with respect to the parameters are not degenerate and the parameters do not affect the covariance.

over, say, point-source amplitude, it is useful to know of the following trick for Gaussian likelihoods:

$$P(\alpha_1 \ldots \alpha_{m-1}|D) = \int \frac{\mathrm{d}A}{(2\pi)^{\frac{m}{2}}||C||^{\frac{1}{2}}} \mathrm{e}^{\left[-\frac{1}{2}[C_i-(\hat{C}_i+AP_i)]\Sigma_{ij}^{-1}[C_j-(\hat{C}_j+AP_j)]\right]}$$

$$\times \frac{1}{\sqrt{2\pi\sigma^2}} \exp\left[-\frac{1}{2}\frac{(A-\hat{A})^2}{\sigma^2}\right], \qquad (5.21)$$

where repeated indices are summed over and $||C||$ denotes the determinant. Here, $A$ is the amplitude of, say, a point-source contribution $P$ to the $C_\ell$ angular power spectrum, $A$ is the $m$-th parameter which we want to marginalize over with a Gaussian prior with variance $\sigma^2$ around $\hat{A}$. The trick is to recognize that this integral can be written as:

$$P(\alpha_1 \ldots \alpha_{m-1}|D) = \int C_0 \exp\left[-\frac{1}{2}C_1 - 2C_2A + C_3A^2\right]\mathrm{d}A, \qquad (5.22)$$

(where $C_{0\ldots3}$ denote constants) and that this kind of integral is evaluated by using the substitution $A \longrightarrow A - C_2/C_3$, giving something proportional to $\exp[-1/2(C_1 - C_2^2/C_3)]$. It is left as an exercise to write the constants explicitly.

### 5.3.4 An example

Let's say you want to constrain Cosmology by studying cluster number counts as a function of redshift. Here we follow the paper of Cash (1979). The observation of a discrete number $N$ of clusters is a Poisson process, the probability of which is given by the product

$$\mathcal{P} = \Pi_{i=1}^N[e_i^{n_i}\exp(-e_i)/n_i!], \qquad (5.23)$$

where $n_i$ is the number of clusters observed in the $i$-th experimental bin and $e_i$ is the expected number in that bin in a given model: $e_i = I(x)\delta x_i$ with $I$ being proportional to the probability distribution. Here $\delta x_i$ can represent an interval in clusters mass and/or redshift. Note: this is a product of Poisson distributions, thus one is assuming that these are independent processes. Clusters may be clustered, so when can this be used?

For unbinned data (or for small bins so that bins have only 0 and 1 counts) we define the quantity:

$$C \equiv -2\ln\mathcal{P} = 2(E - \sum_{i=1}^N \ln I_i), \qquad (5.24)$$

where $E$ is the total expected number of clusters in a given model. The quantity $\Delta C$ between two models with different parameters has a $\chi^2$ distribution! (So all that was said in the $\chi^2$ section applies, even though we started from a highly non-Gaussian distribution.)

## 5.4 Description of random fields

Let's take a break from probabilities and consider a slightly different issue. In comparing the results of theoretical calculations with the observed Universe, it would be meaningless to hope to be able to describe with theory the properties of a particular patch, i.e. to predict the density contrast of the matter $\delta(\boldsymbol{x}) = \delta\rho(x)/\rho$ at any specific point $\boldsymbol{x}$. Instead, it is possible to predict the average statistical properties of the mass distribution.[4] In addition, we consider that the Universe we live in is a random realization of all the possible universes that could have been a realization of the true underlying model (which is known only to Mother Nature). All the possible realizations of this true underlying Universe make up the *ensemble*. In statistical inference one may sometimes want to try to estimate how different our particular realization of the Universe could be from the true underlying one. Thinking back to the example of the urn with colored balls, it would be like considering that the particular urn from which we are drawing the balls is only one possible realization of the true underlying distribution of urns. For example, say that the true distribution has a 50–50 split in red and blue balls but that the urn can have only an odd number of balls. Clearly the exact 50–50 split cannot be realized in one particular urn but it can be realized in the ensemble. . .

Following the *cosmological principle* (e.g. Peebles, 1980), models of the Universe have to be homogeneous on average, therefore, in widely separated regions of the Universe (i.e. independent), the density field must have the same statistical properties.

A crucial assumption of standard cosmology is that the part of the Universe that we can observe is a *fair sample* of the whole. This is closely related to the cosmological principle since it implies that the statistics like the correlation functions have to be considered as averages over the ensemble. But the peculiarity in cosmology is that we have just one Universe, which is just one realization from the ensemble (a quite fictitious one: it is the ensemble of all possible universes). The fair sample hypothesis states that samples from well-separated parts of the Universe are independent realizations of the same physical process, and that, in the observable part of the Universe, there are enough independent samples to be representative of the statistical ensemble. The hypothesis of ergodicity follows: averaging over many realizations is equivalent to averaging over a large (enough) volume. The cosmological field we are interested in, in a given volume, is taken as a realization of the statistical process and, for the hypothesis of ergodicity, averaging over many realizations is equivalent to averaging over a large volume.

Theories can just predict the statistical properties of $\delta(x)$ which, for the cosmological principle, must be a homogeneous and isotropic random field, and our observable Universe is a random realization from the ensemble.

In cosmology, the scalar field $\delta(x)$ is enough to specify the initial fluctuation field, and – we ultimately hope – also the present-day distribution of galaxies and matter. Here lies one of the big challenges of modern cosmology (see e.g. Martínez and Saar, 2002).

A fundamental problem in the analysis of the cosmic structures is to find the appropriate tools to provide information on the distribution of the density fluctuations, on their initial conditions and subsequent evolution. Here we concentrate on power spectra and correlation functions.

### 5.4.1 Gaussian random fields

Gaussian random fields are crucially important in cosmology, for different reasons: first of all it is possible to describe their statistical properties analytically, but also there are strong theoretical motivations, namely inflation, to assume that the primordial fluctuations that gave rise to the present-day cosmological structures follow a Gaussian distribution. Without resorting to inflation, for the central limit theorem, Gaussianity results from a superposition of a large number of random processes.

[4] A very similar approach is taken in statistical mechanics.

The distribution of density fluctuations $\delta$ defined as[5] $\delta = \delta\rho/\rho$ cannot be exactly Gaussian because the field has to satisfy the constraint $\delta > -1$; however if the amplitude of the fluctuations is small enough, this can be a good approximation. This seems indeed to be the case: by looking at the CMB anisotropies we can probe fluctuations when their statistical distribution should have been close to its primordial one; possible deviations from Gaussianity of the primordial density field are small.

If $\delta$ is a Gaussian random field with an average value of 0, its probability distribution is given by:

$$P_n(\delta_1, \cdots, \delta_n) = \frac{\sqrt{\det \mathbf{C}^{-1}}}{(2\pi)^{n/2}} \exp\left[-\frac{1}{2}\delta^T \mathbf{C}^{-1}\delta\right], \qquad (5.25)$$

where $\delta$ is a vector made by the $\delta_i$, $\mathbf{C}^{-1}$ denotes the inverse of the correlation matrix the elements of which are $\mathbf{C}_{ij} = \langle \delta_i \delta_j \rangle$.

An important property of Gaussian random fields is that the Fourier transform of a Gaussian field is still Gaussian. The phases of the Fourier modes are random and the real and imaginary part of the coefficients have Gaussian distribution and are mutually independent.

Let us denote the real and imaginary part of $\delta_k$ by $Re\delta_k$ and $Im\delta_k$ respectively. Their joint probability distribution is the bivariate Gaussian:

$$P(Re\delta_k, Im\delta_k)dRe\delta_k dIm\delta_k = \frac{1}{2\pi\sigma_k^2} \exp\left[-\frac{Re\delta_k^2 + Im\delta_k^2}{2\sigma_k^2}\right] dRe\delta_k dIm\delta_k, \qquad (5.26)$$

where $\sigma_k^2$ is the variance in $Re\delta_k$ and $Im\delta_k$, which in the case of isotropy depends only on the magnitude of $k$. Equation (5.26) can be rewritten in terms of the amplitude $|\delta_k|$ and the phase $\phi_k$ as:

$$P(|\delta_k|, \phi_k)d|\delta_k|d\phi_k = \frac{1}{2\pi\sigma_k^2} \exp\left[-\frac{|\delta_k|^2}{2\sigma_k^2}\right] |\delta_k|d|\delta_k|d\phi_k; \qquad (5.27)$$

that is, $|\delta_k|$ follows a Rayleigh distribution.

From this it follows that the probability that the amplitude is above a certain threshold $X$ is:

$$P(|\delta_k|^2 > X) = \int_{\sqrt{X}}^{\infty} \frac{1}{\sigma_k^2} \exp\left[-\frac{|\delta_k|^2}{2\sigma_k^2}\right] |\delta_k|d|\delta_k| = \exp\left[-\frac{X}{\langle|\delta_k|^2\rangle}\right], \qquad (5.28)$$

which is an exponential distribution.

The fact that the phases of a Gaussian field are random, implies that the two-point correlation function (or the power spectrum) completely specifies the field.

*P.S.* If your advisor now asks you to generate a Gaussian random field you know how to do it. If you are not familiar with Fourier transforms see the next section.

However, the observed fluctuation field is not Gaussian. The observed galaxy distribution is highly non-Gaussian mainly due to gravitational instability. To completely specify a non-Gaussian distribution, higher-order correlation functions are needed;[6] conversely deviations from Gaussian behavior can be characterized by the higher-order statistics of the distribution.

---

[5] Note that $\langle \delta \rangle = 0$

[6] For "non pathological" distributions. For a discussion see e.g. Kendall and Stuart (1977).

## 5.4.2 Basic tools

The Fourier transform of the (fractional) overdensity field $\delta$ is defined as:

$$\delta_{\boldsymbol{k}} = A \int \mathrm{d}^3 r \delta(\boldsymbol{r}) \exp[-i\boldsymbol{k} \cdot \boldsymbol{r}], \qquad (5.29)$$

with inverse

$$\delta(\boldsymbol{r}) = B \int \mathrm{d}^3 k \delta_{\boldsymbol{k}} \exp[i\boldsymbol{k} \cdot \boldsymbol{r}], \qquad (5.30)$$

and the Dirac delta is then given by

$$\delta^D(\boldsymbol{k}) = BA \int \mathrm{d}^3 r \exp[\pm i\boldsymbol{k} \cdot \boldsymbol{r}]. \qquad (5.31)$$

Here I chose the convention $A = 1$, $B = 1/(2\pi)^3$, but always beware of the Fourier transform (FT) conventions.

The *two-point correlation function* (or correlation function) is defined as:

$$\xi(x) = \langle \delta(\boldsymbol{r})\delta(\boldsymbol{r}+\boldsymbol{x}) \rangle = \int < \delta_{\boldsymbol{k}}\delta_{\boldsymbol{k}'} > \exp[i\boldsymbol{k} \cdot \boldsymbol{r}] \exp[i\boldsymbol{k} \cdot (\boldsymbol{r}+\boldsymbol{x})] \mathrm{d}^3 k \mathrm{d}^3 k'. \qquad (5.32)$$

Due to isotropy, $\xi(|x|)$ is only a function of the distance and not of the orientation. Note that in some cases when isotropy is broken one may want to keep the orientation information (see e.g. redshift space distortions, which affect clustering only along the line of sight).

The definition of the power spectrum $P(k)$ follows:

$$< \delta_{\boldsymbol{k}}\delta_{\boldsymbol{k}'}^* >= (2\pi)^3 P(k)\delta^D(\boldsymbol{k}+\boldsymbol{k}'); \qquad (5.33)$$

again for isotropy $P(k)$ depends only on the modulus of the $k$-vector, although in special cases where isotropy is broken one may want to keep the direction information. Since $\delta(\boldsymbol{r})$ is real, we have that $\delta_{\boldsymbol{k}}^* = \delta_{-\boldsymbol{k}}$, so

$$< \delta_{\boldsymbol{k}}\delta_{\boldsymbol{k}'}^* >= (2\pi)^3 \int \mathrm{d}^3 x \xi(x) \exp[-i\boldsymbol{k} \cdot \boldsymbol{x}]\delta^D(\boldsymbol{k}-\boldsymbol{k}'). \qquad (5.34)$$

The power spectrum and the correlation function are Fourier transform pairs:

$$\xi(x) = \frac{1}{(2\pi)^3} \int P(k) \exp[i\boldsymbol{k} \cdot \boldsymbol{r}] \mathrm{d}^3 k, \qquad (5.35)$$

$$P(k) = \int \xi(x) \exp[-i\boldsymbol{k} \cdot \boldsymbol{x}] \mathrm{d}^3 x. \qquad (5.36)$$

At this stage the same amount of information is enclosed in $P(k)$ as in $\xi(x)$. From here the variance is

$$\sigma^2 =< \delta^2(x) >= \xi(0) = \frac{1}{(2\pi)^3} \int P(k) \mathrm{d}^3 k, \qquad (5.37)$$

or better

$$\sigma^2 = \int \Delta^2(k) \mathrm{d}\ln k, \quad \text{where} \quad \Delta^2(k) = \frac{1}{(2\pi)^3} k^3 P(k), \qquad (5.38)$$

and the quantity $\Delta^2(k)$ is independent from the FT convention used.

Now the question is: on what scale is this variance defined? The answer is: in practice one needs to use filters. The density field is convolved with a filter (smoothing) function. There are two typical choices:

$$f = \frac{1}{(2\pi)^{3/2} R_G^3} \exp\left[-\frac{1}{2}\frac{x^2}{R_G^2}\right] \quad \text{Gaussian} \rightarrow f_k = \exp[-k^2 R_G^2/2] \qquad (5.39)$$

$$f = \frac{1}{(4\pi) R_T^3}\Theta(x/R_T) \quad \text{Top hat} \rightarrow f_k = \frac{3}{(kR_T)^3}[\sin(kR_T) - kR_T\cos(kR_T)] \qquad (5.40)$$

roughly $R_T \simeq \sqrt{5}R_G$. Remember: *Convolution in real space is a multiplication in Fourier space; multiplication in real space is a convolution in Fourier space.*

--- Exercise ---

(14) Consider a multivariate Gaussian distribution:

$$P(\delta_1, \ldots, \delta_n) = \frac{1}{(2\pi)^{n/2}(\det \mathbf{C})^{1/2}} \exp[-\frac{1}{2}\delta^T \mathbf{C}^{-1}\delta]$$

where $C_{ij} = <\delta_i \delta_j>$ is the covariance. Show that if $\delta_i$ are Fourier modes then $C_{ij}$ is diagonal. This is an ideal case, of course, but this is telling us that for Gaussian fields the different $k$ modes are independent! This is always a nice feature.

(15) Another question for you: if you start off with a Gaussian distribution (say from inflation) and then leave this Gaussian field $\delta$ to evolve under gravity, will it remain Gaussian forever? Hint: think about present-time Universe, and think about the dark matter density at, say, the center of a big galaxy and in a large void.

### 5.4.2.1 The importance of the power spectrum

The structure of the Universe on large scales is largely dominated by the force of gravity (which we think we know well) and no too much by complex mechanisms (baryonic physics, galaxy formation etc.) – or at least that's the hope. Theory (see chapter 2 on inflation) give us a prediction for the primordial power spectrum:

$$P(k) = A\left(\frac{k}{k_0}\right)^n. \qquad (5.41)$$

The *spectral index*, $n$, is often taken to be a constant and the power spectrum is a power law power spectrum. However there are theoretical motivations to generalize this to

$$P(k) = A\left(\frac{k}{k_0}\right)^{n(k_0) + \frac{1}{2}\frac{dn}{d\ln k}\ln(k/k_0)} \qquad (5.42)$$

as a sort of Taylor expansion of $n(k)$ around the pivot point $k_0$; $dn/d\ln k$ is called the *running of the spectral index*.

Note that different authors often use different choices of $k_0$ (sometimes the same author in the same paper uses different choices...) so things may get confused...so let's report explicitly the conversions:

$$A(k_1) = A(k_0)\left(\frac{k_1}{k_0}\right)^{n(k_0) + 1/2(dn/d\ln k)\ln(k_1/k_0)}. \qquad (5.43)$$

—————————————————————————————— Exercises ——————————————————————————————

(16) Prove the equations above.
(17) Show that given the above definition of the running of the spectral index, $n(k) = n(k_0) + (dn/d\ln k)\ln(k/k_0)$.

_____

It can be shown that as long as linear theory applies – and only gravity is at play – $\delta \ll 1$, different Fourier modes evolve independently and the Gaussian field remains Gaussian. In addition, $P(k)$ changes only in amplitude and not in shape except in the radiation- to matter-dominated era and when there are baryon–photon interactions and baryon–dark matter interactions (see chapter 3 on CMB in this book). In detail, this is described by linear perturbation growth and by the "transfer function."

### 5.4.3 Examples of "real world" issues

Say that now you go and try to measure a $P(k)$ from a realistic galaxy catalog. What are the real-world effects you may find? We have mentioned before redshift space distortions. Here we concentrate on other effects that are more general (and not so specific to large-scale structure analysis).

### 5.4.3.1 Discrete Fourier transform

In the real world when you go and take the FT of your survey or even of your simulation box you will be using something like a fast Fourier transform code (FFT) which is a discrete Fourier transform (DFT).

If your box has a side of size $L$, even if $\delta(r)$ in the box is continuous, $\delta_k$ will be discrete. The k-modes sampled will be given by

$$\boldsymbol{k} = \left(\frac{2\pi}{L}\right)(i, j, k) \quad \text{where} \quad \Delta_k = \frac{2\pi}{L}. \tag{5.44}$$

The discrete Fourier transform is obtained by placing the $\delta(x)$ on a lattice of $N^3$ grid points with spacing $L/N$. Then:

$$\delta_k^{\text{DFT}} = \frac{1}{N^3}\sum_r \exp[-i\boldsymbol{k} \cdot \boldsymbol{r}]\delta(\boldsymbol{r}) \tag{5.45}$$

$$\delta^{\text{DFT}}(\boldsymbol{r}) = \sum_k \exp[i\boldsymbol{k} \cdot \boldsymbol{r}]\delta_k^{\text{DFT}} \tag{5.46}$$

*Beware of the mapping between r and k, some routines use a weird wrapping!*

There are different ways of placing galaxies (or particles in your simulation) on a grid: nearest grid point, cloud in cell, triangular shaped cloud, etc. For each of these *remember*(!) then to deconvolve the resulting $P(k)$ for their effect. Note that

$$\delta_k \sim \left(\frac{\Delta x}{2\pi}\right)^3 N^3 \delta_k^{\text{DFT}} \simeq \frac{1}{\Delta k^3}\delta_k^{\text{DFT}} \tag{5.47}$$

and thus

$$P(k) \simeq \frac{<|\delta^{\text{DFT}}|^2>}{(\Delta k)^3}, \quad \text{since} \quad \delta^D(k) \simeq \frac{\delta^K}{(\Delta k)^3}. \tag{5.48}$$

The discretization introduces several effects:

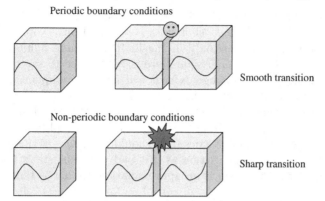

Periodic boundary conditions

Smooth transition

Non-periodic boundary conditions

Sharp transition

FIGURE 5.2. The importance of periodic boundary conditions

**Nyquist frequency.** The Nyquist frequency, $k_{\mathrm{Ny}} = \frac{2\pi}{L}\frac{N}{2}$, is that of a mode which is sampled by two grid points. Higher frequencies cannot be properly sampled and give aliasing (spurious transfer of power) effects. You should always work at $k < k_{\mathrm{Ny}}$. There is also a minimum $k$ (largest possible scale) that your finite box can test: $k_{\min} > 2\pi/L$. This is one of the – many – reasons why one needs ever larger $N$-body simulations...

In addition DFT assumes periodic boundary conditions; if you do not have periodic boundary conditions (Fig. 5.2) then this also introduces aliasing.

### 5.4.3.2 *Window, selection function, masks, etc.*

**Selection function.** Galaxy surveys are usually magnitude limited, which means that as you look further away you start missing some galaxies. The selection function tells you the probability for a galaxy at a given distance (or redshift $z$) to enter the survey. It is a multiplicative effect along the line of sight in real space.

**Window or mask.** You can never observe a perfect (or even better infinite) squared box of the Universe and in CMB studies you can never have a perfect full-sky map (we live in a galaxy . . . ). The mask ("sky cut" in CMB jargon) is a function that usually takes a value of 0 or 1 and is defined on the plane of the sky (i.e. it is constant along the same line of sight). The mask is also a real-space multiplication effect. In addition sometimes in CMB studies different pixels may need to be weighted differently, and the mask is an extreme example of this where the weights are either 0 or 1. Also this operation is a real-space multiplication effect.

Let's recall that a multiplication in real space [where $W(\boldsymbol{x})$ denotes the effects of window and selection functions]

$$\delta^{\mathrm{true}}(\boldsymbol{x}) \longrightarrow \delta^{\mathrm{obs}}(\boldsymbol{x}) = \delta^{\mathrm{true}}(\boldsymbol{x})W(\boldsymbol{x}) \tag{5.49}$$

is a convolution in Fourier space:

$$\delta^{\mathrm{true}}(\boldsymbol{k}) \longrightarrow \delta^{\mathrm{obs}}(\boldsymbol{k}) = \delta^{\mathrm{true}}(\boldsymbol{k}) * W(\boldsymbol{k}) \tag{5.50}$$

the sharper $W(\boldsymbol{r})$ is the messier and delocalized is $W(\boldsymbol{k})$. As a result it will couple different k-modes even if the underlying ones were not correlated!

**Discreteness.** While the dark matter distribution is almost a continuous one, the galaxy distribution is discrete. We usually assume that the galaxy distribution is a sampling of the dark matter distribution. The discreteness effect gives the galaxy distribution a Poisson contribution (also called "shot noise" contribution). Note that the Poisson contribution is non-Gaussian: it is only in the limit of a large number of objects (or of modes) that it approximates Gaussian. Here it will suffice to say that as long as a galaxy number density is high enough (which will need to be quantified and checked for any practical application) and we have enough modes, we say that we will have a superposition of our random field [say the dark matter one characterized by its $P(k)$] plus a white-noise contribution coming from the discreteness the amplitude of which depends on the average number density of galaxies (and should go to zero as this goes to infinity), and we treat this additional contribution as if it had the same statistical properties as the underlying density field (which is an approximation). What is the shot noise effect on the correlation properties?

Following Peebles (1980) we recognize that our random field is now given by

$$f(\boldsymbol{x}) = n(\boldsymbol{x}) = \bar{n}[1 + \delta(\boldsymbol{x})] = \sum_i \delta^D(\boldsymbol{x} - \boldsymbol{x}_i) \tag{5.51}$$

where $\bar{n}$ denotes average number of galaxies: $\bar{n} = <\sum_i \delta^D(\boldsymbol{x} - \boldsymbol{x}_i)>$. Then, as when introducing the Poisson distribution, we divide the volume in infinitesimal volume elements $\delta V$ so that their occupation can only be 0 or 1. For each of these volumes the probability of getting a galaxy is $\delta P = \rho(\boldsymbol{x})\delta V$, the probability of getting no galaxy is $\delta P = 1 - \rho(\boldsymbol{x})\delta V$ and $<n_i> = <n_i^2> = \bar{n}\delta V$. We then obtain a double stochastic process with one level of randomness coming from the underlying random field and one level coming from the Poisson sampling. The correlation function is obtained as:

$$\left\langle \sum_{ij} \delta^D(\boldsymbol{r}_1 - \boldsymbol{r}_i)\delta^D(\boldsymbol{r}_2 - \boldsymbol{r}_j) \right\rangle = \bar{n}^2(1 + \xi_{12}) + n\delta^D(\boldsymbol{r}_1 - \boldsymbol{r}_2) \tag{5.52}$$

thus

$$<n_1 n_2> = \bar{n}^2[1 + <\delta_1\delta_2>^d] \quad \text{where} \quad <\delta_1\delta_2>^d = \xi(x_{12}) + \frac{1}{\bar{n}}\delta^D(\boldsymbol{r}_1 - \boldsymbol{r}_2) \tag{5.53}$$

and in Fourier space

$$<\delta_{k_1}\delta_{k_2}>^d = (2\pi)^3 \left(P(k) + \frac{1}{\bar{n}}\right) \delta^D(\boldsymbol{k}_1 + \boldsymbol{k}_2). \tag{5.54}$$

This is not a complete surprise: the power spectrum of a superposition of two independent processes is the sum of the two power spectra.

### 5.4.3.3 Pros and cons of $\xi(r)$ and $P(k)$

Let us briefly recap the pros and cons of working with power spectra or correlation functions.

**Power spectra.**
  *Pros*: Direct connection to theory. Modes are uncorrelated (in the ideal case). The average density ends up in $P(k = 0)$ which is usually discarded, so no accurate knowledge of the mean density is needed. There is a clear distinction between linear and non-linear scales. Smoothing is not a problem (just a multiplication).
  *Cons:* Window and selection functions act as complicated convolutions, introducing mode coupling! (This is a serious issue).
**Correlation function.**
  *Pros:* No problem with window and selection function.

*Cons:* Scales are correlated. Covariance calculation is a real challenge even in the ideal case. We need to know mean densities very well. There is no clear distinction between linear and non-linear scales. There is no direct correspondence to theory.

### 5.4.3.4 ... And for CMB?

If we can observe the full sky the CMB temperature fluctuation field can be nicely expanded in spherical harmonics:

$$\Delta T(\hat{n}) = \sum_{\ell > 0} \sum_{m=-\ell}^{\ell} a_{\ell m} Y_{\ell m}(\hat{n}), \tag{5.55}$$

where

$$a_{\ell m} = \int d\Omega_n \Delta T(\hat{n}) Y_{\ell m}^*(\hat{n}). \tag{5.56}$$

and thus

$$< |a_{\ell m}|^2 > = \langle a_{\ell m} a_{\ell' m'}^* \rangle = \delta_{\ell \ell'} \delta_{m m'} C_\ell, \tag{5.57}$$

where $C_\ell$ is the angular power spectrum and

$$C_\ell = \frac{1}{(2\ell + 1)} \sum_{m=-\ell}^{\ell} |a_{\ell m}|^2. \tag{5.58}$$

Now what happens in the presence of real-world effects such as a sky cut? Analogously to the real-space case:

$$\tilde{a}_{\ell m} = \int d\Omega_n \Delta T(\hat{n}) W(\hat{n}) Y_{\ell m}^*(\hat{n}), \tag{5.59}$$

where $W(\hat{n})$ is a position-dependent weight that in particular is set to 0 on the sky cut. As any CMB observation gets pixelized this is

$$\tilde{a}_{\ell m} = \Omega_p \sum_p \Delta T(p) W(p) Y_{\ell m}^*(p), \tag{5.60}$$

where $p$ runs over the pixels and $\Omega_p$ denotes the solid angle subtended by the pixel. Clearly this can be a problem (a nasty convolution), but let us initially ignore the problem and carry on.

The pseudo-$C_\ell$s (Hivon *et al.*, 2002) are defined as:

$$\tilde{C}_\ell = \frac{1}{(2\ell + 1)} \sum_{m=-\ell}^{\ell} |\tilde{a}_{\ell m}|^2. \tag{5.61}$$

Clearly $\tilde{C}_\ell \neq C_\ell$ but

$$\langle \tilde{C}_\ell \rangle = \sum_\ell G_{\ell \ell'} \langle C_{\ell'} \rangle, \tag{5.62}$$

where $\langle \rangle$ denotes the ensemble average.

We notice already two things: as expected the effect of the mask is to couple otherwise uncorrelated modes. In large-scale structure studies people usually stop here: convolve the theory with the various real-world effects including the mask and compare that to the observed quantities. In CMB usually we go beyond this step and try to deconvolve the real-world effects.

First of all note that

$$G_{\ell_1\ell_2} = \frac{2\ell_2+1}{4\pi}\sum_{\ell_3}(2\ell_3+1)W_{\ell_3}\begin{pmatrix}\ell_1\ell_2\ell_3\\0\ 0\ 0\end{pmatrix}^2, \quad (5.63)$$

where

$$W_\ell = \frac{1}{2\ell+1}\sum_m |W_{\ell m}|^2 \quad \text{and} \quad W_{\ell m} = \int d\Omega_n W(\hat{n})Y^*_{\ell m}(\hat{n}). \quad (5.64)$$

So if you are good enough to be able to invert $G$ and you can say that $<C_\ell>$ is the $C_\ell$ you want then

$$C_\ell = \sum_{\ell'} G^{-1}_{\ell\ell'}\widetilde{C}_{\ell'}. \quad (5.65)$$

Not for all experiments it is viable (possible) to do this last step.

In addition to this, the instrument has other effects such as *noise* and a finite *beam*.

### 5.4.3.5 Noise and beams

Instrumental noise and the finite resolution of any experiment affect the measured $C_\ell$. The effect of the noise is easily found: the instrumental noise is an independent random process with a Gaussian distribution superposed to the temperature field. In $a_{\ell m}$ space $a_{\ell m} \longrightarrow a^{\text{signal}}_{\ell m} + a^{\text{noise}}_{\ell m}$.

While $\langle a^{\text{noise}}_{\ell m}\rangle = 0$, in the power spectrum this gives rise to the so-called noise bias:

$$C^{\text{measured}}_\ell = C^{\text{signal}}_\ell + C^{\text{noise}}_\ell, \quad (5.66)$$

where $C^{\text{noise}}_\ell = \sum_m |a^{\text{noise}}_{\ell m}|^2/(2\ell+1)$. As the expectation value of $C^{\text{noise}}_\ell$ is non-zero, this is a *biased estimator*.

Note that the noise bias disappears if one computes the so-called cross $C_\ell$ obtained as a cross correlation between different, uncorrelated, detectors (say detector a and b) as $\langle a^{\text{noise,a}}_{\ell m}a^{\text{noise,b}}_{\ell m}\rangle = 0$. One is however not getting something for nothing: when one computes the covariance (or the associated error) for auto and for cross correlation $C_\ell$ (exercise!) the covariance is the same and includes the extra contribution of the noise. It is only that the cross-$C_\ell$ are *unbiased* estimators.

Every experiment sees the CMB with a finite resolution given by the experimental beam (similar concept to the point spread function for optical astronomy). The observed temperature field is smoothed on the beam scales. Smoothing is a convolution in real space:

$$T_i = \int d\Omega'_n T(\hat{n})b(|\hat{n} - \hat{n}'|) \quad (5.67)$$

where we have considered a symmetric beam for simplicity. The beam is often well approximated by a Gaussian of a given full width at half maximum (FWHM). Remember that $\sigma_{\text{b}} = 0.425\text{FWHM}$.

Thus in harmonic space the beam effect is a multiplication:

$$C^{\text{measured}}_\ell = C^{\text{sky}}_\ell e^{-\ell^2\sigma^2_{\text{b}}}, \quad (5.68)$$

and in the presence of instrumental noise

$$C^{\text{measured}}_\ell = C^{\text{sky}}_\ell e^{-\ell^2\sigma^2_{\text{b}}} + C^{\text{noise}}_\ell. \quad (5.69)$$

Of course, one can always deconvolve for the effects of the beam to obtain an estimate of $C_\ell^{\text{measured}}$ as close as possible to $C_\ell^{\text{sky}}$:

$$C_\ell^{\text{measured}'} = C_\ell^{\text{sky}} + C_\ell^{\text{noise}} e^{\ell^2 \sigma_b^2}. \tag{5.70}$$

That is why it is often said that the effective noise "blows up" at high $\ell$ (small scales) and why it is important to know the beam(s) well.

─────────────────────── Exercises ───────────────────────

(18) What happens if you use cross-$C_\ell$s?

(19) What happens to $C_\ell^{\text{measured}'}$ if the beam is poorly reconstructed?

Note that the signal to noise of a CMB map depends on the pixel size (by smoothing the map and making larger pixels the noise per pixel will decrease as $\sqrt{\Omega_{\text{pix}}}$, $\Omega_{\text{pix}}$ being the new pixel solid angle), on the integration time $\sigma_{\text{pix}} = s/\sqrt{t}$ (where $s$ is the detector sensitivity and $t$ the time spent on a given pixel), and on the number of detectors $\sigma_{\text{pix}} = s/\sqrt{M}$ (where $M$ is the number of detectors).

To compare maps of different beam sizes it is useful to have a noise measure that is independent of $\Omega_{\text{pix}}$: $w = (\sigma_{\text{pix}}^2 \Omega_{\text{pix}})^{-1}$.

─────────────────────── Exercise ───────────────────────

(20) Compute the expression for $C_\ell^{\text{noise}}$ given:
  $t =$ observing time,
  $s =$ detector sensitivity (in $\mu\text{K s}^{-\frac{1}{2}}$),
  $n =$ number of detectors,
  $N =$ number of pixels,
  $f_{\text{sky}} =$ fraction of the sky observed.
  Assume uniform noise and observing time uniformly distributed. You may find Knox (1995) very useful.

### 5.4.3.6 Aside: higher-order correlations

From what we have learned so far we can conclude that the power spectrum (or the correlation function) completely characterizes the statistical properties of the density field if it is Gaussian. But what if it is not?

Higher-order correlations are defined as: $\langle \delta_1, \ldots, \delta_m \rangle$ where the "deltas" can be in real space giving the correlation function or in Fourier space giving power spectra.

At this stage, it is useful to present here *Wick's theorem* (or cumulant expansion theorem). The correlation of order $m$ can in general be written as the sum of products of irreducible (*connected*) correlations of order $\ell$ for $\ell = 1 \ldots m$. For example for order three we obtain:

$$\begin{aligned}
\langle \delta_1 \delta_2 \delta_3 \rangle_f &= \langle \delta_1 \rangle \langle \delta_2 \rangle \langle \delta_3 \rangle \\
&+ \langle \delta_1 \rangle \langle \delta_2 \delta_3 \rangle + (3 \text{ cyc. terms}) \\
&+ \langle \delta_1 \delta_2 \delta_3 \rangle,
\end{aligned} \tag{5.71}$$

and for order six (but for a distribution of zero mean):

$$< \delta_1 \ldots \delta_6 >_f = \langle \delta_1 \delta_2 \rangle \langle \delta_3 \delta_4 \rangle \langle \delta_5 \delta_6 \rangle + \ldots (15 \text{ terms})$$
$$+ \langle \delta_1 \delta_2 \rangle \langle \delta_3 \delta_4 \delta_5 \delta_6 \rangle + \ldots (15 \text{ terms})$$
$$+ \langle \delta_1 \delta_2 \delta_3 \rangle \langle \delta_4 \delta_5 \delta_6 \rangle + \ldots (10 \text{ terms})$$
$$+ \langle \delta_1 \ldots \delta_6 \rangle \tag{5.72}$$

For computing covariances of power spectra, it is useful to be familiar with the above expansion of order four.

## 5.5 More on likelihoods

While the CMB temperature distribution is Gaussian (or very close to Gaussian) the $C_\ell$ distribution is not. At high $\ell$ the central limit theorem will ensure that the likelihood is well approximated by a Gaussian but at low $\ell$ this is not the case:

$$\mathcal{L}(T|C_\ell^{\text{th}}) \propto \frac{\exp[-(T S^{-1} T)/2]}{\sqrt{\det(S)}}, \tag{5.73}$$

where $T$ denotes a vector of the temperature map, $C_\ell^{\text{th}}$ denotes the $C_\ell$ given by a theoretical model (e.g. a cosmological parameters set) and $S_{ij}$ is the signal covariance:

$$S_{ij} = \sum_\ell \frac{(2\ell + 1)}{4\pi} C_\ell^{\text{th}} P_\ell(\hat{n}_i \cdot \hat{n}_j) \tag{5.74}$$

and $P_\ell$ denote the Legendre polynomials. If we then expand $T$ in spherical harmonics we obtain:

$$\mathcal{L}(T|C_\ell^{\text{th}}) \propto \frac{\exp[-1/2|a_{\ell m}|^2/C_\ell^{\text{th}}]}{\sqrt{C_\ell^{\text{th}}}}. \tag{5.75}$$

Isotropy means that we can sum over $m's$ thus:

$$-2\ln\mathcal{L} = \sum_\ell (2\ell + 1)\left[ \ln\left(\frac{C_\ell^{\text{th}}}{C_\ell^{\text{data}}}\right) + \left(\frac{C_\ell^{\text{data}}}{C_\ell^{\text{th}}}\right) - 1\right] \tag{5.76}$$

where $C_\ell^{\text{data}} = \sum_m |a_{\ell m}|^2/(2\ell + 1)$.

──────────────────────── Exercises ────────────────────────

(21) Show that for an experiment with (Gaussian) noise the expression is the same but with the substitution $C_\ell^{\text{th}} \longrightarrow C_\ell^{\text{th}} + \mathcal{N}_\ell$ with $\mathcal{N}$ denoting the power spectrum of the noise.

(22) Show that for a partial-sky experiment (that covers a fraction of sky $f_{\text{sky}}$) you can approximately write:

$$\ln\mathcal{L} \longrightarrow f_{\text{sky}} \ln\mathcal{L}.$$

Hint: think about how the number of independent modes scales with the sky area.

─────────────────────────────────────────────────────────

As an aside, you could ask: "But what do I do with polarization data?". Well, if the $a_{\ell m}^T$ are Gaussianly distributed, the $a_{\ell m}^E$ and $a_{\ell m}^B$ will also be. So we can generalize the approach above using a vector $(a_{\ell m}^T a_{\ell m}^E a_{\ell m}^B)$. Let us consider a full-sky, ideal experiment. Start by writing down the covariance, follow the same steps as above and show that:

$$-2\ln\mathcal{L} = \sum_\ell (2\ell+1)\left\{\ln\left(\frac{C_\ell^{BB}}{\hat{C}_\ell^{BB}}\right) + \ln\left(\frac{C_\ell^{TT}C_\ell^{EE}-(C_\ell^{TE})^2}{\hat{C}_\ell^{TT}\hat{C}_\ell^{EE}-(\hat{C}_\ell^{TE})^2}\right)\right.$$
$$\left. + \frac{\hat{C}_\ell^{TT}C_\ell^{EE}+C_\ell^{TT}\hat{C}_\ell^{EE}-2\hat{C}_\ell^{TE}C_\ell^{TE}}{C_\ell^{TT}C_\ell^{EE}-(C_\ell^{TE})^2} + \frac{\hat{C}_\ell^{BB}}{C_\ell^{BB}} - 3\right\}, \quad (5.77)$$

where $C_\ell$ denotes $C_\ell^{\text{th}}$ and $\hat{C}_\ell$ denotes $C_\ell^{\text{data}}$.

It is easy to show that for a noisy experiment $C_\ell^{XY} \longrightarrow C_\ell^{XY} + \mathcal{N}_\ell^{XY}$, where $\mathcal{N}_\ell$ denotes the noise power spectrum and $X,Y = \{T,E,B\}$.

———————————— Exercise ————————————

(23) Generalize the above to partial-sky coverage: for added complication take $f_{\text{sky}}^{TT} \neq f_{\text{sky}}^{EE} \neq f_{\text{sky}}^{BB}$. This is often the case as the sky cut for polarization may be different from that of temperature (the foregrounds are different) and in general the cut (or the weighting) for $B$ may need to be larger than that for $E$.

———————————————————————————————

Following Verde *et al.* (2003) let us now expand in Taylor series Eq. (5.76) around its maximum by writing $\hat{C}_\ell = C_\ell^{\text{th}}(1+\epsilon)$. For a single multipole $\ell$,

$$-2\ln\mathcal{L}_\ell = (2\ell+1)[\epsilon - \ln(1+\epsilon)] \simeq (2\ell+1)\left(\frac{\epsilon^2}{2} - \frac{\epsilon^3}{3} + \mathcal{O}(\epsilon^4)\right). \quad (5.78)$$

We note that the Gaussian likelihood approximation is equivalent to the above expression truncated at $\epsilon^2$: $-2\ln\mathcal{L}_{\text{Gauss},\ell} \propto (2\ell+1)/2[(\hat{C}_\ell - C_\ell^{\text{th}})/C_\ell^{\text{th}}]^2 \simeq (2\ell+1)\epsilon^2/2$.

Also widely used for CMB studies is the *log-normal likelihood* for the equal variance approximation (Bond *et al.*, 1998):

$$-2\ln\mathcal{L}_{\text{LN}}' = \frac{(2\ell+1)}{2}\left[\ln\left(\frac{\hat{C}_\ell}{C_\ell^{\text{th}}}\right)\right]^2 \simeq (2\ell+1)\left(\frac{\epsilon^2}{2} - \frac{\epsilon^3}{2}\right). \quad (5.79)$$

Thus our approximation of likelihood function is given by the form,

$$\ln\mathcal{L} = \frac{1}{3}\ln\mathcal{L}_{\text{Gauss}} + \frac{2}{3}\ln\mathcal{L}_{\text{LN}}', \quad (5.80)$$

where

$$\ln\mathcal{L}_{\text{Gauss}} \propto -\frac{1}{2}\sum_{\ell\ell'}(C_\ell^{\text{th}} - \hat{C}_\ell)Q_{\ell\ell'}(C_{\ell'}^{\text{th}} - \hat{C}_{\ell'}), \quad (5.81)$$

and

$$\ln\mathcal{L}_{\text{LN}} = -\frac{1}{2}\sum_{\ell\ell'}(z_\ell^{\text{th}} - \hat{z}_\ell)Q_{\ell\ell'}(z_{\ell'}^{\text{th}} - \hat{z}_{\ell'}), \quad (5.82)$$

where $z_\ell^{\text{th}} = \ln(C_\ell^{\text{th}} + \mathcal{N}_\ell)$, $\hat{z}_\ell = \ln(\hat{C}_\ell + \mathcal{N}_\ell)$ and $Q_{\ell\ell'}$ is the local transformation of the curvature matrix $Q$ to the lognormal variables $z_\ell$,

$$\mathcal{Q}_{\ell\ell'} = (C_\ell^{\text{th}} + \mathcal{N}_\ell)Q_{\ell\ell'}(\hat{C}_{\ell'}^{\text{th}} + \mathcal{N}_{\ell'}). \quad (5.83)$$

The curvature matrix is the inverse of the covariance matrix evaluated at the maximum likelihood. However we do not want to adopt the "equal variance approximation," so $Q$ for us will be in inverse of the covariance matrix.

The elements of the covariance matrix, even for a ideal full-sky experiment can be written as:

$$\mathbf{C}_{\ell\ell} = 2\frac{(\mathcal{C}_\ell^{\text{th}})^2}{2\ell+1}. \tag{5.84}$$

In the absence of noise the covariance is non-zero: this is the *cosmic variance*.

Note that for the second WMAP release (WMAP3; Page *et al.*, 2007; Spergel *et al.*, 2007), at low $\ell$ the likelihood is computed directly from the maps $\mathbf{m}$. The standard likelihood is given by

$$L(\mathbf{m}|S)\mathrm{d}\mathbf{m} = \frac{\exp\left[-\frac{1}{2}\mathbf{m}^{\mathrm{T}}(S+N)^{-1}\mathbf{m}\right]}{|S+N|^{1/2}}\frac{\mathrm{d}\mathbf{m}}{(2\pi)^{3n_{\mathrm{p}}/2}}, \tag{5.85}$$

where $\mathbf{m}$ is the data vector containing the temperature map, $\mathbf{T}$, as well as the polarization maps, $\mathbf{Q}$, and $\mathbf{U}$, $n_{\mathrm{p}}$ is the number of pixels of each map, and $S$ and $N$ are the signal and noise covariance matrix $(3n_{\mathrm{p}} \times 3n_{\mathrm{p}})$, respectively. As the temperature data are completely dominated by the signal at such low multipoles, noise in temperature may be ignored. This simplifies the form of likelihood as

$$L(\mathbf{m}|S)\mathrm{d}\mathbf{m} = \frac{\exp\left[-\frac{1}{2}\tilde{\mathbf{m}}^{\mathrm{T}}(\tilde{S}_{\mathrm{P}}+N_{\mathrm{P}})^{-1}\tilde{\mathbf{m}}\right]}{|\tilde{S}_{\mathrm{P}}+N_{\mathrm{P}}|^{1/2}}\frac{\mathrm{d}\tilde{\mathbf{m}}}{(2\pi)^{n_{\mathrm{p}}}}\frac{\exp\left(-\frac{1}{2}\mathbf{T}^{\mathrm{T}}S_T^{-1}\mathbf{T}\right)}{|S_T|^{1/2}}\frac{\mathrm{d}\mathbf{T}}{(2\pi)^{n_{\mathrm{p}}/2}}, \tag{5.86}$$

where $S_T$ is the temperature signal matrix $(n_{\mathrm{p}} \times n_{\mathrm{p}})$, the new polarization data vector, $\tilde{\mathbf{m}} = (\tilde{Q}_{\mathrm{p}},\ \tilde{U}_{\mathrm{p}})$ and $\tilde{S}_{\mathrm{P}}$ and $N_{\mathrm{P}}$ are the signal and noise matrices, respectively, for the new polarization vector with the size of $2n_{\mathrm{p}} \times 2n_{\mathrm{p}}$.

At the time of writing, in CMB parameter estimates for $\ell <$ 2000, the likelihood calculation is the bottleneck of the analysis.

## 5.6 Monte Carlo methods

### 5.6.1 Monte Carlo error estimation

Let's go back to the issue of parameter estimation and error calculation. Here is the conceptual interpretation of what it means for an experiment to measure some parameters (say cosmological parameters). There is some underlying true set of parameters $\boldsymbol{\alpha}_{\text{true}}$ that are only known to Mother Nature but not to the experimenter. The true parameters are statistically realized in the observable Universe and random measurement errors are then included when the observable Universe gets measured. This "realization" gives the measured data set $\mathcal{D}_0$. Only $\mathcal{D}_0$ are accessible to the observer (you). Then you go and do what you have to do to estimate the parameters and their errors (chi-square, likelihood, etc. . . ) and get $\boldsymbol{\alpha}_0$. Note that $\mathcal{D}_0$ are not a unique realization of the true model given by $\boldsymbol{\alpha}_{\text{true}}$ : there could be infinitely many other realizations as *hypothetical data sets*, which could have been the measured one: $\mathcal{D}_1, \mathcal{D}_2, \ldots$, each of them with a slightly different fitted parameters $\boldsymbol{\alpha}_1, \boldsymbol{\alpha}_2, \ldots$ The parameter $\boldsymbol{\alpha}_0$ is one set drawn from this distribution. The hypothetical ensemble of universes described by $\boldsymbol{\alpha}_i$ is called an *ensemble*, and one expects that the expectation value $\langle\boldsymbol{\alpha}_i\rangle = \boldsymbol{\alpha}_{\text{true}}$. If we knew the distribution of $\boldsymbol{\alpha}_i - \boldsymbol{\alpha}_{\text{true}}$ we would know everything we need about the uncertainties in our measurement $\boldsymbol{\alpha}_0$. The goal is to infer the distribution of $\boldsymbol{\alpha}_i - \boldsymbol{\alpha}_{\text{true}}$ without knowing $\boldsymbol{\alpha}_{\text{true}}$.

Here's what we do: we say that hopefully $\boldsymbol{\alpha}_0$ is not too wrong and we consider a fictitious world where $\boldsymbol{\alpha}_0$ is the true one. So it would not be such a big mistake to take the probability distribution of $\boldsymbol{\alpha}_0 - \boldsymbol{\alpha}_i$ to be that of $\boldsymbol{\alpha}_{\text{true}} - \boldsymbol{\alpha}_i$. In many cases we know how to simulate $\alpha_0 - \alpha_i$ and so we can simulate many synthetic realization of "worlds" where $\alpha_0$ is the true underlying model. Then we mimic the observation process of these fictitious universes replicating all the observational errors and effects and from each of these universes estimate the parameters. If we simulate enough of them, from $\alpha_i^S - \alpha_0$ we will be able to map the desired multidimensional probability distribution.

With the advent of fast computers this technique has become increasingly widespread. As long as you believe you know the underlying distribution and that you believe you can mimic the observation replicating all the observational effects this technique is extremely powerful and, I would say, indispensable.

### 5.6.2 Monte Carlo Markov chains

When dealing with high-dimensional likelihoods (i.e. many parameters) the process of mapping the likelihood (or the posterior) surface can become very expensive. For example for CMB studies the models considered have from 6 to 11 or more parameters. Every model evaluation even with a fast code such as CAMB can take up to minutes per iteration. A grid-based likelihood analysis would require prohibitive amounts of central processing unit (CPU) time. For example, a coarse grid ($\sim 20$ grid points per dimension) with six parameters requires $\sim 6.4 \times 10^7$ evaluations of the power spectra. At 1.6 seconds per evaluation, the calculation would take $\sim 1200$ days. Christensen & Meyer (2001) proposed using Markov chain Monte Carlo (MCMC) to investigate the likelihood space. This approach has become the standard tool for CMB analyses; the MCMC method simulates posterior distributions. In particular, sampling of the posterior distribution $\mathcal{P}(\alpha|x)$, of a set of parameters $\alpha$ given event $x$, obtained via Bayes' theorem is simulated:

$$\mathcal{P}(\alpha|x) = \frac{\mathcal{P}(x|\alpha)\mathcal{P}(\alpha)}{\int \mathcal{P}(x|\alpha)\mathcal{P}(\alpha)\mathrm{d}\alpha}, \tag{5.87}$$

where $\mathcal{P}(x|\alpha)$ is the likelihood of event $x$ given the model parameters $\alpha$ and $\mathcal{P}(\alpha)$ is the prior probability density; $\alpha$ denotes a set of cosmological parameters (e.g. for the standard, flat $\Lambda$CDM model these could be the cold dark matter density parameter $\Omega_c$, the baryon density parameter $\Omega_b$, the spectral slope $n_s$, the Hubble constant – in units of 100 km s$^{-1}$ Mpc$^{-1}$ – $h$, the optical depth $\tau$ and the power spectrum amplitude $A$), and event $x$ will be the set of observed $\widehat{C}_\ell$. The MCMC generates random draws (i.e. simulations) from the posterior distribution that are a "fair" sample of the likelihood surface. From this sample, we can estimate all of the quantities of interest about the posterior distribution (mean, variance, confidence levels). The MCMC method scales approximately linearly with the number of parameters, thus allowing us to perform likelihood analysis in a reasonable period of time.

A properly derived and implemented MCMC draws from the joint posterior density $\mathcal{P}(\alpha|x)$ once it has converged to the stationary distribution. The primary consideration in implementing MCMC is determining when the chain has *converged*. After an initial *"burn-in"* period, all further samples can be thought of as coming from the stationary distribution. In other words the chain has no dependence on the starting location.

Another fundamental problem of inference from Markov chains is that there are always areas of the target distribution that have not been covered by a finite chain. If the MCMC is run for a very long time, the ergodicity of the Markov chain guarantees that eventually the chain will cover all the target distribution, but in the short term the simulations cannot tell us about areas where they have not been. It is thus crucial that the chain achieves good *"mixing"*. If the Markov chain does not move rapidly throughout the support of the target distribution because of poor *mixing*, it might take a prohibitive

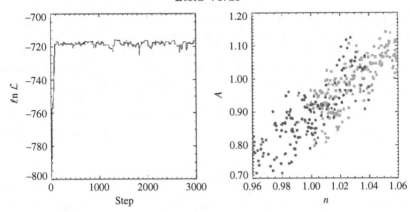

FIGURE 5.3. Unconverged Markov chains. The left panel shows a *trace plot* of the likelihood values versus iteration number for one MCMC (these are the first 3000 steps from the run). Note the burn in for the first 100 steps. In the right panel, the light-grey dots are points of the chain in the $(n, A)$ plane after discarding the burn-in. Dark dots are from another MCMC for the same data set and the same model. It is clear that, although the trace plot may appear to indicate that the chain has converged, it has not fully explored the likelihood surface. Using either of these two chains at this stage will give incorrect results for the best-fit cosmological parameters and their errors. From Verde *et al.* (2003).

amount of time for the chain to fully explore the likelihood surface. Thus it is important to have a convergence criterion and a mixing diagnostic. Plots of the sampled MCMC parameters or likelihood values versus iteration number are commonly used to provide such criteria (left panel of Fig. 5.3). However, samples from a chain are typically serially correlated; very high auto-correlation leads to little movement of the chain and thus makes the chain "appear" to have converged. For a more detailed discussion see Gilks *et al.* (1996). Using a MCMC that has not fully explored the likelihood surface for determining cosmological parameters will yield *wrong* results (see right panel of Fig. 5.3).

### 5.6.3 Markov chains in practice

Here are the necessary steps to run a simple MCMC for the CMB temperature power spectrum. It is straightforward to generalize these instructions to include the temperature–polarization power spectrum and other data sets. The MCMC is essentially a random walk in parameter space, where the probability of being at any position in the space is proportional to the posterior probability.

(1) Start with a set of cosmological parameters $\{\alpha_1\}$, compute the $\mathcal{C}_\ell^1$ and the likelihood $\mathcal{L}_1 = \mathcal{L}(\mathcal{C}_\ell^{1\text{th}}|\widehat{\mathcal{C}_\ell})$.

(2) Take a random step in parameter space to obtain a new set of cosmological parameters $\{\alpha_2\}$. The probability distribution of the step is taken to be Gaussian in each direction $i$ with RMS given by $\sigma_i$. We will refer below to $\sigma_i$ as the "step size." The choice of the step size is important to optimize the chain efficiency (see discussion below)

(3) Compute the $\mathcal{C}_\ell^{2\text{th}}$ for the new set of cosmological parameters and their likelihood $\mathcal{L}_2$.

(4a) If $\mathcal{L}_2/\mathcal{L}_1 \geq 1$, "take the step," i.e. save the new set of cosmological parameters $\{\alpha_2\}$ as part of the chain, then go to step 2 after the substitution $\{\alpha_1\} \longrightarrow \{\alpha_2\}$.

(4b) If $\mathcal{L}_2/\mathcal{L}_1 < 1$, draw a random number $x$ from a uniform distribution from 0 to 1. If $x \geq \mathcal{L}_2/\mathcal{L}_1$ "do not take the step," i.e. save the parameter set $\{\alpha_1\}$ as part of the chain and return to step 2. If $x < \mathcal{L}_2/\mathcal{L}_1$, " take the step," i.e. do as in 4a).

(5) For each cosmological model run four chains starting at randomly chosen, well-separated points in parameter space. When the convergence criterion is satisfied and the chains have enough points to provide reasonable samples from the a-posteriori distributions (i.e. enough points to be able to reconstruct the 1- and 2-$\sigma$ levels of the marginalized likelihood for all the parameters) stop the chains.

It is clear that the MCMC approach is easily generalized to compute the joint likelihood of WMAP data with other data sets.

### 5.6.4 Improving MCMC efficiency

The efficiency of MCMC can be seriously compromised if there are degeneracies among parameters. A typical example is the degeneracy between $\Omega_m$ and $H_0$ for a flat cosmology or that between $\Omega_m$ and $\Omega_\Lambda$ for a non-flat case (see e.g. Spergel *et al.*, 2007).

The Markov chain efficiency can be improved in different ways. Here we report the simplest one.

**Reparameterization.** We describe below the method we use to ensure convergence and good mixing. Degeneracies and poor parameter choices slow the rate of convergence and mixing of the Markov chain. There is one near-exact degeneracy (the geometric degeneracy) and several approximate degeneracies in the parameters describing the CMB power spectrum (Bond *et al.*, 1994; Bond and Efstathiou, 1984). The numerical effects of these degeneracies are reduced by finding a combination of cosmological parameters (e.g., $\Omega_c$, $\Omega_b$, $h$, etc.) that have essentially orthogonal effects on the angular power spectrum. The use of such parameter combinations removes or reduces degeneracies in the MCMC and hence speeds up convergence and improves mixing, because the chain does not have to spend time exploring degeneracy directions. Kosowsky *et al.* (2002) and Jiménez *et al.* (2003) introduced a set of reparameterizations to do just this. In addition, these new parameters reflect the underlying physical effects determining the form of the CMB power spectrum (we will refer to these as physical parameters). This leads to particularly intuitive and transparent parameter dependencies of the CMB power spectrum.

For the six parameter $\Lambda$CDM model these "normal" or "physical" parameters are: the physical energy densities of cold dark matter, $\omega_c \equiv \Omega_c h^2$, and baryons, $\omega_b \equiv \Omega_b h^2$, the characteristic angular scale of the acoustic peaks,

$$\theta_A = \frac{r_s(a_{\text{dec}})}{D_A(a_{\text{dec}})}, \tag{5.88}$$

where $a_{\text{dec}}$ is the scale factor at decoupling,

$$r_s(a_{\text{dec}}) = \frac{c}{H_0\sqrt{3}} \times \tag{5.89}$$

$$\int_0^{a_{\text{dec}}} \frac{dx}{\left[\left(1 + \frac{3\Omega_b}{4\Omega_\gamma}\right)\left((1-\Omega)x^2 + \Omega_\Lambda x^{1-3w} + \Omega_m x + \Omega_{\text{rad}}\right)\right]^{\frac{1}{2}}}$$

is the sound horizon at decoupling, and

$$D_A(a_{\text{dec}}) = \frac{c}{H_0} \frac{S_\kappa(r_{\text{dec}})}{\sqrt{|\Omega - 1|}}, \tag{5.90}$$

where

$$r(a_{\text{dec}}) = |\Omega - 1| \int_{a_{\text{dec}}}^1 \frac{dx}{\left[(1-\Omega)x^2 + \Omega_\Lambda x^{1-3w} + \Omega_m x + \Omega_{\text{rad}}\right]^{-\frac{1}{2}}} \tag{5.91}$$

and $S_\kappa(r)$ as usual coincides with the argument if the curvature $\kappa$ is 0, is a sin function for $\Omega > 1$ and a sinh function otherwise. Here $H_0$ denotes the Hubble constant and $c$ is the speed of light, $\Omega_m = \Omega_c + \Omega_b$, $\Omega_\Lambda$ denotes the dark energy density parameters, $w$ is the equation of state of the dark energy component, $\Omega = \Omega_m + \Omega_\Lambda$ and the radiation density parameter $\Omega_{rad} = \Omega_\gamma + \Omega_\nu$, where $\Omega_\gamma$ and $\Omega_\nu$ are the photon and neutrino density parameters respectively. For reionization sometimes the parameter $\mathcal{Z} \equiv \exp(-2\tau)$ is used, where $\tau$ denotes the optical depth to the last scattering surface (not the decoupling surface).

These reparameterizations are useful because the degeneracies are non-linear, that is they are not well described by ellipses in parameter space. For degeneracies that are well approximated by ellipses in parameter space it is possible to find the best reparameterization automatically. This is what the code CoSMoMC does (Lewis and Bridle, 2002; see http://cosmologist.info/cosmomc/ and the notes at the end of the chapter). To be more precise it computes the parameter covariance matrix from which the axes of the multidimensional degeneracy ellipse can be found. Then it performs a rotation and rescaling of the coordinates (i.e. the parameters) to transform the degeneracy ellipse in an azimuthally symmetric contour. See discussion at `http://cosmologist.info/notes/CosmoMC.pdf` for more information. This technique can improve the MCMC efficiency up to a factor of order 10.

**Step size optimization.** The choice of the step size in the Markov chain is crucial to improve the chain efficiency and speed up convergence. If the step size is too big, the acceptance rate will be very small; if the step size is too small the acceptance rate will be high but the chain will exhibit poor mixing. Both situations will lead to slow convergence.

### 5.6.5 Convergence and mixing

Before we even start this section: *thou shalt always use a convergence and mixing criterion when running MCMCs.*

Let's illustrate here the method proposed by Gelman and Rubin (1992) as an example. They advocate comparing several sequences drawn from different starting points and checking to see that they are indistinguishable. This method not only tests convergence but can also diagnose poor mixing. Let us consider $M$ chains starting at well-separated points in parameter space; each has $2N$ elements, of which we consider only the last $N$: $\{y_i^j\}$ where $i = 1,\ldots,N$ and $j = 1,\ldots,M$, i.e. $y$ denotes a chain element (a point in parameter space); the index $i$ runs over the elements in a chain; the index $j$ runs over the different chains. We define the mean of the chain

$$\bar{y}^j = \frac{1}{N} \sum_{i=1}^{N} y_i^j, \tag{5.92}$$

and the mean of the distribution

$$\bar{y} = \frac{1}{NM} \sum_{ij=1}^{NM} y_i^j. \tag{5.93}$$

We then define the variance between chains as

$$B_n = \frac{1}{M-1} \sum_{j=1}^{M} (\bar{y}^j - \bar{y})^2, \tag{5.94}$$

and the variance within a chain as

$$W = \frac{1}{M(N-1)} \sum_{ij} (y_i^j - \bar{y}^j)^2. \tag{5.95}$$

The quantity

$$\hat{R} = \frac{\frac{N-1}{N} W + B_n \left(1 + \frac{1}{M}\right)}{W} \tag{5.96}$$

is the ratio of two estimates of the variance in the target distribution: the numerator is an estimate of the variance that is unbiased if the distribution is stationary, but is otherwise an overestimate. The denominator is an underestimate of the variance of the target distribution if the individual sequences did not have time to converge.

The convergence of the Markov chain is then monitored by recording the quantity $\hat{R}$ for all the parameters and running the simulations until the values for $\hat{R}$ are always $< 1.03$. Needless to say that the CosMoMC package offers several convergence and mixing diagnostic tools as part of the "getdist" routine.

---
— Exercise —

(24) Question: how does the MCMC sample the prior if all one actually computes is the likelihood?

---

### 5.6.6 MCMC output analysis

Now that you have your multiple chains and the convergence criterion says they are converged what do you do? First discard *burn in* and merge the chains. Since the MCMC passes objective tests for convergence and mixing, the density of points in parameter space is proportional to the posterior probability of the parameters. (Note that CosMoMC saves repeated steps as the same entry in the file but with a weight equal to the repetitions: the MCMC gives to each point in parameter space a "weight" proportional to the number of steps the chain has spent at that particular location.) The marginalized distribution is obtained by projecting the MCMC points. This is a great advantage compared to the grid-based approach where multidimensional integrals would have to be performed. The MCMC basically performs a Monte Carlo integration. The density of points in the $n$-dimensional space is proportional to the posterior, and best-fit parameters and multidimensional confidence levels can be found as illustrated in Section 5.3.1.

Note that the global maximum likelihood value for the parameters does not necessarily coincide with the expectation value of their marginalized distribution if the likelihood surface is not a multivariate Gaussian.

A virtue of the MCMC method is that the addition of extra data sets in the joint analysis can efficiently be done with minimal computational effort from the MCMC output if the inclusion of extra data sets does not require the introduction of extra parameters nor drive the parameters significantly away from the current best fit. If the likelihood surface for a subset of parameters from an external (independent) data set is known, or if a prior needs to be added a posteriori, the joint posterior surface can be obtained by multiplying the new probability distribution with the posterior distribution of the MCMC output. To be more precise: as the density of point (i.e. the weight) is directly proportional to the posterior, then this is achieved by multiplying the weight by the new probability distribution. The CosMoMC package already includes this facility.

## 5.7 Fisher matrix

What if you wanted to forecast how well a future experiment can do? There is the expensive but more accurate way and the cheap and quick way (but often less accurate). The expensive way is in the same spirit of Monte Carlo simulations discussed earlier: simulate the observations and estimate the parameters, repeating the same procedure applied to the real data. However, often you want to have a much quicker way to forecasts parameter uncertainties, especially if you need to compare quickly many different experimental designs. This technique is the Fisher matrix approach.

### 5.7.1 Fisher matrix approach

The question of how accurately one can measure model parameters from a given data set (without simulating the data set) was answered more than 70 years ago by Fisher (1935). Suppose your data set is given by $m$ real numbers $x_1, \ldots, x_m$ [arranged in a vector (they can be CMB map pixels, $P(k)$ of galaxies, etc.) $\boldsymbol{x}$ is a random variable with a probability distribution which depends in some way on the vector of model parameters $\boldsymbol{\alpha}$.] The Fisher information matrix is defined as:

$$F_{ij} = \left\langle \frac{\partial^2 L}{\partial \alpha_i \partial \alpha_j} \right\rangle, \tag{5.97}$$

where $L = -\ln \mathcal{L}$. In practice you will choose a fiducial model and compute the above at the fiducial model. In the one parameter case let's note that if the likelihood is Gaussian then $L = 1/2(\alpha - \alpha_0)/\sigma_\alpha^2$ where $\alpha_0$ is the value that maximizes the likelihood and sigma is the error on the parameter $\alpha$. Thus the second derivative with respect to $\alpha$ of $L$ is $1/\sigma_\alpha^2$ as long as $\sigma_\alpha$ does not depend on $\alpha$. In general we can expand $L$ in a Taylor series around its maximum (or the fiducial model). There by definition the first derivative of $L$ with respect to the parameters is 0.

$$\Delta L = \frac{1}{2} \frac{d^2 L}{d\alpha^2} (\alpha - \alpha_0)^2. \tag{5.98}$$

When $2\Delta L = 1$, which in the case when we can identify it with $\Delta \chi^2$ corresponds to 68% (or 1-$\sigma$), then $1/\sqrt{d^2 L/d\alpha^2}$ is the 1 sigma displacement of $\alpha$ from $\alpha_0$.

The generalization to many variables is beyond our scope here (see Kendall and Stuart, 1977); let's just say that an estimate of the covariance for the parameters is given by:

$$\sigma_{\alpha_i, \alpha_j}^2 \geq (\mathbf{F}^{-1})_{ij}. \tag{5.99}$$

From here it follows that if all other parameters are kept fixed

$$\sigma_{\alpha_i} \geq \sqrt{\frac{1}{F_{ii}}} \tag{5.100}$$

(i.e. the reciprocal of the square root of the diagonal element $ii$ of the Fisher matrix.)

But if all other parameters are estimated from the data as well then the marginalized error is

$$\sigma_\alpha = (\mathbf{F}^{-1})_{ii}^{1/2} \tag{5.101}$$

(i.e. square root of the element $ii$ of the inverse of the Fisher matrix – perform a matrix inversion here!)

In general, say that you have five parameters and that you want to plot the joint two-dimensional contours for parameters 2 and 4 marginalized over all other parameters 1, 3, 5. Then you invert $F_{ij}$, take the minor 22, 24, 42, 24 and invert it back.

The resulting matrix, let's call it Q, describes a Gaussian two-dimensional likelihood surface in the parameters 2 and 4 or, in other words, the chi-square surface for parameters 2,4 – marginalized over all other parameters – can be described by the equation $\tilde{\chi}^2 = \sum_{kq}(\alpha_k - \alpha_k^{\text{fiducial}})Q_{kq}^{-1}(\alpha_q - \alpha_q^{\text{fiducial}})$.

From this equation, getting the errors corresponds to finding the quadratic equation solution $\tilde{\chi}^2 = \Delta\chi^2$. For correspondence between $\Delta\chi^2$ and the confidence region see Section 5.3.1. If you want to make plots, the equation for the elliptical boundary for the joint confidence region in the subspace of parameters of interest is: $\Delta = \delta\alpha Q^{-1}\delta\alpha$.

Note that in many applications the likelihood for the data is assumed to be Gaussian and the data covariance is assumed not to depend on the parameters. For example for the application to CMB the Fisher matrix is often computed as:

$$F_{ij} = \sum_\ell \frac{(2\ell + 1)}{2} \frac{\frac{\partial C_\ell}{\partial \alpha_i} \frac{\partial C_\ell}{\partial \alpha_j}}{(C_\ell + \mathcal{N}_\ell e^{\sigma^2\ell^2})^2} \tag{5.102}$$

This corresponds to a Gaussian approximation to likelihood, which is good in the high-$\ell$, noise-dominated regime. In this regime the central limit theorem ensures a Gaussian likelihood and the cosmic variance contribution to the covariance is negligible; thus the covariance does not depend on the cosmological parameters. However at low $\ell$ in the cosmic-variance dominated-regime, the Gaussian likelihood approximation can overestimate the errors and introduce biases in the recovered parameters. In this case it is therefore preferable to go for the more numerically intensive option of computing the exact form for the likelihood (Eq. 5.77).

Before concluding we report here a useful identity:

$$2L = \ln\det(C) + (x - y)_i C_{ij}^{-1}(x - y)_j = \text{Tr}[\ln C + C^{-1}D] \tag{5.103}$$

where $C$ stands for the covariance matrix of the data, repeated indices are summed over, $x$ denotes the data, $y$ the model fitting the data and $D$ is the data matrix defined as $(x - y)(x - y)^{\text{T}}$. We have used the identity $\ln\det(C) = \text{Tr}\ln C$.

## 5.8 Conclusions

Whether you are a theorist or an experimentalist in cosmology, these days you cannot ignore the fact that to make the most of your data, statistical techniques need to be employed, and used correctly. Incorrect treatment of the data will lead to nonsensical results. A given data set can reveal a lot about the Universe, but there will always things that are beyond the statistical power of the data set. It is crucial to recognize that. I hope I have given you the basis to be able to learn this for yourself. When in doubt always remember: treat your data with respect.

## 5.9 Questions

Here are some of the questions (and their answers ...).

Q: What about the forgotten $\mathcal{P}(D)$ in Bayes' theorem?
A: $\mathcal{P}(D)$ becomes important when one wants to compare different cosmological models (not just different parameter values within the same cosmological model). There is a somewhat extensive literature in cosmology alone on this: "Bayesian evidence" is used to do this "model selection". Bayesian evidence calculation can be, in many cases, numerically intensive.

Q: What do you do when the likelihood surface is very complicated, for example is multi-peaked?

A: Luckily enough, in CMB analysis this is almost never the case (exceptions being, for example, where sharp oscillations are present in the $C_\ell^{\mathrm{th}}$ as happens in trans-Planckian models). In cases where the likelihood surface is very complicated and additional peaks can't be suppressed by a motivated prior, there are several options: (a) if $m$ is the number of parameters, report the confidence levels by considering only regions with $m$-dimensional likelihoods above a threshold before marginalizing (thus local maxima will be included in the confidence contours if significant enough) or (b) simply marginalize as described here, but expect that the marginalized peaks will not coincide with the $m$-dimensional peaks. The most challenging issue when the likelihood surface is complicated is having the MCMC to fully explore the surface. Techniques such as Hamiltonian Monte Carlo are very powerful in these cases; see e.g. Hajian (2007) and Taylor *et al.* (2008).

Q: The Fisher matrix approach is very interesting, is there anything I should look out for?

A: The Fisher matrix approach assumes that the parameters log-likelihood surface can be quadratically approximated around the maximum. This may or may not be a good approximation. It is always a good idea to check at least in a reference case whether this approach has significantly underestimated the errors. In many Fisher matrix implementations the likelihood for the data is assumed to be Gaussian and the data covariance is assumed not to depend on the parameters. While this approximation greatly simplifies the calculations (*Exercise:* show why this is the case), it may significantly fail to estimate the size of the errors. In addition, the Fisher matrix calculation often requires numerical evaluation of second derivatives. Numerical derivatives always need a lot of care and attention: you have been warned.

Q: Does CosmoMC use the sampling described here?

A: The recipe reported here is the so called Metropolis–Hasting algorithm. The CosmoMC package also offers other sampling algorithms: slice sampling, or a split between "slow" and "fast" parameters or the learn–propose option where chains automatically adjust the proposal distribution by repeatedly computing the parameter covariance matrix "on the fly". These techniques can greatly improve the MCMC "efficiency". See for example http://cosmologist.info/notes/CosmoMC.pdf or Gilks *et al.* (1996) for more details.

*Acknowledgments*

I am indebited to A. Taylor for his lectures on statistics for beginners given at the Royal Observatory, Edinburgh in 1997. Also a lot of my understanding of the basics come from Ned Wright's "Journal Club in statistics" – see http://www.astro.ucla.edu/~wright/intro.html. A lot of the techniques reported here were developed and/or tested as part of the analysis of WMAP data between 2002 and 2006. Last, but not least, I would like to thank the organizers of the XIX Canary Islands Winter School, for a very stimulating school.

**Notes**

Two tutorial sessions were organized as part of the Winter School.[7] You may want to try to repeat the same steps.

[7] *Editors' note*: These two tutorials were coordinated by Raúl Jiménez and José Alberto Rubiño-Martín.

The goals of the first tutorial were:

- Download and install Healpix (Górski *et al.*, 2005; see also http://healpix. jpl.nasa.gov). Make sure you have a Fortran 90 compiler installed, and possibly also IDL software installed (see http://www.ittvis.com/Product Services/IDL.aspx).
- The LAMBDA site (see http://lamda.gsfc.nasa.gov) contains a lot of extremely useful information: browse the page.
- Find the on-line calculator for the theory $C_\ell$ given some cosmological parameters, generate a set of $C_\ell$ and save them as a "fits" file.
- Browse the help pages of Healpix and find out what it does.
- In particular the routine "synfast" enables one to generate a map from a set of theory $C_\ell$.
- The routine "anafast" enables one to compute $C_\ell$ from a map.
- Using "synfast", generate two maps with two different random number generators for the $C_\ell$ you generated above. Select a beam of FWHM of, say, 30′, but for now do not impose any galaxy cut and do not add noise. Compare the maps. To do that use "mollview". Why are they different?
- Using "anafast" compute the $C_\ell$ for both maps.
- Using your favorite plotting routine (the IDL part of Healpix offers utilities to do that) plot the original $C_\ell$ and the two realizations. Why do they differ?
- Deconvolve for the beam, and make sure you are plotting the correct units, the correct factors of $\ell(\ell+1)$ etc. Why do the power spectra still differ?
- Try to compute the $C_\ell$ for $\ell_{max} = 3000$ and for a non-flat model with the software code CAMB (see http://camb.info). It takes much longer than for a simple, flat LCDM model. A code like CMBwarp (Jiménez *et al.*, 2003) offers a shortcut. Try it by downloading it at:
  `http://www.ice.csic.es/personal/jimenez/CMBwarp/`.
  Keep in mind however that this offers a fast-fitting routine for a fine grid of precomputed $C_\ell$, it is not a Boltzmann code and thus is valid only within the quoted limits!

The goals of the second tutorial were:

- Download and install the COSMOMC package (and read the instructions).
- Remember that the WMAP data and the likelihood need to be downloaded separately from the LAMBDA site. Do this following the instructions.
- Take a look at the params.ini file. In here you set up the MCMC.
- Set a chain to run.
- Get familiar with the getdist program and distparams.ini file. This program checks convergence for you. Compute a basic parameter estimation from converged chains.
- Download from LAMBDA a set of chains, the suggested one was for the flat, quintessence model for WMAP data only.
- Plot the marginalized probability distribution for the parameter $w$.
- **The challenge**: apply now a Hubble constant prior to this chain, take the HST key project (Freedman *et al.*, 2001) constraint of $H_0 = 72\pm8$ km s$^{-1}$ Mpc$^{-1}$ and assume it has a Gaussian distribution. This procedure is called "importance sampling."

## REFERENCES

BOND, J. R. and EFSTATHIOU, G. (1984). *Ashophys. J. Lett.*, **285**, L45.

BOND, J. R., CRITTENDEN, R., DAVIS, R. L., EFSTATHIOU, G. and STEINHARDT, P. J. (1994). *Phys. Rev. Lett.*, **72**, 13.

BOND, J. R., JAFFE, A. H. and KNOX, L. (1998). *Phys. Rev. D* **57**, 2117.

CASH, W. (1979). *Astrophys. J.* **228** 939.

CHRISTENSEN, N. and MEYER, R. (2001). *Phys. Rev. D*, **64**, 022001.

FISHER, R. A. (1935). *J. R. Stat. Soc.*, **98**, 39.

FREEDMAN, W. L., MADORE, B F., GIBSON, B. K. *et al.* (2001). *Astrophys. J.* **553**, 47.

GELMAN, A. and RUBIN, D. (1992). *Stat. Sci.*, **7**, 457.

GILKS, W. R., RICHARDSON, S. and SPIEGELHALTER, D. J. (1996). *Markov Chain Monte Carlo in Practice*. Chapman and Hall.

GÓRSKI, K. M., HIVON, E., BANDAY, A. J. *et al.* (2005). *Astrophys. J.* **622** 759.

HAJIAN, A. (2007). *Phys. Rev. D*, **75**, 083525.

HIVON, E., GÓRSKI, K. M., NETTERFIELD, C. B. *et al.* (2002). *Astrophys. J.* **567**, 2.

JIMÉNEZ, R., VERDE, L., PEIRIS, H. and KOSOWSKY, A. (2003). *Phys. Rev. D* **70**, 3005.

KENDALL, M. and STUART, A. (1977). *The Advanced Theory of Statistics*, 4th edn. Griffin.

KNOX, L. (1995). *Phys. Rev. D* **52** , 4307.

KOSOWSKY, A., MILOSAVLJEVIC, M. and JIMENEZ, R. (2002). *Phys. Rev. D* **66**, 63007.

LEWIS, A. and BRIDLE, S. (2002). *Phys. Rev. D*, **66**, 103511.

MARTÍNEZ, V. and SAAR, E. (2002). *Statistics of the Galaxy Distribution*, Chapman & Hall/CRC Press.

PAGE, L., HINSHAW, G., KOMATSU, E. *et al.* (2007). *Astrophys. J. Suppl. Ser.*, **170** 335.

PEEBLES, P. J. E. (1980). *The Large-scale Structure of the Universe*, Princeton University Press.

PRESS, W. H., TEUKOLSKY, S. A., VETTERLING, W. T. and FLANNERY, B. P. (1992). *Numerical Recipes in FORTRAN. The Art of Scientific Computing*. Cambridge University Press.

SPERGEL, D. N., VERDE, L., PEIRIS, H. V. *et al.* (2003). *Astrophys. J. Suppl. Ser.* **148** 175.

SPERGEL, D. N., BEAN, R., DORÉ, O. *et al.* (2007). *Astrophys. J. Suppl. Ser.* **170**, 377.

TAYLOR, J. F., ASHDOWN, M. A. J. and HOBSON, M. P. (2008). *Mor. Not. R. Astron. Soc.* **389**, 1284.

VERDE, L., PEIRIS, H. V., SPERGEL, D. N. *et al.* (2003). *Astrophys. J. Suppl. Ser.* **148**, 195.

WALL, J. V. and JENKINS, C. R. (2003). *Practical Statistics for Astronomers*. Cambridge University Press.

# 6. Gaussianity

ENRIQUE MARTÍNEZ-GONZÁLEZ

## Abstract

In this chapter the present status of the Gaussianity studies of the CMB anisotropies is reviewed, including physical effects producing non-Gaussianity, methods to test it and observational constraints.

## 6.1 Introduction

The study of the CMB temperature and polarization anisotropy has had an essential role in establishing the standard cosmological model: a flat $\Lambda$CDM model composed of baryonic matter ($\approx 4\%$), dark matter ($\approx 24\%$) and dark energy ($\approx 72\%$), with nearly scale-invariant adiabatic fluctuations in the energy density. The information extracted from the CMB to derive the cosmological model has been mostly based on the CMB anisotropy power spectrum. However, much more information is still waiting to be extracted from the statistics of the CMB beyond the second-order moment.

The study of the Gaussianity of the CMB anisotropies has in recent years become a very relevant topic in the CMB field. The main reason for this is the availability of high sensitivity and high-resolution maps provided by the new generation of CMB experiments, with which the Gaussian character of the anisotropies predicted by the standard inflationary paradigm can be tested at a high accuracy. The CMB anisotropies form an ideal data set for testing the Gaussianity of the primordial energy-density fluctuations since the dominant physical effects producing the anisotropies involve just linear physics, and Gaussianity is preserved under linear transformations. Another approach to test it is by studying the large-scale structure of the Universe as described by the galaxy distribution. The density perturbations in the linear regime should also be a good representation of the initial conditions. However, galaxy formation is a very non-linear process involving complex physical effects which very much complicate the analysis.

Different models of the early Universe have been proposed to account naturally for the early stages of its history and, thus, not to have to rely on ad hoc initial conditions. They are based on different theories, such as string theory or M theory, and include, or not, an inflationary phase (an example of the latter being the Ekpyrotic cosmology; Khoury et al., 2001). Most of those models predict very specific properties about the probability distribution of the CMB anisotropies, which in many cases imply deviations from Gaussianity with amplitudes within reach of present or near-future experiments. Some of those models are already constrained by present data and many more are expected to be disproved in the coming years, specially with the launch of the Planck satellite at the beginning of 2009.

The analysis of the angular distribution of the CMB anisotropies is a formidable task with profound consequences on our understanding of the Universe. Before any meaningful result can be achieved, it is, however, crucial to control at a high level of precision all possible systematics that can be introduced by the experiment and the pipeline process used to reduce the data. On the other hand, there are several Galactic and extragalactic emissions in the microwave band which blur the CMB signal. Disentangling the cosmic signal from the others is also crucial and requires a very good knowledge of the astrophysical emissions. A lot of observational and theoretical effort has been dedicated to that aim in recent years. Below we summarize the present status of the most important aspects of the analysis of Gaussianity and the results achieved.

## 6.2 The isotropic Gaussian random field (IGRF)

One of the most robust predictions of the standard inflationary model is that the CMB anisotropies should be well represented by an isotropic Gaussian random field (IGRF) on the celestial sphere. This is a very powerful prediction to test the standard model. There are many statistical quantities for an IGRF that can be derived analytically, which greatly facilitates the Gaussianity test of the CMB data. However, a disadvantage may come from the central limit theorem that implies that the sum of several independent non-Gaussian distributions tends to a more Gaussian one. As we will see below, this complicates the Gaussianity analysis since the observational data are composed of several contributions which can be either intrinsic or extrinsic to the CMB anisotropies.

### 6.2.1 Definition

A random field defined on a given support is said to be Gaussian if for any $N$ points of the support $x_1, \ldots, x_n$ the values of the random field $y_1, \ldots, y_n$ follow a multinormal distribution

$$f(y_1, \ldots, y_n) = \frac{1}{(2\pi)^{n/2}|M|^{\frac{1}{2}}} \exp\left(-\frac{1}{2}\sum_{ij}(y_i - <y_i>)M_{ij}^{-1}(y_j - <y_j>)\right) \quad (6.1)$$

where $\mathbf{M}$ is the covariance matrix defined as $M_{ij} = <(y_i - <y_i>)(y_j - <y_j>)>$. Thus the n-pdf (probability density function), and also all the moments, is given in terms of just the first two moments.

In the case of the temperature anisotropies of the CMB the mean value is set to zero, and thus the standard model predicts a Gaussian pdf characterized by only the second moment. The support is in this case the two-dimensional sphere. Thus the temperature anisotropies can be expanded in terms of the spherical harmonic coefficients,

$$\Delta(\mathbf{n}) \equiv \frac{\Delta T}{T}(\mathbf{n}) = \sum_{lm} a_{lm}Y_{lm}(\mathbf{n}) \quad (6.2)$$

where $Y_{lm}(\mathbf{n})$ are the spherical harmonic functions for direction $\mathbf{n}$ and $a_{lm}$ are the spherical harmonic coefficients, which for the standard model are Gaussian distributed.

If the field is isotropic then the two-point correlation function only depends on the modulus of the difference of the two directions. In harmonics space, isotropy translates in that the harmonic coefficients are uncorrelated

$$< a_{lm}a_{l'm'}^* > = C_l\delta_{ll'}\delta_{mm'}, \quad (6.3)$$

where $\delta_{ij}$ is the Kronecker delta. The parameter $C_l$ is the temperature power spectrum, which for a realization can be estimated as $(2l+1)^{-1}\sum_m |a_{lm}|^2$.

### 6.2.2 Properties

The IGRF is one of the best-studied random fields and many of its properties have been thoroughly analyzed. One of the most remarkable properties is that the expectation of any even combination of the field $\Delta(\mathbf{n})$ can be given in terms of the second moment, the two-point correlation function, and the expectation of any odd combination is zero; i.e. the n-point correlation functions for n odd are zero. The same property translates to the spherical harmonics space, where the expectation of even combinations of the coefficients $a_{lm}$ can be expressed in terms of the power spectrum $C_l$; and the expectation of odd combinations, e.g. the bispectrum, is zero. More generally, a very useful characteristic of an IGRF is that the expectations of many statistical quantities can be calculated (semi)analytically. This is the case for the morphological descriptors, number, shape and correlation of peaks (maxima), scalars on the sphere, . . . .

The Minkowski functionals are useful descriptors of the morphology of point sets or smoothed fields in spaces with arbitrary dimension $d$ (see e.g. Adler, 1981). As stated in (Schmalzing and Gorski, 1998), under a few simple requirements any morphological descriptor can be written as a linear combination of $d+1$ Minkowski functionals. For the sphere there are therefore three; namely, total length of the contour $\mathcal{C}(\nu)$ of the excursion set, total area $A(\nu)$ of the excursion set and the genus $G(\nu)$ above a threshold $\nu$. Their average value per unit area for a IGRF can be simply given by

$$< \mathcal{C}(\nu) >= \frac{1}{8\theta_c} \exp\left(-\frac{\nu^2}{2}\right), \tag{6.4}$$

$$< A(\nu) >= \frac{1}{2} - \frac{2}{\sqrt{\pi}} \int_0^{\nu/\sqrt{2}} \exp\left(-x^2\right) \mathrm{d}x, \tag{6.5}$$

$$< G(\nu) > \frac{1}{(2\pi)^{3/2}\theta_c^2} \nu \exp\left(-\frac{\nu^2}{2}\right), \tag{6.6}$$

where $\theta_c = \left(-\frac{C(0)}{C''(0)}\right)^{1/2}$ is the coherence angle of the random field which is defined by the ratio of the two-point correlation function $C(\theta)$ to its second derivative at zero lag (see Fig. 6.1).

Properties of peaks in a two-dimensional IGRF have been studied by Longuet-Higgins (1957), Bond and Efstathiou (1987) and Barreiro *et al.* (1997). The local description of a peak involves the second derivative of the field along the two principal directions. The curvature radii are defined in the usual way from the second derivative of the temperature anisotropies $\Delta$ at the position of the maximum: $R_1 = [-\Delta_1''(\max)/2\sigma_0]^{-1/2}$ $R_2 = [-\Delta_2''(\max)/2\sigma_0]^{-1/2}$, where $\sigma_0$ is the anisotropy rms and $\Delta_i''$ is the second derivative along the principal direction $i$. The two invariant quantities, Gaussian curvature $\kappa$ and eccentricity $\epsilon$ can be constructed from them:

$$\kappa = \frac{1}{R_1 R_2}, \quad \epsilon = \left[1 - \left(\frac{R_2}{R_1}\right)^2\right]^{1/2}. \tag{6.7}$$

It is straightforward to obtain the number of peaks on the celestial sphere $N(\kappa, \epsilon, \nu)\, \mathrm{d}\kappa\, \mathrm{d}\epsilon\, \mathrm{d}\nu$ with Gaussian curvature, eccentricity and threshold between $(\kappa, \kappa + \mathrm{d}\kappa)$, $(\epsilon, \epsilon + \mathrm{d}\epsilon)$ and $(\nu, \nu + \mathrm{d}\nu)$, in terms of the two spectral parameters $\gamma$ and $\theta_*$ that characterize the cosmological model:

$$\gamma = \frac{\sigma_1^2}{\sigma_0\sigma_2}, \quad \theta_* = 2^{\frac{1}{2}} \frac{\sigma_1}{\sigma_2}, \tag{6.8}$$

$$\sigma_0^2 = C(0), \quad \sigma_1^2 = -2C''(0), \quad \sigma_2^2 = \frac{8}{3}C^{(iv)}(0). \tag{6.9}$$

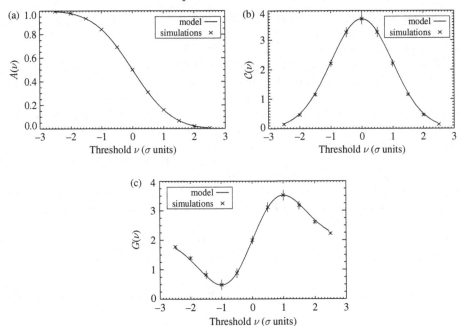

FIGURE 6.1. Expected value of the Minkowski functionals (a) area, (b) contour length and (c) genus for an IGRF obtained from Eqs. 6.4, 6.5, 6.6. Also plotted are the average values (asterisk) and error bars of the Minkowski functionals obtained with 1000 Gaussian simulations. Adapted from Curto *et al.* (2007).

The number of peaks above a threshold $\nu$, $N(\nu)$, can be calculated from the previous differential quantity by integrating over the whole parameter space for $\kappa$ and $\epsilon$ and over the interval $(\nu, \infty)$ for $\nu$:

$$
\begin{aligned}
N(\nu) = N_T \left(\frac{6}{\pi}\right)^{\frac{1}{2}} \exp(-\nu^2/2) & \left\{\gamma^2(\nu^2 - 1)\left[1 - \frac{1}{2}\mathrm{erfc}(\gamma\nu s)\right]\right. \\
& + \nu\gamma(1 - \gamma^2)\frac{s}{\pi^{1/2}}\exp(-\gamma^2\nu^2 s^2) \\
& \left. + t\left[1 - \frac{1}{2}\mathrm{erfc}(\gamma\nu s t)\right]\exp(-\gamma^2\nu^2 t^2)\right\},
\end{aligned}
\tag{6.10}
$$

where

$$
s = [2(1 - \gamma^2)]^{-\frac{1}{2}}, \quad t = (3 - 2\gamma^2)^{-\frac{1}{2}},
\tag{6.11}
$$

and $N_T = (3^{\frac{1}{2}}\theta_*^2)^{-1}$ is the total number of peaks on the whole celestial sphere.

Another interesting quantity is the distribution of the curvature of the peaks. The pdf of peaks with inverse of the Gaussian curvature $L \equiv \kappa^{-1}$ between $(L, L + \mathrm{d}L)$ above the threshold $\nu$, $p(L)$, can be also obtained from $N(\kappa, \epsilon, \nu)\,\mathrm{d}\kappa\,\mathrm{d}\epsilon\,\mathrm{d}\nu$:

$$
\begin{aligned}
p(L) = \left(\frac{6}{\pi}\right)^{\frac{1}{2}} & (2\gamma\theta_c^2)^4 t L^{-5} \exp[(2\gamma\theta_c^2)^2 L^{-2}] \\
& \times \int_\nu^\infty \mathrm{d}\nu \exp(-3t^2\nu^2/2)\mathrm{erfc}\left[\frac{s}{t}(2\gamma\theta_c^2 L^{-1} - \gamma\nu t^2)\right].
\end{aligned}
\tag{6.12}
$$

The distribution of eccentricities can be calculated in a similar manner. The pdf of peaks with eccentricity between $(\epsilon, \epsilon + d\epsilon)$ above a threshold $\nu$, $p(\epsilon)$, can be obtained after a straightforward calculation:

$$
p(\epsilon) = \frac{32\sqrt{6}}{\pi} \epsilon^3 \frac{1 - \epsilon^2}{(2 - \epsilon^2)^5}
$$

$$
\times \int_\nu^\infty d\nu \exp(\nu^2/2) \left\{ (H\pi)^{\frac{1}{2}} \exp(-G) \left[ 1 - \operatorname{erfc}(H^{\frac{1}{2}}\gamma\nu s)/2 \right] \right.
$$

$$
\times \left[ 3H^2(1 - \gamma^2)^2 + 6H^3\gamma^2(1 - \gamma^2)\nu^2 + (H\gamma\nu)^4 \right]
$$

$$
\left. + \exp(-s^2\gamma^2\nu^2)s \left[ 5H^3\gamma(1 - \gamma^2)^2\nu + H^4(\gamma\nu)^3(1 - \gamma^2) \right] \right\}, \qquad (6.13)
$$

where $H = (2 - \epsilon^2)^2/[(3 - 2\gamma^2)\epsilon^4 + 4(1 - \epsilon^2)]$ and $G = H(\gamma\nu\epsilon^2)^2/(2 - \epsilon^2)^2$.

Scalar quantities can be constructed from the derivatives of the CMB field on the sphere. A single scalar can be constructed in terms of the ordinary derivative of the field $\Delta_{,i}$. Only two independent scalars can be obtained from the second covariant derivatives on the sphere, $\Delta_{;ij}$. Following Monteserín *et al.* (2005), many scalar quantities can be defined from the first and second derivatives of the field, associated with the Hessian matrix, the distortion, the gradient and the curvature. However, all except three are correlated. For testing Gaussianity, it is convenient to use normalized scalars for which the dependence of the scalars on the power spectrum has been removed. Here, as an example, we focus on three independent normalized scalars, namely the Laplacian, the fractional anisotropy and the square of the modulus of the gradient. The first two scalars have been proved to be very efficient as detectors of non-Gaussianity (Monteserín *et al.*, 2006). The third scalar, the square of the modulus of the gradient, is the only scalar from the list given in that paper which is independent from all the others. It depends only on the first derivatives of the field. The Laplacian and the fractional anisotropy depend only on second derivatives and can be defined in terms of the eigenvalues $\lambda_1$, $\lambda_2$, of the negative Hessian matrix $\mathbf{A}$ of the field: $\mathbf{A} = (-T_{;ij})$. The eigenvalues, i.e. the negative second derivatives along the two principal directions, can be written as a function of the covariant second derivatives of the field $\Delta(\mathbf{n})$:

$$
\lambda_1 = -\frac{1}{2} \left[ \left( \Delta^{;i}{}_i \right) - \sqrt{ \left( \Delta^{;i}{}_i \right)^2 - 2 \left( \Delta^{;i}{}_i \Delta^{;j}{}_j - \Delta^{;j}{}_i \Delta^{;i}{}_j \right) } \right], \qquad (6.14)
$$

$$
\lambda_2 = -\frac{1}{2} \left[ \left( \Delta^{;i}{}_i \right) + \sqrt{ \left( \Delta^{;i}{}_i \right)^2 - 2 \left( \Delta^{;i}{}_i \Delta^{;j}{}_j - \Delta^{;j}{}_i \Delta^{;i}{}_j \right) } \right]. \qquad (6.15)
$$

We will assume $\lambda_1 \geq \lambda_2$. Considering the values of $\lambda_1$ and $\lambda_2$, three types of points can be distinguished (see e.g. Doré *et al.*, 2003): hill (both positive), lake (both negative) and saddle (one positive and one negative).

The normalized Laplacian, or trace of the Hessian matrix, $\bar{\lambda}_+$, is defined in terms of the eigenvalues as:

$$
\bar{\lambda}_+ = \frac{\lambda_1 + \lambda_2}{\sigma_2}, \quad -\infty < \bar{\lambda}_+ < \infty. \qquad (6.16)
$$

Since the Laplacian is given by linear transformations of the CMB temperature fluctuation field $\Delta$, if the field is Gaussian then its one-point pdf is also Gaussian:

$$
p(\bar{\lambda}_+) = \frac{1}{\sqrt{2\pi}} \exp\left( -\frac{\bar{\lambda}_+^2}{2} \right). \qquad (6.17)
$$

The fractional anisotropy (Basser and Pierpaoli, 1996) has been used in other fields such as in the analysis of medical images. The normalized quantity $\bar{f}_a$ is defined as:

$$\bar{f}_a = \frac{1}{\sqrt{2}} \frac{\bar{\lambda}_1^2 - \bar{\lambda}_2^2}{\sqrt{\bar{\lambda}_1^2 + \bar{\lambda}_2^2}}, \tag{6.18}$$

where $\bar{\lambda}_1$ and $\bar{\lambda}_2$ are the normalized eigenvalues given by:

$$\begin{pmatrix} \bar{\lambda}_1 \\ \bar{\lambda}_2 \end{pmatrix} = \frac{1}{2} \begin{pmatrix} \frac{1}{\sigma_2} + \frac{1}{\sqrt{\sigma_2^2 - 2\sigma_1^2}} & \frac{1}{\sigma_2} - \frac{1}{\sqrt{\sigma_2^2 - 2\sigma_1^2}} \\ \frac{1}{\sigma_2} - \frac{1}{\sqrt{\sigma_2^2 - 2\sigma_1^2}} & \frac{1}{\sigma_2} + \frac{1}{\sqrt{\sigma_2^2 - 2\sigma_1^2}} \end{pmatrix} \begin{pmatrix} \lambda_1 \\ \lambda_2 \end{pmatrix}. \tag{6.19}$$

The pdf of the normalized fractional anisotropy is given by:

$$p(\bar{f}_a) = \frac{2\bar{f}_a}{(1 - \bar{f}_a^2)^{\frac{1}{2}}(1 + \bar{f}_a^2)^{\frac{3}{2}}}, 0 < \bar{f}_a < 1. \tag{6.20}$$

The normalized square of the modulus of the gradient $\bar{g}$, which depends only on the first derivatives of the field, is defined as:

$$\bar{g} = \frac{|\nabla \Delta|^2}{\sigma_1^2} = \frac{\Delta^{,i}\Delta_{,i}}{\sigma_1^2}, \tag{6.21}$$

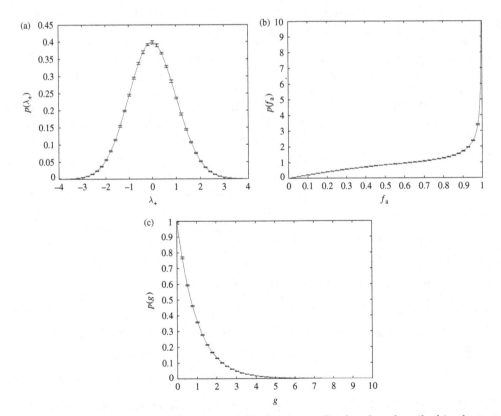

FIGURE 6.2. Probability density functions of the three normalized scalars described in the text: (a) the Laplacian [Eq. (6.17)], (b) the fractional anisotropy [Eq. (6.20)] and (c) the square of the modulus of the gradient [Eq. (6.23)]. The average values and error bars of 1000 IGRF simulations are also represented. Adapted from Monteserin *et al.* (2005).

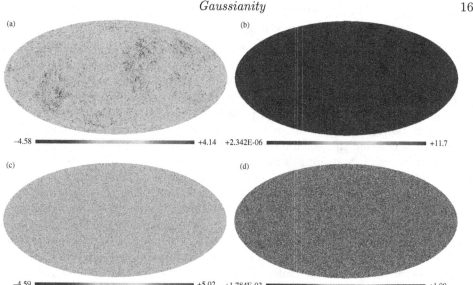

FIGURE 6.3. Maps of the three normalized scalars described in the text (a) normalized temperature, (b) normalized gradient, (c) normalized Laplacian and (d) of the corresponding normalized temperature for a realization of the IGRF. Adapted from Monteserin *et al.* (2005).

where $\sigma_1^2$ is the dispersion of the unnormalized square of the modulus of the gradient and accounts for the normalization factor. In terms of the derivatives of the field with respect to the spherical coordinates $(\theta, \phi)$, $\bar{g}$ takes the form:

$$\bar{g} = \frac{1}{\sigma_1^2} \left[ \left( \frac{\partial \Delta}{\partial \theta} \right)^2 + \frac{1}{\sin^2 \theta} \left( \frac{\partial \Delta}{\partial \phi} \right)^2 \right]. \tag{6.22}$$

Taking into account that the square of the gradient modulus is given by the addition of two independent squared Gaussian variables, its pdf follows a $\chi_2^2$ distribution. Since we consider the normalized quantity $\bar{g}$, then its mean and dispersion are equal to one and its distribution takes the simple form:

$$p(\bar{g}) = \exp(-\bar{g}), \quad 0 < \bar{D}_g < \infty. \tag{6.23}$$

The pdfs of the three normalized scalars can be seen in Fig. 6.2. Maps of the same scalars for a random realization of a IGRF are shown in Fig. 6.3.

## 6.3 Physical effects producing deviations from the standard IGRF

There are a number of physical effects which may produce deviations from the standard IGRF. They are normally related to secondary anisotropies produced by photon scattering, gravitational effects generated by the non-linear evolution of the matter density, variations from standard inflation in the early Universe, topological defects, non-standard geometry and topology of the Universe or primordial magnetic fields. Below we summarize some relevant aspects of the most studied effects.

### 6.3.1 Secondary anisotropies

On the last scattering surface CMB anisotropies are generated via the gravitational potential (Sachs–Wolfe effect, Sachs and Wolfe, 1967) and the physics of the baryon–photon plasma. These anisotropies are usually referred to as *primary anisotropies*. After matter–radiation decoupling when the temperature drops below $\approx 3000\,\text{K}$, new anisotropies are generated during the trip made by the CMB photons to reach us. These *secondary anisotropies* can be originated via the gravitational redshift suffered by the photons when crossing the gravitational potentials produced by the large-scale matter distribution, or by the scattering of the microwave photons with the ionized matter after the reionization epoch ($z \lesssim 10$). The gravitational potential can produce two types of effect: the gravitational redshift suffered by the photons when they cross the varying potential wells formed by the matter evolution (Rees–Sciama effect, Rees and Sciama, 1968; Martínez-González and Sanz, 1990), and the lensing effect produced by the same gravitational potentials which bends their trajectory (see e.g. Bartelmann and Schneider, 2001, for a review). The CMB power spectrum produced by the Rees–Sciama effect is subdominant on all angular scales (Sanz *et al.*, 1996; Hu and Dodelson, 2002), and the expected three-point correlations have an amplitude much below the cosmic variance (Mollerach *et al.*, 1995). The lensing effect produces a smoothing of the acoustic peaks in the CMB power spectrum (Seljak, 1996; Martínez-González *et al.*, 1997). It also induces high-order correlations in the CMB temperature and polarization fields (Hu, 2000).

The scattering of the CMB photons with the ionized medium produces a randomization of the directions of the rescattered photons, implying a suppression of the primary CMB anisotropies whose resulting power spectrum depends on the Thomson optical depth $\tau$ as $\exp(-2\tau)$. In addition, new anisotropies are generated by the scattering with the free electrons moving along the line of sight which produces a Doppler effect. This effect is strongly suppressed due to the cancellation of velocities along the line of sight. However the Doppler effect can survive cancellation if it is modulated by either density or ionization fluctuations. Its amplitude is given by:

$$\frac{\Delta T}{T} = -\sigma_T \int \mathrm{d}t\, e^{-\tau(\theta,t)} n_e(\theta,t) v_r(\theta,t). \tag{6.24}$$

In this formula $\sigma_T$ is the Thomson cross-section, $n_e$ the electron density and $v_r$ the line of sight velocity of the electrons. The density $n_e$ can be further expressed as $n_e(\theta,t) = \bar{n}_e(t)[1 + \delta(\theta,t) + \delta_i(\theta,t)]$ where $\bar{n}_e(t)$ is the mean electron density, $\delta$ the density fluctuation and $\delta_i$ the ionization fraction. Two second-order effects generating secondary anisotropies appear in this formula. The first one is the Doppler effect modulated by the density variation and known as the *Ostriker and Vishniac effect* in the linear regime (Ostriker and Vishniac, 1986; Vishniac, 1986). The second effect is the Doppler effect modulated by the ionization fraction and is usually referred to as *patchy reionization* (see e.g. Aghanim *et al.*, 1996). Since they are of second order, both effects produce non-Gaussian perturbations in the CMB. However, their amplitude is much smaller than the thermal and kinetic *Sunyaev–Zeldovich effects (SZ; 1972)* generated in the non-linear regime at lower redshifts and produced by the scattering of hot, ionized gas associated with collapsed structures. In addition to temperature anisotropies, due to the primary quadrupole moment reionization also produces polarization.

For more details about secondary anisotropies the reader is referred to e.g. Hu and Dodelson (2002) and Aghanim and Puget (2006).

### 6.3.2 Non-standard models of the early Universe

### 6.3.2.1 Non-standard inflationary models

In the standard, single-field, slow-roll inflationary model, the dominant linear effects in the evolution of density fluctuations preserve the initial Gaussian distribution produced by the quantum fluctuations of the inflaton. On the contrary, second-order effects perturb the original Gaussianity of the fluctuations, although resulting in a negligible effect in the standard model. In non-standard inflationary models, primordial non-Gaussianity should be added to the second-order effects associated to the evolution of density fluctuations after inflation finishes. In any case, it is common to characterize phenomenologically the deviations from Gaussianity by introducing the $f_{NL}$ parameter in the gravitational potential (Komatsu and Spergel, 2001):

$$\phi = \phi_L + f_{NL}\phi_L^2, \tag{6.25}$$

where $\phi_L$ is the gravitational potential at the linear order and it is distributed following a Gaussian random field. In general the *non-linear coupling parameter* $f_{NL}$ is a function of the distance vectors and the product is really a convolution. However, an effective $f_{NL}$ which accounts for those complexities can still be used. A detailed study of the values of $f_{NL}$ for several non-standard inflationary models, including multifield inflation, the curvaton scenario, inhomogeneous reheating and the Dirac–Born–Infeld (DBI) inflation inspired by string theory, can be found in Bartolo *et al.*, 2004 (see also Matarrese, this volume). Here it is worth noticing that sometimes the specific $f_{NL}$ parameter given by Eq. 6.25 is called the *local* non-linear coupling parameter $f_{NL}^{local}$, referring to the fact that $\phi$ is obtained from $\phi_L$ at the same position in space. In addition to $f_{NL}^{local}$, which contains mainly information of the squeezed configurations (those with two large and similar wavevectors and the other small), the *equilateral* non-linear coupling parameter $f_{NL}^{equil}$ is used to characterize equilateral configurations of the bispectrum for which the lengths of the three wavevectors are equal in Fourier space. The two non-linear coupling parameters suppose a fair representation of a large class of models. For instance, $f_{NL}^{local}$ can be generated in curvaton and reheating scenarios whereas $f_{NL}^{equil}$ can be produced in DBI inflation within the context of String theory.

In the standard single-field, slow-roll inflation the effective $f_{NL}$ is dominated by second-order gravitational corrections leading to values of order unity. These low values require very sensitive measurements to be detected and are not even within reach of the Planck mission. Non-standard inflationary models generally predict larger values, some of them already constrained by WMAP. A positive detection of $f_{NL} \gtrsim 10$ would rule out the standard inflationary models.

For an ideal experiment with white noise, no foreground residuals and no Galactic and point-source mask, it has been shown that the optimal estimator for $f_{NL}$ is based on a bispectrum test constructed from a cubic combination of appropriately filtered temperature and polarization maps (Komatsu *et al.*, 2005; Creminelli *et al.*, 2006; Yadav *et al.*, 2007). An extension to this estimator to deal with data under realistic experimental conditions has been made by Yadav *et al.*, 2008. Several observational constraints on $f_{NL}$ have been derived from different experiments. The first ones were obtained with COBE–DMR (Komatsu *et al.*, 2002; Cayón *et al.*, 2003). Further constraints have been derived with other experiments like VSA (Smith *et al.*, 2004), BOOMERanG (de Troia *et al.*, 2007) and Archeops (Curto *et al.*, 2007, 2008). However, the best limits have been derived with the various releases of the WMAP data (Komatsu *et al.*, 2003, 2009; Spergel *et al.*, 2007) representing an improvement of at least an order of magnitude over previous ones (see below).

## 6.3.2.2 Topological defects

In standard theories of particle physics the fundamental forces of nature unify progressively when the energy scale exceeds certain thresholds. These unified theories imply that the Universe went through several phase transitions during the early stages of its evolution. At energies above $10^2$ GeV, the electromagnetism and the weak nuclear force are merged into the electroweak force. At higher energies of the order of $10^{15}$ GeV it is believed that the electroweak force unifies with the strong nuclear force, a process usually called grand unified theory (GUT). At even higher energies it is speculated that it is possible to merge gravity with the other three interactions.

Unified theories are based on symmetry. When the symmetry is broken spontaneously via the Higgs mechanism, topological defects generically appear (Kibble, 1976; for a more pedagogical discussion about topological defects see Vilenkin and Shellard, 1994). Depending on the dimensionality of the symmetry which is broken, a type of topological defect is formed. When a discrete symmetry is broken domain walls form, in a two-dimensional symmetry breaking cosmic strings appear, monopoles form when the symmetry breaking is three-dimensional and textures appear when it is in four or more dimensions. In the 1980s topological defects were considered as an alternative scenario to inflationary quantum fluctuations for the process of structure formation. Cosmic microwave background observations showed that the former scenario could only play a subdominant role as the source of cosmic structure. Moreover, due to the catastrophic effects that the presence of domain walls and monopoles would have on our Universe only the existence of cosmic strings and textures is usually considered.

Much effort has been made to study the cosmological consequences of cosmic strings. It is generally believed that the best strategy to detect them is by searching for their imprint on CMB maps. The non-Gaussian character of the density field of strings produce line discontinuities in the CMB anisotropies at arcmin angular scales as a consequence of the metric deformation around the strings (the Kaiser–Stebbins effect; Kaiser and Stebbins, 1984). However, at larger angular scales a Gaussian distribution emerges from the central limit theorem.

Cosmic textures were first studied in detail by Turok (1989). They are much more diffuse than the other defects which are localized at a point (monopole), on a line (cosmic string) or a surface (domain wall). Contrary to the others, they are unstable and consist of twisted configurations of fields which collapse and unwind. Each texture creates a time-varying gravitational potential which produces a red- or blue-shift to the CMB photons passing through such a region. Thus, textures generate hot and cold spots on the CMB anisotropy maps whose amplitude is set by the symmetry-breaking energy scale. The shape of the spots is approximately spherically symmetric and an approximated analytical formula has been derived by Turok and Spergel (1990). Recently, a very cold spot detected in the WMAP temperature map has been found to be consistent with the effect produced by a texture (Cruz *et al.*, 2007). If confirmed, this result will have outstanding consequences in our understanding of the Universe (see Section 6.6.6.2 for more details on this finding).

### 6.3.3 Non-standard geometry and topology

The geometry of the observable Universe is believed to be well approximated on large scales by a Friedmann–Lemaître–Robertson–Walker (FLRW) metric characterized by a homogeneous and isotropic space-time. Support for the homogeneity of the Universe is provided by the largest surveys of the galaxy distribution [e.g. SDSS DR6 (Ademan-McCarthy *et al.*, 2008), 2dFGRS (Colless *et al.*, 2001)] and by the smallness of the CMB temperature fluctuations. The cosmological principle, stating that the Universe is homogeneous and isotropic on large scales, follows from those observations and the assumption that we are not located at a special place in the Universe (the Copernican principle).

However, recent observations of the CMB as measured by the WMAP satellite might question the cosmological principle. They are the large-scale power asymmetry found between the two hemispheres of a reference frame close to that of the ecliptic one (Eriksen *et al.*, 2004a; Hansen *et al.*, 2004), the planarity and alignment of the low multipoles (de Oliveira-Costa *et al.*, 2004; Land and Magueijo, 2005a), the non-Gaussian cold spot present at scales of $\approx 10°$ (Vielva *et al.*, 2004; Cruz *et al.*, 2005) and the alignment of local structures also at similar scales (Wiaux *et al.*, 2006; Vielva *et al.*, 2007; see Section 6.6 for more details). These results may imply the existence of privileged directions in the CMB map and thus motivates the study of alternatives to the standard FLRW metric.

An interesting class of alternative models is that for which the metric is homogeneous but anisotropic. They are known as Bianchi models and are classified according to their space-time properties. Their predictions for the CMB anisotropy were studied in Barrow *et al.* (1985). Since the signatures left by those models appear on large angular scales, they have been already constrained by the COBE–DMR experimental data (Martínez-González and Sanz, 1995; Bunn *et al.*, 1996; Kogut *et al.*, 1997). One particularly interesting case is Bianchi $VII_h$ which experiences anisotropic expansion and global rotation. These two properties produce a characteristic pattern in the CMB in the form of a spiral pattern and spots. This model has been recently used to account for some of the large-scale WMAP anomalies mentioned above (Jaffe *et al.*, 2005, 2006a,b; Bridges *et al.*, 2007; see Section 6.6.6.1 for more details). An important problem with this model comparison approach is that the CMB anisotropies are not computed in a self-consistent way but are assumed to be the sum of two independent components: an isotropic one produced by the energy density fluctuations which is assumed to be the same than for the FLRW model, and an anisotropic component which is the deterministic effect produced by the anisotropic model.

A different source of anisotropic features in the CMB sky is the global topology of the Universe. The local character of the general theory of Relativity does not theoretically constrain it. If the topology of the Universe is non-trivial (i.e. multiconnected, meaning that there is not a unique way to connect two points by geodesics) then CMB photons originated from the same location on the last scattering surface can be observed in different directions. This effect manifests itself in the CMB sky as anisotropic patterns, correlated (matched) circles (Cornish *et al.*, 2004) or, more generically, as deviations from a IGRF (for a discussion on different topologies and tests developed to detect them see the reviews by Lachièze-Rey and Luminet, 1995 and Levin, 2002). An additional consequence of a multiconnected Universe is the lack of fluctuations in the CMB above the wavelength corresponding to the size of the Universe. This property has led several authors to suggest that the low quadrupole and the alignment of the low multipoles measured by WMAP might be evidence of a non-trivial topology (Luminet *et al.*, 2003; Cresswell *et al.*, 2006). Constraints on the topology of the Universe started with the COBE–DMR data (de Oliveira-Costa, 1996; Roukema, 2000; Rocha *et al.*, 2004) and followed with the WMAP data (Inoue and Sugiyama, 2003; Kunz *et al.*, 2006; Phillips and Kogut, 2006; Niarchou and Jaffe, 2007; Kunz *et al.*, 2008). All those analyses concluded that the WMAP data do not show any evidence of multiconnected universes.

### 6.3.4  Primordial magnetic fields

In recent years there has been an increasing interest in studying the consequences that the possible existence of primordial magnetic fields might have on the CMB. Magnetic fields of order of a few µG have been measured in a wide range of astrophysical structures, from individual galaxies (Han and Wielebinski, 2002) to galaxy clusters (Kronberg, 2004). It is also widely believed that they are present in superclusters. Although the origin of those magnetic fields is still unclear and their existence does not necessarily imply a primordial origin, studying the interplay between magnetic fields and CMB is justified by the important consequences that it may have on the CMB temperature and polarization anisotropy, and also on the distortion of the blackbody spectrum (for reviews on this topic the reader is referred to Giovannini, 2004; Durrer, 2007). In particular, the presence of magnetic fields introduce non-Gaussianity in the CMB anisotropy since its amplitude depends on the square of the magnetic field intensity.

Two very different cases can be considered for the primordial magnetic field: a uniform field and a stochastic one. The former breaks the spatial isotropy of the background geometry by introducing shear through an anisotropic stress. This leads to the well-known homogeneous and anisotropic Bianchi models (Barrow *et al.*, 1985). Since those models also imply an anisotropic CMB field, a uniform magnetic field generates phase correlations between different $a_{lm}$. The uniform magnetic field has been constrained with the CMB quadrupole and also with the phase correlations, implying comparable constraints on the magnetic field intensity of a few nG (Barrow *et al.*, 1997; Durrer *et al.*, 1998; Chen *et al.*, 2004).

The stochastic magnetic field is a more realistic scenario which can be generated during inflation. In this case the isotropy of the background geometry is preserved. Temperature and polarization anisotropies are generated through the vector and tensor modes associated with the magnetic field energy-momentum tensor. Allowed amplitudes of the magnetic field intensity of about several nG can produce a potentially observable $B$-mode polarization for nearly scale-invariant spectra (Lewis, 2004). This signal could be distinguished from the one generated by the inflationary gravitational wave background by its non-Gaussian character. In addition Faraday rotation induces $B$-mode polarization from the ordinary $E$-mode with the characteristic $\nu^{-2}$ dependence.

Therefore, the presence of a primordial magnetic field can leave unambiguous imprints in the CMB anisotropy that would allow its identification with the sensitive data expected from the coming experiments.

## 6.4  Methods to test Gaussianity

Testing the Gaussianity of CMB data is not an easy task. In principle it consists in just proving the properties of the IGRF that we discussed in Section 6.2.2: isotropy and multinormality. The CMB data represent a single realization of the underlying random field which for the standard model is nearly Gaussian and isotropic. In practice the analysis is complicated by the characteristics of the experiment which need to be known very precisely: calibration uncertainties, instrumental noise (white and 1/f) which is normally anisotropic in pixel space depending on the scanning strategy, beam response (usually close to Gaussian), data processing, etc. And by foreground contamination which demands certain previous cleaning operations in the data, requiring the masking of certain areas where foregrounds are very intense and leaving some amount of residuals in the rest of the surveyed area. The result of cleaning depends on our a-priori knowledge of the physical properties of the foregrounds and the component separation method used for their removal. These ingredients must be considered in the analysis by performing simulations accounting for them.

Unless one is interested in the compatibility of the data with a specific alternative model for which an optimal method may be found (as specific non-standard inflation, geometry or topology), there are infinite ways in which a random field can deviate from the IGRF one. Methods to test Gaussianity can be classified by the type of property that they try to probe. Typical examples are cumulants and n-point correlation functions in real space (the former should vanish and the latter either vanish for the odd order or can be expressed in terms of two-point correlation functions for the even order), moments in spherical harmonics space (bispectrum, trispectrum, ...), or moments in other spaces to which the data is transformed by linear operations which preserve Gaussianity: filters, wavelets, signal-to-noise eigenvectors, etc. Other approaches may test different properties of the CMB random field, like the morphology of the data using the Minkowski functionals or the geometry of the data using scalar quantities constructed by the first and second covariant derivatives as, for example, the local curvature (see Section 6.2.2). These and other methods are described in more detail in Barreiro (2007) and Martínez-González (2008) and references therein. As an example of the typical statistical procedure followed in the analysis, below we focus on a few statistical methods which have often been used in the literature based on the Minkowski functionals, bispectrum or wavelets.

As we have described in Section 6.2.2, for the excursion set above a given threshold $\nu$, there are three Minkowski functionals on the sphere: total contour length $\mathcal{C}(\nu)$, total area $A(\nu)$ and the genus $G(\nu)$ (see Schmalzing and Gorski, 1998). In the case of an IGRF their expected values follow simple analytical expressions as a function of $\nu$ – Eqs. (6.4), (6.5) and (6.6). However, simulations are needed to account for the experimental characteristics, basically noise, beam response and mask. As it can be shown with simulations, the 1-pdf for any Minkowski functional at each $\nu$ follows a nice bell-shape distribution, implying the natural choice of a generalized $\chi^2$ as the appropriate statistical test to be used in this case to combine all the information. More specifically, considering $n_{\mathrm{th}}$ different thresholds we can define a $3n_{\mathrm{th}}$ vector $\mathbf{v}$,

$$\mathbf{v} = (A(\nu_1), \ldots, A(\nu_{n_{\mathrm{th}}}), \mathcal{C}(\nu_1), \ldots, \mathcal{C}(\nu_{n_{\mathrm{th}}}), \mathcal{G}(\nu_1), \ldots, \mathcal{G}(\nu_{n_{\mathrm{th}}})). \qquad (6.26)$$

The generalized $\chi^2$ statistic to test the Gaussianity of a data map can then be constructed as

$$\chi^2 = \sum_{i,j=1}^{3n_{\mathrm{th}}} (v_i - <v_i>) \, C_{ij}^{-1} \, (v_j - <v_j>) \qquad (6.27)$$

where $<>$ is the expected value for the Gaussian case and $C_{ij}$ is the covariance matrix, $C_{ij} = <v_i v_j> - <v_i><v_j>$, both of them usually constructed with simulations. For testing the compatibility of the data with a non-Gaussian model (e.g. a non-standard inflationary model characterized by the $f_{\mathrm{NL}}$ parameter) we simply have to use the corresponding expected value and covariance matrix for the Minkowski functionals at different thresholds in Eq. (6.27). If the deviations from Gaussianity are small (as in the case of the $f_{\mathrm{NL}}$ models with $f_{\mathrm{NL}} \lesssim 1000$) the covariance matrix can be well approximated by that of the Gaussian case. More information is added to the analysis by considering $n_{\mathrm{res}}$ different resolutions of a given data map. Including this extra information simply increases the vectors and covariance matrix present in the $\chi^2$ expression to a dimension $3n_{\mathrm{th}}n_{\mathrm{res}}$. Examples of applications of this method to different data sets can be found in Komatsu et al. (2003) for the one-year WMAP data, de Troia et al. (2007) for the BOOMERanG three-year data and Curto et al. (2008) for the Archeops data. Recently, perturbative formulae of the Minkowski functionals as a function of $f_{NL}$ have been derived for the $f_{NL}$ models (Hikage *et al.*, 2006). The results of applying them to the three-year WMAP data show constraints on $f_{NL}$ very similar to those obtained with the optimal bispectrum (Hikage *et al.*, 2008).

Generically, non-standard models of inflation produce small deviations of Gaussianity which are more prominent in the three-point correlation function or equivalently, its harmonic transform the bispectrum $B_{l_1 l_2 l_3}^{m_1 m_2 m_3} \equiv < a_{l_1 m_1} a_{l_2 m_2} a_{l_3 m_3} >$. The trispectrum, the harmonic transform of the four-point correlation function, can also play an important role in discriminating inflationary models since some models do not produce any bispectra but produce significant trispectra, or produce similar amplitudes of the bispectra but very different trispectra (e.g. DBI inflation, Huang and Shiu, 2006, or new ekpyrotic cosmology, Buchbinder *et al.*, 2008). Here we briefly describe the bispectrum; for more details on it and the trispectrum see Komatsu (2001). The average bispectrum, $B_{l_1 l_2 l_3}$, is the rotationally invariant third-order moment of spherical harmonic coefficients and is given by the following expression (see Hu, 2001; and Bartolo *et al.*, 2004 for more details on this and higher-order moments):

$$B_{l_1 l_2 l_3} = \sum_{m_1 m_2 m_3} \begin{pmatrix} l_1 & l_2 & l_3 \\ m_1 & m_2 & m_3 \end{pmatrix} \langle a_{l_1 m_1} a_{l_2 m_2} a_{l_3 m_3} \rangle, \qquad (6.28)$$

where the symbol within the parentheses is the Wigner-3j symbol. The bispectrum must satisfy the selection rules. Rotational invariance of the three-point correlation function implies that the bispectrum can be written as

$$B_{l_1 l_2 l_3}^{m_1 m_2 m_3} = G_{l_1 l_2 l_3}^{m_1 m_2 m_3} b_{l_1 l_2 l_3}, \qquad (6.29)$$

where $G_{l_1 l_2 l_3}^{m_1 m_2 m_3}$ is the Gaunt factor and $b_{l_1 l_2 l_3}$ is a real symmetric function of $l_1$, $l_2$ and $l_3$ called the *reduced bispectrum*. By substituting Eq. (6.29) into Eq. (6.28) it is straightforward to obtain the following relation between the averaged bispectrum and the reduced bispectrum

$$B_{l_1 l_2 l_3} = \sqrt{\frac{(2l_1 + 1)(2l_2 + 1)(2l_3 + 1)}{4\pi}} \begin{pmatrix} l_1 & l_2 & l_3 \\ 0 & 0 & 0 \end{pmatrix} b_{l_1 l_2 l_3}. \qquad (6.30)$$

Therefore under rotational invariance the reduced bispectrum contains all physical information of the bispectrum. In particular $b_{l_1 l_2 l_3}$ is fully identified by the inflationary models characterized by the non-linear coupling parameter $f_{NL}$. Thus these models can be optimally tested by using the averaged bispectrum. The computation of the full averaged bispectrum scales as $N_{pix}^{5/2}$, where $N_{pix}$ is the number of pixels, and is already not feasible for the WMAP data. Komatsu *et al.* (2005) solved this problem by constructing a cubic statistic, the KSW estimator, that combines the triangle configurations of the bispectrum optimally for determining $f_{NL}^{local}$, with its computation scaling as $N_{pix}^{3/2}$. The extension of this estimator to $f_{NL}^{equil}$, and also including polarization, has been made in Creminelli *et al.* (2006) and Yadav *et al.* (2007). The strongest constraints on the non-linear coupling parameter to date have been obtained by applying the KSW estimator to the WMAP data (see next section).

Wavelets are compensated filters which allow one to extract information which is localized in both real and harmonic space. In particular wavelets may provide information on the position and scale of different features in astrophysical images. They can be more sensitive than classical methods [for a review on wavelets see Jones (2008) and for applications to the CMB see Vielva (2007)]. One example which has been used many times in cosmological applications is the Mexican hat wavelet defined as the Laplacian of a

Gaussian function. For CMB analyses the extension of an Euclidean wavelet to the sphere can be made with an inverse stereographic projection (Antoine and Vandergheynst, 1998). By this procedure the spherical Mexican hat wavelet can be constructed preserving dilation and compensation (Martínez-González *et al.*, 2002), and was first applied to CMB analyses by Cayón *et al.* (2001) using COBE–DMR data. Also it can be made directional, i.e. sensitive not only to the scale but also to the orientation of a feature, by simply considering different widths along the two axes of the original two-dimensional Gaussian. A very useful property to study directional properties of structures in an image is the so called steerability. In general, a directional or non-axisymmetric filter is said to be *steerable* if any rotation about itself can be expressed as a finite linear combination of non-rotated basis filters. This concept has been recently extended to the sphere by Wiaux *et al.* (2005). An important consequence of the steerability property is that it makes the wavelet analysis very efficient computationally (see Section 6.6.4). An interesting spherical steerable wavelet is the second Gaussian derivative one which may be rotated in terms of three basis wavelets: the second derivative in direction $x$, the second derivative in direction $y$, and the cross-derivative. This wavelet provides information on the three local morphological measures of orientation, signed intensity (amplitude at the orientation which maximizes the absolute value of the coefficient) and elongation. It has been recently applied to the CMB analysis to test global isotropy and Gaussianity (Wiaux *et al.*, 2006; Vielva *et al.*, 2007; Wiaux *et al.*, 2008; see Section 6.6.4).

## 6.5 Constraints from observations

Previous to any Gaussianity study, one major problem in the analysis of the observational data has been the separation of the different Galactic and extragalactic components from the CMB itself. A key property to distinguish them is the frequency dependence of the specific emission in the microwave range. Thus, the intensity of the Galactic synchrotron and free–free emissions decreases with frequency as a power law with approximated spectral indexes $-3$ and $-2$, respectively. On the contrary, the intensity of the Galactic thermal dust increases with frequency following approximately a grey-body spectrum with emissivity proportional to $\nu^2$ and temperature $T_D \approx 10 - 20$ K. Extragalactic sources emitting in the microwave band are typically radio galaxies and infrared (IR) galaxies, with their emission dominating at frequencies $\lesssim 100$ GHz for the former (synchrotron-like) and at higher ones (dust-like) for the latter. An additional extragalactic emission comes from galaxy clusters through the SZ effect, as was discussed in Section 6.3.1. The spatial distribution of the different foregrounds is also very different from the CMB one, differing very much from that of an IGRF. Thus, the Galactic foreground emissions dominate at large angular scales with the power spectrum of fluctuations decaying approximately as $C_l \propto l^{-3}$. On the contrary the extragalactic foregrounds dominate at the smallest angular scales and appear as point sources for typical CMB experiments. (For more details on the properties of foregrounds see contributions by Davies and Partridge to this volume). All these properties have to be exploited in order to best disentangle the foreground emissions from the CMB one. Indeed, the numerous component separation methods already developed take advantage of our knowledge of those foreground properties to obtain a clean CMB map (see Barreiro, 2008 and Delabrouille and Cardoso, 2008 for further reading on this topic).

The Gaussianity of the CMB signal has been studied with data measured by many different experiments. The first systematic analysis was carried out with the all-sky COBE–DMR satellite (Smoot *et al.*, 1992; Bennett *et al.*, 1994). Other analyses were based on data covering a fraction of the sky and obtained from experiments onboard stratospheric balloons, like QMAP, MAXIMA, BOOMERanG or Archeops, and from ground-based ones like Saskatoon, QMASK or VSA. The result of most of those Gaussianity studies was a systematic compatibility with the standard IGRF. Deviations were also claimed in a few cases which were later proved to be due either to systematics or an incomplete analysis. Upper limits on the $f_{NL}^{local}$ parameter were derived of approximately a few thousands (Komatsu *et al.*, 2002; Cayón *et al.*, 2003). For a more detailed discussion on those analyses see Martínez-González (2008). A significant improvement has been recently achieved with measurements by BOOMERanG and Archeops, lowering the upper limit to $f_{NL} \lesssim 1000$ at the $2\sigma$ confidence limits (de Troia *et al.*, 2007; Curto *et al.*, 2008). It is worth mentioning a deviation from the IGRF found in one of the VSA fields, the Corona Borealis supercluster (Génova-Santos *et al.*, 2005; Rubiño-Martín *et al.*, 2006). The deviation consists in a strong and resolved negative spot, of $\approx -250\,\mu K$ and angular size of $\approx 20$ arcmin, which is not associated with any of the clusters of that supercluster. An SZ effect produced by a diffuse, extended warm/hot gas distribution has been suggested as a possible explanation (Génova-Santos *et al.*, 2008). This hypothesis, if confirmed, would be of relevance for providing the location of the missing baryons in the local Universe.

The precision with which the Gaussianity of the CMB can be tested has been strongly improved with the quality data measured with the WMAP satellite (Bennett *et al.*, 2003a; Hinshaw *et al.*, 2007, 2009). The WMAP team has provided "cleaned" CMB maps at the $Q$, $V$ and $W$ frequency bands for the five-year data collected by the experiment, and masks covering a region around the Galactic plane and a catalog of radio sources where the foreground emission cannot be removed at the required sensitivity. By a noise-weighting combination of the cleaned maps at the cosmological frequencies and applying a conservative mask the WMAP team has performed a Gaussianity study of the data based on the Minkowski functionals and the optimal bispectrum. For both quantities the one, three and five-year WMAP data releases have been found to be compatible with the IGRF (Komatsu *et al.*, 2003; Spergel *et al.*, 2007; Komatsu *et al.*, 2009). Stringent limits have also been derived on the local and equilateral $f_{NL}$ parameter using an optimal bispectrum-like quantity: $-9 < f_{NL}^{local} < 111$ and $-151 < f_{NL}^{equil} < 253$ at the $2\sigma$ confidence level.

## 6.6 WMAP anomalies

The WMAP team found the WMAP data consistent with the IGRF using the Minkowski functionals and the bispectrum. Subsequently, many works have tested the Gaussianity of the WMAP data in many different ways. Some of them have found agreement with Gaussianity whereas others found significant deviations of the IGRF. Examples of the former are analyses based on three-point correlation analysis (Gaztañaga and Wagg, 2003; Chen and Szapudi, 2005), integrated bispectrum (Cabella *et al.*, 2006), real-space statistics (Bartosz, 2008) or isotropy analyses based on the bipolar power spectrum (Hajian *et al.*, 2005; Hajian and Souradeep, 2006). Examples of the latter are analyses based on phase correlations (Chiang *et al.*, 2003), the genus (Park, 2004), isotropic wavelets (Mukherjee and Wang, 2004; Vielva *et al.*, 2004; Cruz *et al.*, 2005), 1-pdf (Monteserín *et al.*, 2007), isotropy analyses based on local n-point correlations (Eriksen *et al.*, 2004a), local curvature (Hansen *et al.*, 2004), multipole vectors (Copi *et al.*, 2004, 2007) or directional wavelets (McEwen *et al.*, 2005; Wiaux *et al.*, 2006, 2008; Vielva *et al.*, 2007). However, some of the analyses have been performed on the whole-sky internal linear combination (ILC) map (Bennett *et al.*, 2003b) or similar maps (e.g. Tegmark *et al.*, 2003) which are known to suffer from Galactic contamination and, as stated by the WMAP team itself, should not be used for cosmological analyses. From here on, we will concentrate on the most relevant works based on WMAP maps where a certain region around the Galactic plane has been masked as well as several hundred of extragalactic sources. The deviations or "anomalies" reported have been detected using other statistical quantities different from the ones originally used by the WMAP team. They are the north–south asymmetry (Eriksen *et al.*, 2004a; Hansen *et al.*, 2004), alignment of the low multipoles (de Oliveira-Costa, 2004; Land and Magueijo, 2005a), the cold spot (Vielva *et al.*, 2004, Cruz *et al.*, 2005, 2007), non-Gaussian features detected with directional wavelets (McEwen *et al.*, 2005, 2008), alignment of CMB structures (Wiaux *et al.*, 2006; Vielva *et al.*, 2007) low variance (Monteserín *et al.*, 2007). Below we describe these anomalies and discuss several relevant aspects such as significance, origin, etc.

### 6.6.1 North–south asymmetry

North–south asymmetries in ecliptic coordinates have been observed in the WMAP CMB maps using several local quantities: power spectrum and two and three-point correlation functions (Eriksen *et al.*, 2004a, 2005), Minkowski functionals (Eriksen *et al.*, 2004b), local power spectrum (Hansen *et al.*, 2004; Donoghue and Donoghue, 2005), local bispectra (Land and Magueijo, 2005b) and local curvature (Hansen *et al.*, 2004). Varying the coordinate system, a maximum asymmetry for the two hemispheres is obtained for a system whose north pole is lying at $(\theta, \phi) = (80°, 57°)$. This direction is close to the north ecliptic pole $(\theta, \phi) = (60°, 96°)$ (see Fig. 6.4). The result of all those works basically indicates a significant lack of power in the north ecliptic hemisphere compared to the south one. The asymmetry has been also confirmed by Bernui *et al.* (2006, 2007) using a new directional indicator, based on separation histograms of pairs of pixels.

The asymmetry remains stable with respect to variations in the Galaxy cut and to the frequency band. Also a similar asymmetry is found in the COBE–DMR map with the axis of maximum asymmetry close to the one found in the WMAP data. Analyses of the possible foreground contamination and known systematics indicate that these do not seem to be the cause of such asymmetry (Eriksen *et al.*, 2004a).

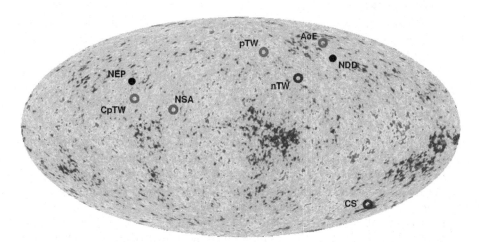

FIGURE 6.4. Directions in the microwave sky derived from the anomalies found in WMAP data: northern direction of the north–south asymmetry (NSA), "axis of evil" (AoE), the cold spot (CS), cluster of positive total weights (CpTW), perpendicular axis to the positive total weights plane (pTW), perpendicular axis to the negative total weights plane (nTW). For reference the north ecliptic pole (NEP) and the northern direction of the CMB dipole (NDD) are also shown.

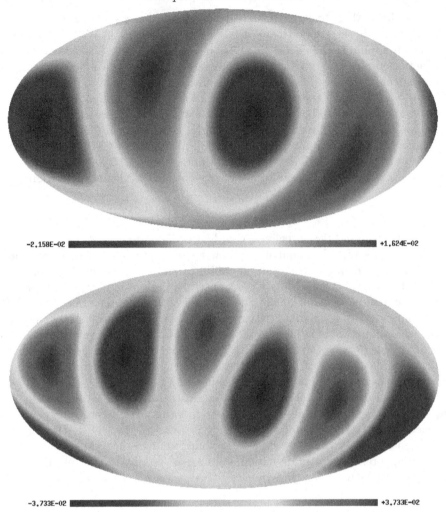

-2.158E-02 ▮▮▮▮▮▮▮▮▮▮▮▮▮▮▮▮▮▮▮▮▮▮▮▮▮▮▮ +1.624E-02

-3.733E-02 ▮▮▮▮▮▮▮▮▮▮▮▮▮▮▮▮▮▮▮▮▮▮▮▮▮▮▮ +3.733E-02

FIGURE 6.5. Maps of the quadrupole and octopole obtained from the WMAP five-year ILC map. The ILC map can be obtained from the LAMBDA web page (http://lambda.gsfc.nasa.gov/).

More recently, the asymmetry has been found again in the three-year WMAP CMB map (Eriksen *et al.*, 2007).

### 6.6.2 Alignment of the low multipoles

The lowest multipoles, especially $l = 2, 3$, of the WMAP data have been found to be anomalously planar and aligned (Copi *et al.*, 2004; de Oliveira-Costa *et al.*, 2004; Katz and Weeks, 2004; Schwartz *et al.*, 2004; Bielewicz *et al.*, 2005 Copi *et al.*, 2007). In Fig. 6.5 the quadrupole and octopole of the ILC map are shown. Both multipoles present maxima and minima following a planar shape, whose perpendicular axis points towards a similar direction called the "axis of evil." The axes of the two multipoles are separated by $\approx 10°$. The probability that the two directions are separated by that angle or less by chance is $\approx 1.5\%$. Further alignments have also been claimed for higher multipoles $l \leq 5$ (Land and Magueijo, 2005a) and $l = 6, 7$ (Freeman *et al.*, 2006). The northern end of the alignment points towards $(\theta, \phi) = (30°, 260°)$, a direction close to the CMB dipole one whose northern end is at $(\theta, \phi) = (42°, 264°)$; see Fig. 6.4.

A problem which appears when trying to estimate the low multipole components is that they are very much affected by the mask. Varying the mask produces significant changes in their amplitude estimates, especially for the quadrupole, implying consequent uncertainties in the determination of their axes. Detailed analyses of this effect tend to weaken the significance of the detection (de Oliveira-Costa and Tegmark, 2006; Land and Magueijo, 2007).

### 6.6.3 The cold spot

A large and prominent cold spot has been observed in the WMAP data which is hard to explain within the standard inflationary scenario. It was detected with the spherical Mexican-hat wavelet (SMHW) as defined in (Martínez-González *et al.*, 2002). The first evidence came from the kurtosis of the wavelet coefficients of the first-year data which showed an excess with respect to the Gaussian hypothesis at a wavelet scale of $\approx 4°$ (corresponding to a structure size of $\approx 10°$; Vielva *et al.*, 2004). A very cold spot at Galactic coordinates $(\theta, \phi) = (-57°, 209°)$ was identified as the possible source of the excess (see Fig. 6.6). An analysis of the area of the spots at different thresholds in the SMHW coefficient map at around 4° proved that indeed the cold spot had a very large area and was the source of the excess of the kurtosis (Cruz *et al.*, 2005).

The cold spot has also been shown to deviate from Gaussianity using the Max and Higher Criticism estimators (Cayón *et al.*, 2005; Cruz *et al.*, 2005). A study of the morphology of the spot with the elliptical MHW on the sphere has found an almost circular shape (Cruz *et al.*, 2006). In the same paper the possible foreground contribution was considered in detail; it was concluded that contributions from the SZ effect and the Galaxy had to be negligible.

The cold spot has been also identified as the most prominent spot in the CMB sky using steerable wavelets (Vielva *et al.*, 2007). In that work two other spots are identified as deviations from the IGRF three-year WMAP best-fitting model. However, the deviation from Gaussianity seen in the kurtosis of the coefficients of that wavelet (Wiaux *et al.*, 2008) cannot be assigned exclusively to those three spots.

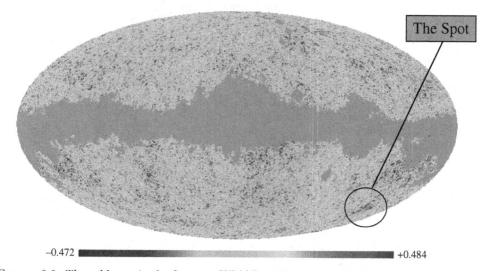

FIGURE 6.6. The cold spot in the five-year WMAP combined map. This map is a noise-weighted combination of the V and W maps given in Hinshaw *et al.* (2009) where the pixels contaminated by Galactic or extragalactic foregrounds have been masked with the KQ75 mask. The WMAP maps and masks can be found at http://lambda.gsfc.nasa.gov/

A number of non-Gaussian features, including the cold spot, have also been detected using the directional wavelets elliptical Mexican hat and Morlet (McEwen *et al.*, 2005, 2008).

All the previous results have been confirmed with the three-year WMAP CMB map (Cruz *et al.*, 2007) and are expected to be almost unaltered for the five-year data.

### 6.6.4 Alignment and signed-intensity of local structures

Some of the previous anomalies imply preferred directions in the sky that, under the assumption of Gaussianity, represent deviations from statistical isotropy. Here we describe a different violation of statistical isotropy based on the alignment of CMB structures. The structures are identified by convolving the CMB map with the steerable wavelet formed by the second Gaussian derivative (Wiaux *et al.*, 2005). For each scale and position in the sky the wavelet identifies the orientation which maximizes the absolute value of the wavelet coefficients. Thus this orientation corresponds to the characteristic orientation of the local feature of the signal. The wavelet coefficient in that specific orientation defines the so-called signed-intensity (Vielva *et al.*, 2007). It should be remarked that these two quantities, local orientation and signed-intensity, are computationally feasible because of the steerability property (see Section 6.4).

Once the local orientation is determined for every position in the sky at a given scale, we can construct the following isotropy test. First, the great circle passing by that position and tangent to the local orientation is identified. Every other pixel which is now crossed by that great circle is considered to be seen by the local orientation of the first position (this procedure is illustrated in Fig. 6.7), with a weight naturally given by the absolute value of the signed-intensity of the first position. Second, we can assign to each pixel in the sky the total weight given by the sum of the weights of all the pixels that see that pixel. This new total-weight signal is even under parity and thus its analysis can

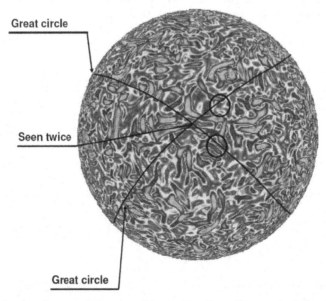

FIGURE 6.7. A simulated map of the signed-intensity where the great circles corresponding to two local features are shown. All the pixels crossed by each great circle are considered to be seen by the corresponding local feature. The position that is crossed by both great circles is thus said to be seen twice. See Vielva *et al.* (2007) for more details.

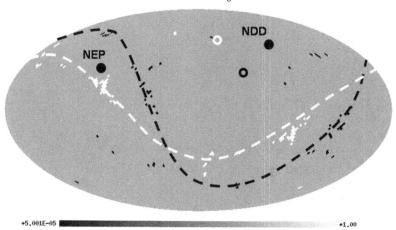

FIGURE 6.8. The highest (lowest) total weights above (below) $3\sigma$ are plotted in white (black). A cluster of positive total weights can be seen close to the NEP. The white (black) dashed line is the best-fit plane to the highest (lowest) total weights. The normal axes defined by both planes point towards directions close to the northern direction of the dipole NDD. Adapted from Vielva *et al.* (2007).

be restricted to one hemisphere of reference. The highest total weights represent the positions towards which the CMB features are predominantly directed while the lowest ones represent the positions predominantly avoided by the CMB features. Of course, for an all-sky IGRF all the pixels should have the same total weight on average.

This isotropy test based on the alignment of local orientations has been applied to the first- (Wiaux *et al.*, 2006) and three-year (Vielva *et al.*, 2007) WMAP data. In both cases a significant violation of isotropy was found at a scale around $10°$. The highest total weights (above $3\sigma$) define an axis located very close to the ecliptic one. The highest and the lowest total weights define two planes whose normal axes are close to the CMB dipole one (Fig 6.8).

Besides the alignment of the local orientations, an analysis of the signed-intensities show three spots at angular scales also around $10°$ containing a total of 39 $1.8°$-pixels whose values have a (signal + noise) probability in the $3\sigma$ tails (see Fig. 6.9). This pattern is similar to the one found with the axisymmetric Mexican hat wavelet (Vielva *et al.*, 2004). The three spots are located in the southern galactic hemisphere confirming the north–south asymmetry. Two of the them are cold, one being identified with the cold spot already detected in Vielva *et al.* (2004).

The two anomalies which appear at similar scales, with global significance levels around 1 per cent, are however quite independent.

### 6.6.5 Low variance

Very recently (Monteserín *et al.*, 2007) the 1-pdf of the WMAP three-year data has been analyzed and an anomalously low value of the variance has been found, compared to the one expected from the WMAP best-fit cosmological model. The result is even more prominent if only the north ecliptic hemisphere is considered (see Fig. 6.10), in agreement with the lack of power found in that hemisphere in previous work (see e.g. Eriksen *et al.*, 2004a). The variance of the CMB signal is obtained by fitting the normalized temperature distribution to a Gaussian of zero mean and unit variance. The significance of the result is around 1% (see Fig. 6.11).

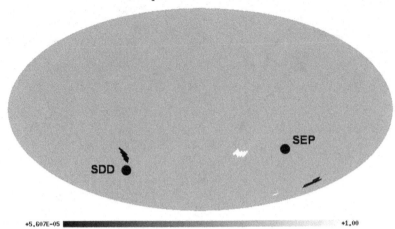

FIGURE 6.9. Map of signed-intensities with probability in the 3σ tails. In black are the pixels with negative values and in white the positive ones. The signed-intensities are grouped around three clusters, one formed by the negative values and other two by the positive ones. Also shown are the positions of the south ecliptic pole (SEP) and the southern direction of the CMB dipole (SDD). Adapted from Vielva *et al.* (2007).

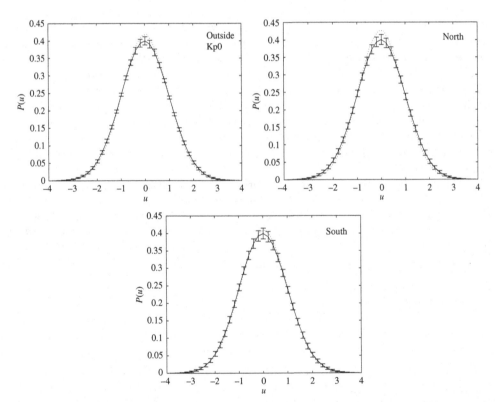

FIGURE 6.10. The histogram of the normalized WMAP data outside the Kp0 mask (dotted line) is compared with the average histogram obtained from 1000 Gaussian simulations (solid line) in the top panel. The error bars represent the dispersion obtained from the simulations. Analogous histograms for the northern and southern ecliptic hemispheres are given in the middle and bottom panels. (Adapted from Monteserín *et al.*, 2006; see paper for more details).

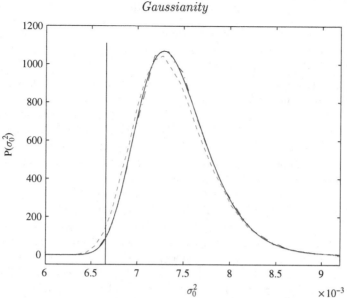

FIGURE 6.11. Theoretical pdf of the CMB variance (calculated following Cayón *et al.*, 1991, solid line) compared to the averaged pdf obtained from 60000 Gaussian simulations of the WMAP best-fit model over the whole sky (dot–dashed line) and using only the pixels outside the Kp0 mask (dashed line). The solid vertical line indicates the value obtained from the three-year WMAP data. See Monteserín *et al.*, 2006 for more details.

In order to find a possible origin for this anomaly the behavior of single-radiometer and single-year data as well as the effect of residual foregrounds and $1/f$ noise have been studied. None of these possibilities can explain the low value of the variance.

Since the largest contribution to the variance comes from the lower multipoles, it is interesting to see if the low quadrupole measured by COBE and WMAP is the cause of the anomalously low variance. Performing the same analysis after subtracting the best-fit quadrupole outside the Kp0 mask the significance of the result is slightly reduced although the variance is still anomalously low.

Beyond the inconsistency found between the best-fit model and the measured variance, one could ask if the latter is consistent with the actual measured power spectrum of the WMAP data. The analysis performed by the same authors shows that a strong discrepancy is indeed found. This last result suggests a possible deviation of the CMB data from the IGRF.

### 6.6.6 Cosmological consequences

Given that neither foreground contamination nor known systematics seem to be causing most of the previous anomalies, there have been a number of attempts to explain the cause of the WMAP anomalies by an intrinsic origin. Among them we mention the Rees–Sciama effect produced by large voids (Tomita, 2005; Inoue and Silk, 2007; Rudnick *et al.*, 2007), inhomogeneous (Adler *et al.*, 2006; Land and Magueijo, 2006) or anisotropic Universes (Jaffe *et al.*, 2005) and cosmic defects (Cruz *et al.*, 2007). Although no further evidence has been found for those explanations most of them still remain as plausible. Below we discuss in more detail two interesting possibilities: the Bianchi model and the cosmic texture.

### 6.6.6.1 Bianchi model

A first interesting attempt to explain the best-studied WMAP anomalies was performed by Jaffe *et al.* (2005, 2006a) who suggested that such features might be produced by an anisotropic Universe, the Bianchi VII$_h$ model. This type of model has a global anisotropic expansion and vorticity that produces geodesic focusing and a spiral pattern in the CMB anisotropy at large angular scales. By fitting the free parameters of that model to the WMAP map the large-scale CMB anisotropies produced by its non-standard geometry are determined. After subtracting that pattern from the WMAP data, the best-studied WMAP anomalies, namely the low-multipole alignments, the north–south asymmetry and the cold spot, were significantly reduced.

This result seemed to suggest that the Universe was not well represented by the homogeneous and isotropic FLRW model and that a perturbation in the form of a Bianchi VII$_h$ was a better representation. However, a more detailed examination of the best-fitted parameters of the Bianchi model when dark energy was included showed values of the dark energy and matter energy density far from the ones measured by many current cosmological tests (Jaffe *et al.*, 2006b; Bridges *et al.*, 2007). Considering other Bianchi models with vorticity and shear, like the Bianchi IX with a closed geometry, does not help since they do not exhibit geodesic focusing or the spiral pattern.

Very recently, Bridges *et al.* (2008) computed the Bayesian evidence of the Bianchi template when the cold spot was not included in the analysis (see below for an alternative interpretation of the cold spot). The result was that the evidence was now significantly reduced, reinforcing the idea that the cold spot was likely to be driving any Bianchi VII$_h$ detection.

### 6.6.6.2 The cold-spot texture

Recently, Cruz *et al.* (2007) performed a Bayesian evidence analysis to test the hypothesis that the cold spot is produced by a cosmic texture. This is a type of topological defect that, as explained in Section 6.3.2.2, in the standard theories of unification of the fundamental forces of nature, is expected to appear in the early Universe. The motivation for considering a texture as the origin of the cold spot is its typical spherical anisotropy pattern left in the CMB and the relatively small number of spots expected at scales of several degrees (see Fig. 6.12).

The test hypothesis considered consisted in the comparison of the following two hypotheses: the null hypothesis $H_0$ for which the cold spot is just a rare fluctuation of the IGRF predicted by the standard inflationary model, and the alternative hypothesis, $H_1$ for which on top of the inflationary Gaussian CMB fluctuations there are non-Gaussian ones as produced by the texture model. The test is applied to a circular area of 40° diameter centered on the cold-spot position. The posterior probability ratio of the two hypotheses is:

$$\rho = \frac{Pr(H_1|\mathbf{D})}{Pr(H_0|\mathbf{D})} = \frac{E_1}{E_0}\frac{Pr(H_1)}{Pr(H_0)}, \tag{6.31}$$

where $\mathbf{D}$ is the data vector, and $E_i$ the evidence, which is the average likelihood $L$ with respect to the priors $\Pi(\Theta_i)$ in the parameters $\Theta_i$ of the hypothesis $H_i$,

$$E_i = Pr(\mathbf{D}|H_i) = \int L_i(\Theta_i|H_i)\Pi(\Theta_i)\mathrm{d}\Theta_i, \tag{6.32}$$

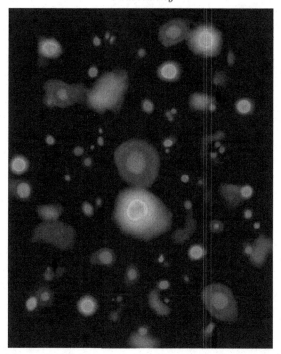

FIGURE 6.12. Image of the high-resolution texture simulation performed by N. Turok and V. Travieso. The distribution of texture spots on the sky is predicted to be scale-invariant. (The image is publicly available at http://www.damtp.cam.ac.uk/cosmos/viz/movies/neil.html.)

and $Pr(H_i)$ is the a-priori probability of hypothesis $H_i$. The a-priori probability ratio is usually set to unity for lack of information, but since in our case the analysis is centered in the cold spot (an a-posteriori selected pixel) it should be given by the sky fraction $f_s$ covered by textures. Given that a scale-invariant distribution of spots is expected and considering only textures above 1° (photon diffusion would smear out textures smaller than that) $f_s = 0.017$. The likelihood function is simply $L \propto \exp(-\chi^2/2)$ where $\chi^2 = (\mathbf{D}-\mathbf{T})'\mathbf{N}^{-1}(\mathbf{D}-\mathbf{T})$ and $\mathbf{N}$ is the CMB + noise covariance matrix. The anisotropy pattern $T$ produced by a texture can be approximated by an analytical spherical profile (Vilenkin and Shellard, 1994) with only two free parameters: the amplitude, which is related to the fundamental symmetry-breaking energy scale, and the angular scale, which is related to the redshift when the texture unwinds.

The result of the analysis is a probability ratio $\rho = 2.5$ for the three-year WMAP data, favoring the texture hypothesis (see Fig 6.13 for the resulting best-fit texture template). This result is slightly increased ($\rho = 2.7$) for the five-year data due to the reduction in the noise amplitude. The texture interpretation helps to alleviate the excess found in the kurtosis, being at the same time compatible with the observed abundance, shape, size and amplitude of the spot. In particular, the symmetry-breaking scale inferred from this analysis, $\phi_0 \approx 8.7 \times 10^{15}$ GeV, is compatible with the upper limit obtained from the CMB power-spectrum analysis (Bevis *et al.*, 2004).

More recently, Cruz *et al.* (2008) have extended the Bayesian evidence analysis to models based on the Rees–Sciama effect produced by voids or on the SZ effect. The result is that, contrary to the texture model, no positive evidence is found for those models.

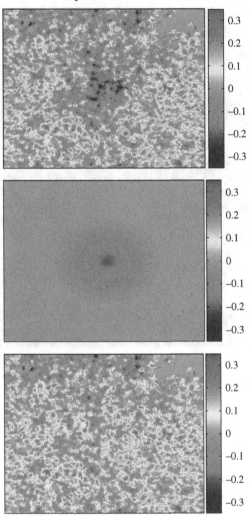

FIGURE 6.13. A 43° × 43° patch centered at Galactic coordinates (−57°, 209°) and obtained from the five-year WMAP combined map given in Fig. 6.6, is shown in the top panel. The best-fit texture template is in the middle and the result of subtracting it from the WMAP map in the bottom. The units shown in the shaded bars are mK. Adapted from Cruz *et al.* (2007); see that paper for more details.

## 6.7 Concluding remarks and future perspectives

The main conclusion derived from the large amount of analyses, involving a variety of methods, performed to test the Gaussianity of the CMB anisotropies, specially those measured by WMAP, is that the standard IGRF prediction is a good representation of their properties as a first approach. Furthermore, a number of significant deviations from the ideal IGRF has also been reported, the origin and interpretation of which are still under debate. Some of them might have to do with foreground residuals or unknown systematics while others could be a hint of new physics with profound implications for our understanding of the Universe. Probably, to answer those questions we will have to wait for new data coming from the advanced experiments being built and expected to be operative in the next years.

Future experiments will shed light on the open questions remaining from the up-to-date analyses of the CMB data, specially on the WMAP anomalies discussed in Section 6.6. In particular, the Planck mission is expected to provide all-sky, high-quality, multifrequency maps in the frequency range 30–900 GHz. The wider frequency range and the higher sensitivity and resolution will allow an improvement in the quality of the resulting CMB map. As a consequence, an improvement in the control of the foreground emission is expected as well as a reduction in the sky area required to be masked. In addition to the temperature, improvements are also expected in the polarization maps which will be provided by Planck, meaning an important complement for probing the nature of the anomalies as well as for testing the different physical interpretations proposed for them. Missions for measuring polarization at the highest sensitivity allowed by present technology, and with the main aim of probing the existence of the gravitational wave background, have recently been proposed to both agencies, the European Space Agency and NASA. As for the temperature, the linear polarization expected to be produced as a consequence of the standard inflationary period of the Universe also possesses properties very close to those of the IGRF studied in Section 6.2. Therefore, extensions of the methods already discussed in Section 6.4 for temperature are also expected to be applied to test the Gaussianity of the future polarization maps.

## Acknowledgments

I thank R. B. Barreiro and P. Vielva for useful comments on the manuscript and M. Cruz, A. Curto and C. Monteserín for helping me with some figures. I acknowledge financial support from the Spanish MEC project ESP2007-68058-C03-02. I also acknowledge the use of LAMBDA, support for which is provided by the NASA Office of Space Science. The work has also used the software package HEALPix (http://www.eso.org/science/healpix) developed by K. M. Gorski, E. F. Hivon, B. D. Wandelt, J. Banday, F. K. Hansen and M. Bartelmann.

## REFERENCES

ADLER, R. J. (1981). *The Geometry of Random Fields*, John Wiley & Sons.

ADLER, R. J., BJORKEN, J. D. AND OVERDUIN, J. M. (2006). gr-qc/0602102

ADEMAN-MCCARTHY, J. K., AGÜERSOS, J. K., ALLAM, S. S. *et al.* (2008). *Astrophys. J. Suppl. Ser.* **175**, 297

AGHANIM, N. and PUGET, J. L., (2006). In *Proceedings of CMB and Physics of the* Early Universe (CMB2006), ed. G. De Zotti *et al.*, SISSA.

AGHANIM, N., DESERT, F. X., PUGET, J. L. and GISPERT, R., 1996. *Astron. Astrophys.* **311**, 1.

ANTOINE, J.-P. and VANDERGHEYNST, P. (1998). *J. Math. Phys.* **39**, 3987.

BARREIRO, R. B. (2007). In *Highlights of Spanish Astrophysics IV. Proceedings of the VII Scientific Meeting of the Spanish Astronomical Society (SEA), Barcelona, September 12-15, 2006* ed. F. Figuras *et al.* Springer.

BARREIRO, R. B. (2008). In *Data Analysis in Cosmology*, eds. V. J. Martínez *et al.* Springer.

BARREIRO, R. B., SANZ, J. L., MARTÍNEZ-GONZÁLEZ, E., CAYÓN, L. and SILK, J. (1997). *Astrophys. J.* **478**, 1.

BARROW, J. D., JUSZKIEWICZ, R. and SONODA, D. H. (1985). *Mon. Not. R. Astron. Soc.* **213**, 917.

BARROW, J. D., FERREIRA, P. G. and SILK J. (1997). *Phys. Rev. Lett.* **78**, 3610.

BARTELMANN, M. and SCHNEIDER, P. (2001). *Phys. Rep.* **340**, 291.

BARTOLO, N., KOMATSU, E., MATARRESE, S. and RIOTTO, A. (2004). *Phys. Rep.* **402**, 103

BARTOSZ, L. (2008). *J. Cosmol. Astropart. Phys.*, **JCAP08**, paper 17.

BASSER, P. J. and PIERPAOLI, C. (1996). *J. Magn. Reson. Med. B* **111**, 209.

BENNETT, C. L., KOGUT, A., HINSHAW, G. *et al.* (1994). *Astrophys. J.* **436**, 443.

BENNETT, C. L., HALPERN, N., HINSHAW, G. *et al.* (2003a). *Astrophys. J. Suppl. Ser.* **148**, 1.

BENNETT, C. L., HILL, R. S., HINSHAW, G. *et al.* (2003b). *Astrophys. J. Suppl. Ser.* **148**, 97.

BERNUI, A., VILLELA, T., WUENSCHE, C. A., LEONARDI, R. and FERREIRA, I. (2006). *Astron. Astrophys.*, **454**, 409.

BERNUI, A., MOTA, B., REBOUCAS, M. J. and TAVAKOL, R. (2007). *Astron. Astrophys.* **464**, 479.

BEVIS, N., HINDMARSH, M. and KUNZ, M. (2004). *Phys. Rev. D* **70**, 043508.

BIELEWICZ, P., ERIKSEN, H. K., BANDAY, A. J., GÒRSKI, K. M. and LILJE, P. B. (2005), *Astrophys. J.* **635**, 750.

BOND, J. R. and EFSTATHIOU, G. (1987). *Mon. Not. R. Astron. Soc.* **226**, 655.

BRIDGES, M., MCEWEN, J. D., LASENBY, A. N. and HOBSON, M. P. (2007). *Mon. Not. R. Astron. Soc.* **377**, 1473

BRIDGES, M., MCEWEN, J. D., CRUZ, M. *et al.* (2008). *Mon. Not. R. Astron. Soc.* **390**, 1372.

BUCHBINDER, E., KHOURY, J. and OVRUT, B. A. (2008). *Phys. Rev. Lett.* **100**(17), 171302.

BUNN, E. F., FERREIRA, P. and SILK, J. (1996). *Phys. Rev. Lett.* **77**, 2883.

CABELLA, P., HANSEN, F. K., LIGUORI, M. *et al.* (2006). *Mon. Not. R. Astron. Soc.* **369**, 819.

CAYÓN, L., MARTÍNEZ-GONZÁLEZ, E. and SANZ, J. L. (1991). *Mon. Not. R. Astron. Soc.* **253**, 599.

CAYÓN, L., SANZ, J. L., MARTINEZ-GONZALEZ, E. *et al.* (2001). *Mon. Not. R. Astron. Soc.* **326**, 1243.

CAYÓN, L., MARTÍNEZ-GONZÁLEZ, E., ARGÜESO, F., BANDAY, A. J. and GORSKI, K. M. (2003). *Mon. Not. R. Astron. Soc.* **339**, 1189.

CAYÓN, L., JIN, J. and TREASTER, A. (2005). *Mon. Not. R. Astron. Soc.* **362**, 826.

CHEN, G., MUKHERJEE, P., KAHNIASHVILI, T. *et al.* (2004). *Astrophys. J.* **611**, 655.

CHEN, G. and SZAPUDI, I. (2005). *Astrophys. J.* **635**, 743.

CHIANG, L.-Y., NASELSKY, P. D., VERKHODANOV, O. V. and WAY, M. J. (2003). *Astrophys. J. Lett.* **590**, 65.

COLLESS, M. M., DALTON, G., MADDOX, S. *et al.* (2001). *Mon. Not. R. Astron. Soc.* **328**, 1039.

COPI, C. J., HUTERER, D. and STARKMAN, G. D. (2004). *Phys. Rev. D* **70**, 043515.

COPI, C. J., HUTERER, D., SCHWARZ, D. and STARKMAN, G. D. (2007). *Phys. Rev. D* **75**, 023507.

CORNISH, N. J., SPERGEL, D. N., STARKMAN, G. D. and KOMATSU, E. (2004). *Phys. Rev. Lett.* **92**, 201302.

CREMINELLI, P., NICOLIS, A., SENATORE, L., TEGMARK, M. and ZALDARRIAGA, M. (2006). *J. Cosmol. Astropart. Phys.* **0605**, 004.

CRESSWELL, J. G., LIDDLE, A. R., MUKHERJEE, P. and RIAZUELO, A. (2006). *Phys. Rev. D* **73**, 041302.

CRUZ, M., MARTÍNEZ-GONZÁLEZ, E., VIELVA, P. and CAYÓN, L. (2005). *Mon. Not. R. Astron. Soc.* **356**, 29.

CRUZ, M., TUCCI, M., MARTÍNEZ-GONZÁLEZ, E. and VIELVA, P. (2006). *Mon. Not. R. Astron. Soc.* **369**, 57.

CRUZ, M., TUROK, N., VIELVA, P., MARTÍNEZ-GONZÁLEZ, E. and HOBSON, M. (2007). *Science* **318**, 1612.

CRUZ, M., MARTÍNEZ-GONZÁLEZ, E., VIELVA, P. *et al.* (2008) *Mon. Not. R. Astron. Soc.*, **390**, 919.

CURTO, A., AUMONT, J., MACÍAS-PÉREZ, J. F. *et al.* (2007). *Astron. Astrophys.* **474**, 23.

CURTO, A., MACÍAS-PÉREZ, J. F., MARTINEZ-GONZÁLEZ, E. *et al.* (2008), *Astron. Astrophys.* **486**, 383

DE OLIVEIRA-COSTA, A., SMOOT, G. F. and STAROBINSKY, A. A. (1996). *Astrophys. J.* **468**, 457

DE OLIVEIRA-COSTA, A., TEGMARK, M., ZALDARRIAGA, M. and HAMILTON, A. (2004). *Phys. Rev. D* **69**, 063516.

DE OLIVEIRA-COSTA, A. and TEGMARK, M. (2006). *Phys. Rev. D* **74**, 023005

DE TROIA, G., ADE, P. A. R., BOCK, J. J. *et al.* (2007). *Astrophys. J. Lett.* **670**, 73.

DELABROUILLE, J. and CARDOSO, J. F. (2008). In *Data Analysis in Cosmology*, eds. V. J. Martínez *et al.* Springer.

DONOGHUE, E. P. and DONOGHUE, J. F. (2005). *Phys. Rev. D* **71**, 043002.

DORÉ, O., COLOMBI, S. and BOUCHET, F. R. (2003). *Mon. Not. R. Astron. Soc.* **344**, 905.

DURRER, R. (2007). *New Astron. Rev.* **51**, 257.

DURRER, R., KAHNIASHVILI, T. and YATES, A. (1998). *Phys. Rev. D* **58**, 123004.

ERIKSEN, H. K., HANSEN, F. K., BANDAY, A. J., GORSKI, K. M. and LILJE, P. B. (2004a). *Astrophys. J.* **605**, 14.

ERIKSEN, H. K., NOVIKOV, D. I., LILJE, P. B., BANDAY, A. J. and GORSKI, K. M. (2004b). *Astrophys. J.* **612**, 64.

ERIKSEN, H. K., BANDAY, A. J., GORSKI, K. M. and LILJE, P. B. (2005). *Astrophys. J.* **622**, 58.

ERIKSEN, H. K., BANDAY, A. J., GORSKI, K. M., HANSEN, F. K. and LILJE, P. B. (2007). *Astrophys. J. Lett.* **660**, L81.

FREEMAN, P. E., GENOVESE, C. R., MILLER, C. J., NICHOL, R. C. and WASSERMAN, L. (2006). *Astrophys. J.* **638**, 1.

GAZTAÑAGA, E. AND WAGG, J. (2003). *Phys. Rev. D* **68**, 021302.

GÉNOVA-SANTOS, R., RUBIÑO-MARTÍN, J. A., REBOLO, R. *et al.* (2005). *Mon. Not. R. Astron. Soc.* **363**, 79.

GÉNOVA-SANTOS, R., RUBIÑO-MARTÍN, J. A., REBOLO, R. *et al.* (2008). *Mon. Not. R. Astron. Soc.* **391**, 1127.

GIOVANNINI, M. (2004). *Int. J. Mod. Phys. D* **13**, 391.

GÓRSKI, K. M., HIVON, E. F., WANDELT, B. D. *et al.* (2005). *Astrophys. J.* **622**, 759.

HAJIAN, A. and SOURADEEP, T. (2006). *Phys. Rev. D* **74**, 123521.

HAJIAN, A., SOURADEEP, T. and CORNISH, N. (2005). *Astrophys. J.* **618**, 63.

HAN, J. and WIELEBINSKI, R. (2002). *Chinese J. Astron. Astrophys.* **2**, 293.

HANSEN, F. K., BANDAY, A. J. and GORSKI, K. M. (2004). *Mon. Not. R. Astron. Soc.* **354**, 641

HANSEN, F. K., CABELLA, P., MARINUCCI, D. and VITTORIO, N. (2004). *Astrophys. J.* **607**, L67

HIKAGE, C., KOMATSU, E. and MATSUBARA, T. (2006). *Astrophys. J.* **653**, 11

HIKAGE, C., MATSUBARA, T., COLES, P. *et al.* (2008). *Mon. Not. R. Astron. Soc.* **389**, 1439.

HINSHAW, G., NOLTA, M. R., BENNETT, C. L., *et al.* (2007). *Astrophys. J. Suppl. Ser.* **170**, 288

HINSHAW, G., WEILAND, J. L., HILL, R. S. *et al.* (2009). *Astrophys. J. Suppl. Sec.* **180**, 225.

HUANG, M.-X. and SHIU, G. (2006). *Phys. Rev. D* **74**, 121301

HU, W. (2000). *Phys. Rev. D* **62**, 043007

HU, W. (2001). *Phys. Rev. D* **64**, 083005

HU, W. and DODELSON, S. (2002). *Annu. Rev. Astron. Astrophys.* **40**, 171

INOUE, K. T. and SILK, J. (2007). *Astrophys. J.* **664**, 650

INOUE, K. T. and SUGIYAMA, N. (2003). *Phys. Rev. D* **67**, 043003

JAFFE, T. R., BANDAY, A. J., ERIKSEN, H. K., GORSKI, K. M. and HANSEN, F. K. (2005). *Astrophys. J. Lett.* **629**, L1.

JAFFE, T. R., BANDAY, A. J., ERIKSEN, H. K., GORSKI, K. M. and HANSEN, F. K. (2006a). *Astron. Astrophys.* **460**, 393.

JAFFE, T. R., HERVIK, S., BANDAY, A. J. and GORSKI, K. M. (2006b). *Astrophys. J.* **644**, 701

JONES, B. J. T. (2008). In *Data Analysis in Cosmology*, eds. V. J. Martínez *et al.* Springer.

KAISER, N. and STEBBINS, A. (1984). *Nature* **310**, 391.

KATZ, G. and WEEKS, J. (2004). *Phys. Rev. D* **70**, 063527.

KHOURY, J., OVRUT, B. A., STEINHARDT, P. J. and TUROK, N. (2001). *Phys. Rev. D* **64**, 123522.

KIBBLE, T. W. B. (1976). *J. Phys. A* **9**, 1387.

KOGUT, A., HINSHAW, G. and BANDAY, A. J. (1997). *Phys. Rev. D* **55**, 1901.

KOMATSU, E. (2001). Ph.D. Thesis, astro-ph/0206039.

KOMATSU, E. and SPERGEL, D. N. (2001). *Phys. Rev. D* **63**, 063002.

KOMATSU, E., WANDELT, B. D., SPERGEL, D. N., BANDAY, A. J. and GORSKI, K. M. (2002). *Astrophys. J.* **566**, 19.

KOMATSU, E., KOGUT, A., NOLTA, M. R. *et al.* (2003). *Astrophys. J. Suppl. Ser.* **148**, 119.

KOMATSU, E., SPERGEL, D. N. and WANDELT, B. D. (2005). *Astrophys. J.* **634**, 14.

KOMATSU, E. DUNKLEY, J., NOLTA, M. R. *et al.* (2009). *Astrophys. J. Suppl. Ser.* **180**, 333.

KRONBERG, P. P. (2004). *J. Korean Astron. Soc.* **37**, 501.

KUNZ, M., AGHANIM, N., CAYÓN, L., FORNI, O., RIAZUELO, A. and UZAN, J. P. (2006). *Phys. Rev. D* **73**, 023511.

KUNZ, M., AGHANIM, N., RIAZUELO, A. and FORNI, O. (2008). *Phys. Rev. D* **77**, 023525.

LACHIÈZE-REY, M. and LUMINET, J.-P. (1995). *Phys. Rep.* **254**, 135.

LAND, K. and MAGUEIJO, J. (2005a). *Phys. Rev. Lett.* **95**, 071301.

LAND, K. and MAGUEIJO, J. (2005b). *Mon. Not. R. Astron. Soc.* **357**, 994.

LAND, K. and MAGUEIJO, J. (2006). *Mon. Not. R. Astron. Soc.* **367**, 1714.

LAND, K. and MAGUEIJO, J. (2007). *Mon. Not. R. Astron. Soc.* **378**, 153.

LEVIN, J. (2002). *Phys. Rep.* **365**, 251.

LEWIS, A. (2004). *Phys. Rev. D* **70**, 043011.

LONGUET-HIGGINS, M. S. (1957). *Phil. Trans. Roy. Soc. London, A* **249**, 321.

LUMINET, J.-P., WEEKS, J. R., RIAZUELO, A., LEHOUCQ, R. and UZAN, J.-P. (2003). *Nature* **425**, 593.

McEWEN, J. D., HOBSON, M. P., LASENBY, A. N. and MORTLOCK, D. J. (2005). *Mon. Not. R. Astron. Soc.* **359**, 1583.

McEWEN, J. D., HOBSON, M. P., LASENBY, A. N. and MORTLOCK, D. J. (2008). *Mon. Not. R. Astron. Soc.* **388**, 659.

MARTÍNEZ-GONZÁLEZ, E., SANZ, J. L. and SILK, J. (1990). *Astrophys. J. Lett.* **335**, L5.

MARTÍNEZ-GONZÁLEZ, E. and SANZ, J. L. (1990). *Mon. Not. R. Astron. Soc.* **247**, 473.

MARTÍNEZ-GONZÁLEZ, E. and SANZ, J. L. (1995). *Astron. Astrophys.* **300**, 346.

MARTÍNEZ-GONZÁLEZ, E., SANZ, J. L. and CAYÓN, L. (1997). *Astrophys. J.* **484**, 1.

MARTÍNEZ-GONZÁLEZ, E., GALLEGOS, J. E., ARGÜESO, F., CAYÓN, L. and SANZ, J. L. (2002). *Mon. Not. R. Astron. Soc.* **336**, 22.

MARTÍNEZ-GONZÁLEZ, E. (2008). In *Data Analysis in Cosmology*, eds. V. J. Martínez *et al.* Springer.

MOLLERACH, S., GANGUI, A., LUCHIN, F. and MATARRESE, S. (1995). *Astrophys. J.* **453**, 1.

MONTESERÍN, C., BARREIRO, R. B., SANZ, J. L. and MARTÍNEZ-GONZÁLEZ, E. (2005). *Mon. Not. R. Astron. Soc.* **360**, 9.

MONTESERÍN, C., BARREIRO, R. B., MARTÍNEZ-GONZÁLEZ, E. and SANZ, J. L. (2006). *Mon. Not. R. Astron. Soc.* **371**, 312.

MONTESERÍN, C., BARREIRO, R. B., VIELVA, P., MARTÍNEZ-GONZÁLEZ, E., HOBSON, M. P. and LASENBY, A. N. (2007). astro-ph/07064289.

MUKHERJEE, P. and WANG, Y. (2004). *Astrophys. J.* **613**, 51.

NIARCHOU, A. and JAFFE, A. (2007). *Phys. Rev. Lett.* **99**, 081302.

OSTRIKER, J. P. and VISHNIAC, E. T. (1986). *Astrophys. J.* **306**, 510.

PARK, C.-G. (2004). *Mon. Not. R. Astron. Soc.* **349**, 313.

PHILLIPS, N. G. and KOGUT, A. (2006). *Astrophys. J.* **645**, 820.

REES, M. J. and SCIAMA, D. W. (1968). *Nature* **517**, 611.

ROCHA, G., CAYÓN, L., BOWEN, R. *et al.* K. M. (2004). *Mon. Not. R. Astron. Soc.* **351**, 769.

ROUKEMA, B. (2000). *Mon. Not. R. Astron. Soc.* **312**, 712.

RUBIÑO-MARTÍN, J. A., ALIAGA, A. M., BARREIRO, R. B. *et al.* (2006). *Mon. Not. R. Astron. Soc.* **369**, 909.

RUDNICK, L., BROWN, S. and WILLIAMS, L. R. (2007). *Astrophys. J.* **671**, 40.

SACHS, R. K. and WOLFE, A. M. (1967). *Astrophys. J.* **147**, 73.

SANZ, J. L., MARTÍNEZ-GONZÁLEZ, E., CAYÓN, L., SILK, J. and SUGIYAMA, N. (1996). *Astrophys. J.* **467**, 485.

SCHMALZING, J. and GORSKI, K. M. (1998). *Mon. Not. R. Astron. Soc.* **297**, 355.

SELJAK, U. (1996). *Astrophys. J.* **463**, 1.

SMITH, S., ROCHA, G., CHALLINOR, A. *et al.* (2004). *Mon. Not. R. Astron. Soc.* **352**, 887.

SMOOT, G. F., BENNETT, C. L., KOGUT, A. *et al.* (1992). *Astrophys. J. Lett.* **396**, L1.

SPERGEL, D. N., BEAN, R., DORÉ, O. *et al.* (2007). *Astrophys. J. Suppl. Ser.* **170**, 377.

SCHWARTZ, D. J., STARKMAN, G. D., HUTERER, D. and COPI, C. J. (2004). *Phys. Rev. Lett.* **93**, 1301.

SUNYAEV, R. A. and ZELDOVICH, Y. B. (1972). *Comm. Astrophys. Space Phys.* **4**, 173.

TEGMARK, M., DE OLIVERIA-COSTA, A. and HAMILTON, A. J. (2003). *Phys. Rev. D* **68**, 123523

TOMITA, K. (2005). *Phys. Rev. D* **72**, 103506

TUROK, N. (1989). *Phys. Rev. Lett.* **63**, 2625

TUROK, N. and SPERGEL, D. (1990). *Phys. Rev. D* **64**, 2736

VILENKIN, A. and SHELLARD, E. P. S. (1994). *Cosmic Strings and Other Topological Defects.* Cambridge University Press.

VIELVA, P., MARTÍNEZ-GONZÁLEZ, E., BARREIRO, R. B., SANZ, J. L. and CAYÓN, L. (2004). *Astrophys. J.* **609**, 22.

VIELVA, P., WIAUX, Y., MARTÍNEZ-GONZÁLEZ, E. and VANDERGHEYNST, P. (2007). *Mon. Not. R. Astron. Soc.* **381**, 932.

VIELVA, P. (2007). *Proc. SPIE*, **6701**, 670119.

VISHNIAC, E. T. (1986). *Astrophys. J.* **322**, 597.

WIAUX, Y., JACQUES, L. and VANDERHEYNST, P. (2005). *Astrophys. J.* **632**, 5.

WIAUX, Y., VIELVA, P., MARTÍNEZ-GONZÁLEZ, E. and VANDERGHEYNST, P. (2006). *Phys. Rev. Lett.* **96**, 151303.

WIAUX, Y., VIELVA, P., BARREIRO, R. B., MARTÍNEZ-GONZÁLEZ, E. and VANDERGHEYNST, P. (2008). *Mon. Not. R. Astron. Soc.* **385**, 939.

YADAV, A. P. S., KOMATSU, E. and WANDELT, B. D. (2007). *Astrophys. J.* **664**, 680.

YADAV, A. P. S., KOMATSU, E., WANDELT, B. D., LIGUORI, M., HANSEN, F. K. and MATARRESE, S. (2008). *Astrophys. J.* **678**, 578.

# 7. Galactic and extragalactic foregrounds

## RODNEY D. DAVIES

### Abstract

In this chapter the four main Galactic foregrounds are described. These include the synchrotron and free–free emissions along with the anomalous emission from dust which are dominant at frequencies below $\sim 100$ GHz. At higher frequencies the thermal (vibrational) emission from dust is the main contributor. In addition to these diffuse components there is a Poissonian component arising from "point" sources such as the radio galaxies, the millimeter-wave dusty galaxies and the Sunyaev–Zeldovich effect in galaxy clusters. There are also foregrounds nearer home arising from the Solar System zodiacal light emission and from the Earth's atmosphere. The physical processes occurring in each are also considered.

## 7.1 Synchrotron and free–free emission

### 7.1.1 Synchrotron emission

Synchrotron radiation is produced when electrons moving at relativistic velocities spiral in magnetic fields. The importance of synchrotron emission in astronomy was first demonstrated in the discovery of strong linear polarization at optical wavelengths in filaments of the Crab Nebula. This gave support to the suggestion by Alfven and Herlofson (1950) that the as-yet unexplained low-frequency emission from the Galaxy was possibly synchrotron emission.

Synchrotron emission has two distinctive features. The first is its linear polarization. The percentage polarization depends primarily on the level of tangling of the magnetic field in the line of sight. The second is its spectral index which can distinguish it from other emission processes. Generally synchrotron emission has a steeper spectrum than free–free emission for example. The synchrotron index mirrors that of the energy spectrum of the relativistic electrons responsible.

The relativistic electrons are the electron component of the cosmic rays which pervade our Galaxy. They are continually being produced in supernova explosions that occur in the plane of the Galaxy and then diffuse into the halo. Some relativistic electrons may be produced in strong interstellar shocks by the Alfven process. Figure 7.1 shows the all-sky 408 MHz map made from observations taken at Jodrell Bank and Bonn in the north and Parkes in the south (Haslam *et al.*, 1982). The major part of the emission seen in this map is synchrotron. Clearly it is Galactic in origin, with substantial symmetry about the Galactic plane. Extragalactic emission is visible in the Magellanic Clouds some 40° below the plane. The extragalactic point sources which are synchrotron emitters have been removed from this map.

Asymmetries are evident in the synchrotron distribution. The most obvious is the stronger emission in the "northern" Galactic plane – the First Quadrant. The strong spur in the northern Galaxy reaching almost to the north galactic pole – the North Polar Spur – is most likely a supernova remnant (SNR). Its centre is about 100 parsecs above the plane and it has a diameter of some 150 parsecs. Many features of this type can be seen extending from the plane. They are all probably the low surface brightness remains of

192

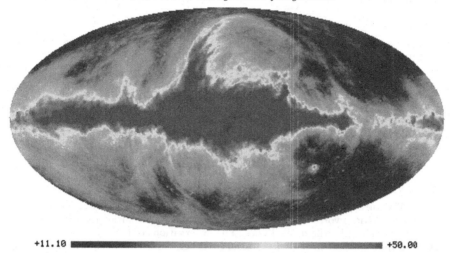

+11.10 +50.00

FIGURE 7.1. The 408 MHz map of the Galaxy plotted in Galactic coordinates. This all-sky map is derived from observations taken in the northern (Bonn and Jodrell Bank) and southern (Parkes) hemispheres (Haslam *et al.*, 1982). (Map available at LAMBDA web site, http://lambda.gsfc.nasa.gov.)

old SNRs which have burst from the plane. Although the majority of the emission in Figure 7.1 is synchrotron in origin at 408 MHz, a fraction comes from free–free emission which becomes more dominant relative to synchrotron at higher frequencies. This is particularly true along the Galactic plane and in the Gould Belt system which extends up to 30° above the plane in the Galactic centre region and below the anticentre (at either end of Figure 7.1).

Supernovae play an important role in supplying the cosmic ray electrons, which produce synchrotron emission in the Galactic magnetic field. Approximately 100 have been identified along the Galactic plane (Green, 2004). They range in size from arcminutes to degrees depending on their age and distance. A rough estimate of the supernova rate in our Galaxy can be made from historical records of supernova ("new star") sightings. The most famous in recent times are those of AD 1054 (the Crab Nebula), AD 1572 (Tycho's star) and AD 1604 (Kepler's star); the most recent of all was in the constellation of Cassiopeia (the Cas A radio source) in about AD 1650, but was not recorded visually despite being only 3 kiloparsecs away in the Perseus spiral arm. It was however behind five magnitudes (a factor of 100) of dust absorption. The historical records suggest that there have been ∼ 10 supernova sightings in the last 2000 years. If we assume that this number represents all the supernovae in a radius of 3 kpc around the Sun, then the total number within the solar circle (i.e. out to 8 kpc from the Galactic centre) would be ∼ seven times this number, indicating a supernova frequency of 1 per ∼ 30 years. This rate is similar to that found in nearby spiral galaxies of the same morphological type (Sbc) as the Galaxy. Such a rate is believed to be sufficient to maintain the cosmic ray electron reservoir of the Galaxy. Supernovae can be classified into two broad types. Type I supernovae have a wider distribution about the Galactic plane; their parent stars are older objects of about the solar mass or slightly more. Type II supernovae are more closely confined to the Galactic plane as would be expected for their more massive red supergiant progenitors. The types can be distinguished optically by the fact that a type II has a rich hydrogen spectrum compared with a type I. In external galaxies type II supernovae are found in spiral arms whereas type I can be anywhere in the Galactic disk; the same applies in our Galaxy.

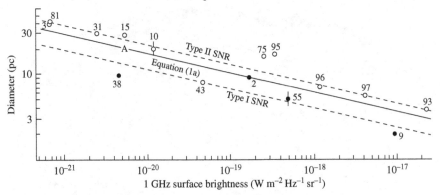

FIGURE 7.2. The surface brightness ($\Sigma$) versus diameter, $D$, plot for Galactic supernova remnants. Individual SNRs evolve upwards to the left. Possible evolutionary tracks for SNRs of types I and II are shown by the dashed lines. The observations of type II supernovae are shown as open circles and type I as filled circles. Taken from fig. 1 in Milne, 1970, with kind permission of CSIRO publishing (http://www.publish.csiro.au/journals).

Supernova remnants can be seen anywhere in the Galaxy at radio wavelengths since there is no absorption at these wavelengths. The 100 or so SNRs that can be clearly distinguished from HII regions by virtue of their different spectral index show a range of surface brightness. It is evident that they evolve from highly compact self-absorbed sources in the first months of life to low surface brightness diffuse objects over timescales of thousands of years. Figure 7.2 is the $\Sigma$–$D$ plot of surface brightness, ($\Sigma$), against diameter, $(D)$, for the assembly of SNRs. The fall of surface brightness with increasing diameter clearly demonstrates the evolution. The Sedov model of an expanding SNR is consistent with Fig. 7.2. As SNRs age beyond the timescales of the oldest (largest) objects in the diagram they will not only become less bright but will break into disconnected filaments as they sweep through regions of varying interstellar density. These are most likely the filaments seen extending from the Galactic plane in Fig. 7.1. Approximately one solar mass of gas is injected into the interstellar medium at the time of a supernova explosion with an energy of $10^{51}$ erg ($= 10^{44}$ Joules) at speeds of $\sim 10\,000\,\mathrm{km\,s^{-1}}$. The expanding shock sweeps up interstellar gas in the Sedov phase (the adiabatic expansion) and produces a shell-like SNR in most cases. A few supernovae, like the Crab Nebula, have continuing activity in their central neutron star. This has the effect of maintaining the flow of relativistic electrons which populate the inner regions of the SNR; such SNRs are designated "plerions". The energy input of the central neutron star in the Crab Nebula is $10^{38}$ erg s$^{-1}$; this energy comes from the slow-down of the pulsar (neutron star).

The cosmic ray electrons which escape from the SNRs will diffuse out into the interstellar medium; those with the highest energy will eventually escape from the Galaxy altogether. It is those relativistic electrons still held in the Galaxy and spiralling in its magnetic field that produce the diffuse radio halo, the foreground to the CMB. Observations of this emission give some of the best evidence of the structure of the magnetic field of the Galaxy. We now consider the synchrotron process. An electron of mass $m$ spiralling in a magnetic field $B$ has a Larmor frequency, $\nu_\mathrm{L}$, given by

$$\nu_\mathrm{L} = \frac{eB}{2\pi mc} = 2.8 \left(\frac{B}{1\,\mathrm{Gauss}}\right)\,\mathrm{MHz}. \qquad (7.1)$$

A relativistic electron has an energy $E = \gamma E_0$ where $E_0 = mc^2$ (0.51 MeV). Its gyro-frequency is accordingly

$$\nu_g = \frac{eB}{2\pi\gamma mc} = \frac{\nu_L}{\gamma}. \qquad (7.2)$$

The emission from such a relativistic electron is in an instantaneous cone angle along its trajectory of $\sim \gamma^{-1}$ steradians. The spectrum of such a single electron consists of the harmonics of the apparent pulse duration ($\sim \nu_a^{-1}$, with $\nu_a \sim \gamma^2 \nu_L$) as seen by an external observer, and has a maximum emission at $\nu_c \sim 3/2\nu_a$. And then in a magnetic field of amplitude $B$ (in Gauss), we have

$$\nu_c = \frac{3eB\gamma^2}{4\pi mc} = 4.2\left(\frac{B}{\text{Gauss}}\right)\gamma^2 \text{ MHz} = 1.6 \times 10^7 \left(\frac{B}{\text{Gauss}}\right)\left(\frac{E}{\text{GeV}}\right)^2 \text{ MHz}. \qquad (7.3)$$

Thus at radio frequencies of 1 GHz and interstellar magnetic fields $\sim 3$–$10$ µG, the Lorentz factor $\gamma$ is $10^4$ to $10^5$, corresponding to energies of 10–100 GeV.

The radius of gyration of a relativistic electron on a magnetic field is given by $R \approx c/(2\pi\nu_g)$. A $10^9$ GeV electron in a magnetic field of 10 µG has a radius of 100 pc, which is comparable to the thickness of the gas and stellar disks of the Galaxy. Accordingly, the 10–100 GeV electrons responsible for the synchrotron emission are locked to the interstellar magnetic fields. They can only escape the field by diffusing on collision with the thermal HI in the Galactic disk. They ultimately make their way into the Galactic halo. The timescale for this diffusion is $\sim 10^9$ years for a 10 GeV electron.

We now consider the emission spectrum of an assembly of relativistic electrons with a spread of energies. This energy spectrum can be expressed as a power law in the form

$$N(E)dE = N_0 E^{-p} dE \qquad (7.4)$$

In combining this expression with the emission spectrum for each energy range it is found that the emission spectrum for the assembly is

$$I(\nu) \propto N_0 B^{(p+1)/2} \nu^{-(p-1)/2} \qquad (7.5)$$

The spectral index of the emission ($\alpha$) is thus directly related to the energy spectrum of the relativistic electrons (cosmic rays) by $\alpha = -(p-1)/2$.

The spectrum of the synchrotron emission from an assembly of relativistic electrons is modified by the energy loss by radiation at different energies. The half-life of an electron of energy $E$ is

$$t_{\frac{1}{2}} = 1/(120B^2E) \text{ years}, \qquad (7.6)$$

where $B$ is given in Gauss and as a consequence high-energy electrons lose energy faster leading to a steeper high-frequency emission spectrum. The high-frequency steepening is $\Delta\alpha = -0.5$.

The final property of synchrotron emission which we will consider is its polarization. Since the acceleration that a relativistic electron suffers is perpendicular to the magnetic field, the emission is linearly polarized in the same plane, The net polarization of the electrons spiralling at different pitch angles and with emission beam-widths at different energies is dependent on the spectral index, $p$, of the electrons. The fractional polarization $\Pi$ is of the form

$$\Pi = \frac{p+1}{p+7/3} \qquad (7.7)$$

A typical value in a region of well-aligned magnetic field is 70%. In a tangled field the polarization of the resultant emission can be much less. Another mechanism which can reduce the fractional linear polarization is Faraday rotation in a magneto-ionic medium such as the interstellar medium. The Faraday rotation angle, $\theta$, is given by

$$\theta[\text{rad}] = 8.1 \times 10^5 \lambda^2 \int n_e \text{B}_{||} \text{d}L \tag{7.8}$$

where $n_e$ is in cm$^{-3}$, $\text{B}_{||}$ is the line-of-sight component of the magnetic field in Gauss and $\lambda$ is in metres. Thus by measuring the position angle of the polarized emission at a range of frequencies, it is possible to determine the intrinsic alignment of the emission and therefore the magnetic direction projected on the sky. It can also show whether the field is directed towards or away from the observer. As a consequence of this Faraday rotation, any synchrotron emission distributed throughout the magneto-ionic medium will experience different rotation from different depths thereby reducing the fractional polarization from the integrated emission depth. Measurements of the linear polarization of the Galaxy at higher frequencies, such as those of the Planck mission where $\nu \geq$ 30 GHz will experience little rotation and so give direct information about magnetic-field geometry. On the other hand, Galactic emission data at $\nu \leq 1$ GHz are seriously affected at intermediate and low latitudes. A combination of data from a range of frequencies can in principle give an indication of field structure with depth. The North Galactic Spur is an example of a well-aligned magnetic field with a polarization of 50%; this field has probably been swept up in an expanding supernova shell.

### 7.1.2 Free–free emission

Free–free emission is equally wide-spread in the Galaxy with a maximum contribution along the Galactic plane. As one may expect, there are significant differences between the emission mechanisms and the Galactic distributions, which enable free–free and synchrotron Galactic components to be separated.

Free–free emission results from collisions between thermal electrons and protons plus other ions in the warm (ionized) interstellar medium. It is a tracer of star formation in the Galaxy. The ultraviolet radiation from newly formed hot stars ionizes the surrounding neutral gas. In all, only a few percent of the diffuse interstellar gas is ionized. Despite this low fraction, the ionized component locks all the gas, both ionized and neutral, to the magnetic field. The locking occurs because the electrons are trapped as they spiral around the magnetic field and the neutral gas in turn is collisionally locked to the ionized gas. As a consequence, the magnetic field energy density and the turbulent gas kinetic energy are of comparable magnitude in the interstellar medium (ISM).

The ionized component of the ISM can be studied not only through its free–free continuum spectrum but also by its spectral line emission. At optical wavelengths the Balmer lines of hydrogen, particularly the H$\alpha$ ($n = 2 \rightarrow 3$) line, are used to measure the surface brightness of the ionized emission over the entire Galaxy. This optical emission will be absorbed in dusty regions, so account must be taken of this fact. The corresponding lines at radio wavelengths, the radio recombination lines (RRLs), although relatively weak, make a crucial contribution on and near the Galactic plane where dust absorption makes it impossible to use H$\alpha$ data.

A full-sky map of the H$\alpha$ emission can now be made by combining the Wisconsin H-Alpha mapper (WHAM) northern Fabry–Perot survey of Haffner *et al.* (2003) with the Southern H-Alpha Sky Survey Atlas (SHASSA) filter survey of Gaustad *et al.* (2001). The WHAM survey has a resolution of 1° while the SHASSA resolution is 6 arcmin but has lost information on scales $\geq 10°$. By smoothing the latter survey to a resolution of 1° and applying a baseline correction of cosecant form at declination $< -30°$, an all-sky

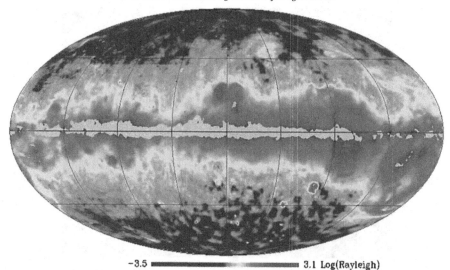

-3.5 ▬▬▬▬▬▬▬▬▬▬ 3.1 Log(Rayleigh)

FIGURE 7.3. The Hα map of the Galaxy from Dickinson *et al.* (2003). The faintest regions have a surface brightness of ∼ 1 Rayleigh, a value well above the sensitivity level of the surveys used. The map has been corrected using a model for dust absorption (see text).

map at this resolution can be constructed (Dickinson *et al.*, 2003). The Hα map has then to be corrected for the dust absorption using the Galactic absorption calibrated against the 100 μm dust emission and taking the well-known reddening curve in the optical range of $A(H\alpha) = 0.81A(V)$ which leads to an absorption at the Hα wavelength of

$$A(H\alpha) = 2.51E(B - V) = 0.0462D^T \text{ magnitudes} \qquad (7.9)$$

where $D^T$ is the 100 μm surface brightness in units of MJy sr$^{-1}$. An absorption of half this value has been used on the assumption that the dust and gas are uniformly mixed along the line of sight. The dust-corrected Hα map expressed in Rayleighs is shown in Figure 7.3. The region of the sky along the Galactic plane where the absorption is greater than 2 magnitudes, and would give an unreliable estimate of Hα emission, is outlined. It can be seen that although the emission is mainly concentrated to the plane, there is also significant Hα emission at intermediate Galactic latitudes. The most striking activity is in the Gould Belt system which includes the Orion region in the southern Galaxy ($l = 180° \pm 30°$) and Sagittarius–Scorpio region in the north ($l = 0° \pm 30°$).

We now consider the physical parameters of free–free emission as applied to the interstellar medium of our Galaxy. The ionized electron gas at an electron temperature $T_e$ at a given frequency, $\nu$, has an optical depth $\tau_\nu$. The brightness temperature of its emission is $T_b$ given by

$$T_b = T_e(1 - e^{-\tau_\nu}) \approx T_e\tau_\nu, \qquad \text{for } \tau_\nu \ll 1. \qquad (7.10)$$

The frequency and electron temperature dependence of the brightness temperature emission can be written as

$$T_b = 8.235 \times 10^{-2} a(\nu, T_e) \left(\frac{T_e}{K}\right)^{-0.35} \left(\frac{\nu}{GHz}\right)^{-2.1} (1 + 0.08) \frac{EM}{cm^{-6} \, pc}. \qquad (7.11)$$

The factor $(1 + 0.08)$ represents the contribution from helium atoms, all of which are assumed to be singly ionized; $a(\nu, T_e)$ is a factor of order unity dependent on $\nu$ and $T_e$; $EM$ is the emission measure which is the integral of the electron density squared along

FIGURE 7.4.  The Rosette Nebula NGC 2237-38-39. The young star cluster, NGC2244, responsible for the ionization can be seen at its centre (from the IAC Image Gallery, http://www.iac.es).

the line of sight ($= \int n_e^2 dl$). It is seen that the temperature spectral index is relatively weak at $-0.35$ while the frequency spectral index is $-2.1$.

Unlike synchrotron emission, free–free emission is unlikely to be highly polarized. The only process which will produce polarization is Thomson scattering of radiation on the ionized gas. This will occur in regions where the brightness (optical depth) gradient is large, as for example around the edges of compact HII regions. Even then, the maximum polarization is $\sim 5$–$10\%$ (Rybicki and Lightman, 1979). Since most HII regions are fairly symmetrical, the integrated linear polarization will largely cancel out. Observations reported by Dickinson *et al.* (2007) indicate values significantly less than $0.6\%$.

Now let us consider individual compact ionized hydrogen objects – HII regions. A typical object in the anticentre region of the Galaxy is the Rosette Nebula, NGC 2237-38-39 (see Fig. 7.4). It is an evolved HII region in which an expanding shell of gas at a distance of 1.1 kpc is moving outwards at a few tens of km s$^{-1}$. The electron gas in the nebula is $\sim 8000$ K, the equilibrium temperature between energy loss by line radiation and energy gain from the ultraviolet radiation of the recently formed central star cluster (NGC 2244). The radio spectrum of the nebula fits the index given above over a wide frequency range as expected for an optically thin ionized region ($\tau_\nu \ll 1$).

Two contrasting spectra are seen in adjacent HII regions on the southern Galactic plane in the survey of Dickinson *et al.* (2007). Spectra of the integrated emission of a radio object at a wavelength $\lambda$ are normally given in terms of the flux density, $S_\lambda$, where

$$S_\lambda = 2k_B T_b \Omega / \lambda^2, \tag{7.12}$$

where $\Omega$ is the solid angle of the source. The two radio sources in question are G291.3–0.7 and G291.6–0.5 which are associated with the nebulosities NGC 3576 and NGC 3603. Figure 7.5 shows a contour plot of the region taken at 31 GHz with the 6.7 arcmin resolution of the cosmic background imager (CBI). It can be seen immediately that G291.6–0.5 is extended relative to the CBI beam. Its deconvolved size is $7.1 \times 7.1$ arcmin$^2$. With these dimensions the HII region will be optically thin over the frequency range plotted except for the lowest frequency (408 MHz). Figure 7.6 shows that this object has a typical optically thin spectrum except at $\nu > 100$ GHz where thermal dust emission is

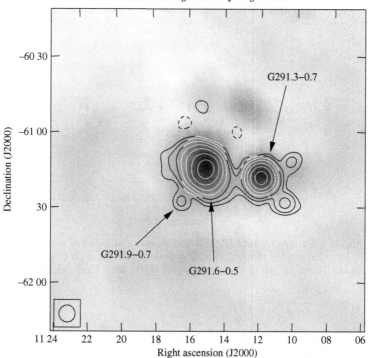

FIGURE 7.5. A 31 GHz map of the G291.6 (NGC 3603) and G291.3 (NGC 3576) region (from fig. 8 in Dickinson *et al.* 2007, with kind permission from Blackwell Publishing).

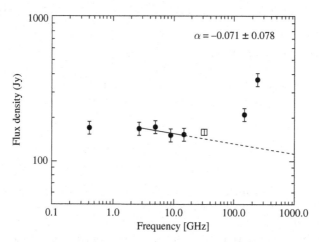

FIGURE 7.6. Spectrum of G291.6–0.5. A power law fit to data in the range 2.7–15 GHz gives $\alpha = -0.071 \pm 0.078$. The spectral index within the CBI band was found to be $\alpha = -0.12$ (indicated by the dashed line). Dust emission dominates at the higher frequencies (from fig. 9 in Dickinson *et al.* 2007, with kind permission from Blackwell Publishing).

evident. By contrast, Figure 7.7 shows that G291.3–0.7 is clearly optically thick at lower frequencies. At 1.4 GHz for example, the flux density corresponds to a $T_b \sim 3000$ K for a source size of 2 arcmin. For $T_e = 7000$ K this is equivalent to $\tau \sim 1$. At lower frequencies (higher optical depth) $T_b \rightarrow T_e$ and the spectral index becomes that of a blackbody ($\alpha = 2.0$).

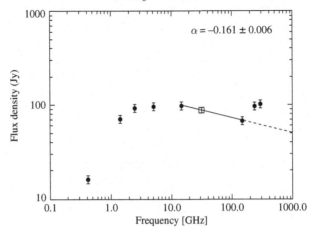

FIGURE 7.7. The 31 GHz spectrum of G291.3–0.7. A spectral index of $\alpha = -0.16$ is plotted as a dashed line. At low frequencies the emission is self-absorbed. At high frequencies dust emission begins to dominate (from fig. 10 in Dickinson *et al.* 2007, with kind permission from Blackwell Publishing).

The great value of the Hα map of the sky is its use as a template for the Galactic free–free emission. The Hα surface brightness, $I_{\mathrm{H}\alpha}$, can be converted into a free–free brightness temperature $T_{\mathrm{b}}$ using the following relation.

$$\frac{T_{\mathrm{b}}^{\mathrm{ff}}}{I_{\mathrm{H}\alpha}} = 8.396 \times 10^3 a(\nu, T_{\mathrm{e}})\, \nu_{\mathrm{GHz}}^{-2.1}\, T_4^{0.667} 10^{0.029/T_4}(1 + 0.08) \tag{7.13}$$

where $T_4$ is the electron temperature, $T_{\mathrm{e}}$, in units of $10\,000$ K. Clearly it is then necessary to assume a value of $T_{\mathrm{e}}$ appropriate to the area of the sky concerned. A large body of RRL data is available for HII regions on and near the Galactic plane. These show a significant increase of $T_{\mathrm{e}}$ with galactocentric distance, $R$. The extensive compilation by Paladini *et al.* (2004) gave the relationship

$$T_{\mathrm{e}} = (4170 \pm 120) + (314 \pm 20)R. \tag{7.14}$$

At $R = R_0$, the Sun's distance, $T_{\mathrm{e}} = (7200 \pm 1200)$ K; the apparently large scatter takes account of the scatter in $T_{\mathrm{e}}$ at a given $R$. It is possible that the $T_{\mathrm{e}}$ of diffuse HII emission may be different from that of HII regions in the plane. There are indications from observation and theory that the diffuse gas may have a higher $T_{\mathrm{e}}$ than in density-bounded HII regions which contain ionizing stars (Wood and Mathis, 2004). Nevertheless, $T_{\mathrm{e}} = 7000$ K would seem to be an acceptable working value.

The expressions above can now be used to evaluate the free–free emission at the frequencies of interest in foreground studies of the CMB. Table 7.1 gives the radio brightness temperature expected for 1.0 Rayleigh of Hα brightness for a gas with $T_{\mathrm{e}} = 7000$ K. The values range from 51.2 mK at 408 MHz to 0.30 μK at 100 GHz. In most of the intermediate-latitude sky at 408 MHz the free–free emission is weaker than the synchrotron emission, except in parts of the Gould Belt system. At frequencies above $\sim 30$ GHz free–free is stronger than synchrotron emission. This situation will be discussed in more detail in later sections.

TABLE 7.1. Free–free brightness temperature per Rayleigh of Hα emission at an electron temperature of 7000 K (Dickinson *et al.*, 2003).

| Frequency [GHz] | Free–Free brightness temperature per unit Rayleigh |
|---|---|
| 0.408 | 51.2 mK |
| 1.420 | 3.79 mK |
| 2.326 | 1.33 mK |
| 10 | 60.9 μK |
| 30 | 5.83 μK |
| 44 | 2.56 μK |
| 70 | 0.94 μK |
| 100 | 0.43 μK |

## 7.2 Dust emission, thermal and "anomalous" (spinning)

### 7.2.1 Thermal dust emission

Dust has two broad-band emission components in the frequency range 10 to 1000 GHz. Anomalous emission is dominant at the lower frequencies while the thermal (vibrational) component is responsible for the higher end. We now consider the properties of the "classical" thermal dust which was the bane of optical astronomers, particularly those interested in extragalactic objects who identified a Zone of Avoidance around the Galactic plane where absorption became too large to make meaningful observations. This situation changed as astronomy moved to include infrared and then far-infrared wavelengths where the absorption was less and the emission rather than absorption became the dominant feature. Interest in the physics of the dust itself also flourished over this time.

The COBE–DIRBE full-sky maps at 100, 140 and 240 μm made with 0.7° resolution have commonly been used as tracers of the thermal component (Kogut *et al.*, 1996). The most sensitive full-sky map of dust emission is the 100 μm data 6 arcmin resolution from the Infrared Astronomical Satellite (IRAS). These data have been re-calibrated using the COBE–DIRBE data and reanalyzed to give reduced artifacts due to zodiacal emission and to remove point sources (Schlegel *et al.*, 1998).

Figure 7.8 shows the dust distribution in the sky derived from IRAS data; the multi-wavelength IRAS observations were used to correct the 100 μm emission to a common temperature of 18.3 K (Finkbeiner *et al.*, 1999). The strongest emission is in a narrow distribution along the Galactic plane. The Gould Belt system is clearly seen at intermediate latitudes as in the case of the Hα emission although the detailed structure is different (see Fig. 7.9). The temperature of the dust varies significantly throughout the interstellar medium. Its value depends on the strength of the local interstellar radiation field (ISRF) and varies between 10 K and 25 K at intermediate latitudes; with most of the dust being at 15–20 K. Within HII regions where the ISRF is high due to the proximity of the central ionizing stars, the dust temperatures may reach 40 K or above.

The dust in galaxies is the end product of stellar evolution. The ejected material from stellar nuclear burning progressively enriches the metal content of the primordial hydrogen and helium. The dust grains which form in the various environments of the ISM are composed of different components, each having different grain sizes and chemical compositions. Figure 7.10 shows the spectrum of a typical sample of interstellar dust (Desert *et al.*, 1990). It is a composite of the three following components.

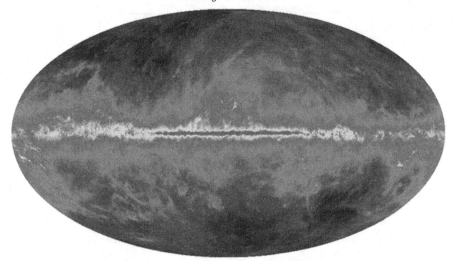

FIGURE 7.8.  The IRAS 100 μm map of the thermal dust distribution in the Galaxy (Finkbeiner *et al.*, 1999). Image available at the LAMBDA web page (http://lambda.gsfc.nasa.gov/).

FIGURE 7.9.  The Galactic centre in Sagittarius at optical wavelengths (United Kingdom Schmidt telescope, UKST). Note the complex filamentary structure in the absorbing dust; structure on all angular scales is evident. The dust absorption is in the Gould Belt system within the local spiral arm – it extends north of the Galactic plane in the Galactic centre direction.

- Grains composed of polycyclic aromatic hydrocarbons (PAHs) containing only 30 to 100 atoms. They have a rich spectrum in the 10 μm wavelength range. Their diameters are $\sim 10$ Å$(= 10^{-3}\mu m)$.
- Very small grains having a radius of 0.001 to 0.01 μm and radiating at 25–60 μm with a peaked spectrum resulting from the very small grain (VSG) size distribution. These are probably not PAHs but are most likely carbon-dominated grains with perhaps a mix of other metals.
- Large grains up to 1 μm in diameter radiating with a blackbody spectrum peaking at 100–200 μm. Silicates appear to be an important component of these large grains; carbon is also needed to explain the optical emission, possibly as a coating to the

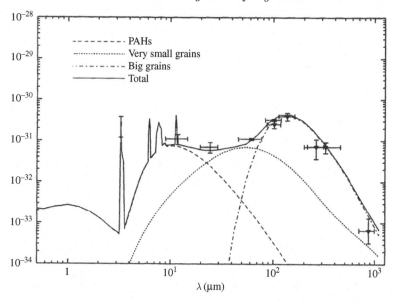

FIGURE 7.10. The composite spectrum of dust in the ISM. Three components are identified. The vertical scale is $\nu F_\nu$ in W per H atom. Taken from Desert *et al.*, 1990, with permission.

silicate. At the long wavelength side of the peak the spectrum falls with a slope between 1.3 and 2.0; a value of 1.7 is often taken as a working estimate, but this does not apply to all dust clouds.

All three components are found in individual dust clouds as well as in the diffuse interstellar medium. The thermal radiation from this dust constitutes a major foreground to the CMB at frequencies $\geq 100$ GHz ($\lambda \leq 3$ mm).

### 7.2.2 Anomalous (spinning) dust emission

The first detections of the dust-correlated radio emission were misinterpreted as being free–free emission on the reasonable assumption that dust is closely coupled to gas, both neutral and ionized, in the ISM. The COBE–differential microwave radiometer (DMR) all-sky maps (Kogut *et al.*, 1996) of radio emission, correlated with the 240 μm dust emission, appeared to have a free–free spectrum between 31 and 53 GHz and were interpreted as such. De Oliveira-Costa *et al.* (1997) interpreted their Saskatoon data in the same way. A deep scan around the north celestial pole at 14.5 and 31.7 GHz by Leitch *et al.* (1997) showed a strong correlation with 100 μm dust and a free–free-like spectrum. However no associated Hα emission at the expected level was evident so a $T_e = 10^6$ K ionized gas was thought to be responsible. These data were just what Draine & Lazarian (1998a,b) needed to substantiate their model of electric dipole emission from very small spinning dust grains. The spinning dust paradigm has been increasingly accepted as the explanation of the anomalous emission.

The distinguishing characteristic of spinning dust emission is the turn-over in the spectrum at low frequencies. The first indication of this turn-over came as a result of combining the Tenerife data at 10 and 15 GHz with the COBE–DMR data (de Olivera-Costa *et al.*, 1997) which showed a peak at $\sim 20$ GHz in the radio emission per unit of 100 μm dust emission. This spectrum was representative of dust in the 5000 deg$^2$ of the Tenerife intermediate latitude survey. This result was more clearly confirmed when the

*Rodney D. Davies*

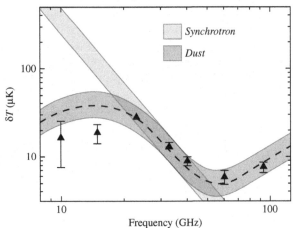

FIGURE 7.11. The dust-correlated radio emission seen in the Tenerife 10 and 15 GHz survey of 5000 deg$^2$ at intermediate and high latitudes compared with WMAP data (from de Oliveira-Costa *et al.* 2004). The vertical scale is in μK per MJy sr$^{-1}$ at 100 μm plotted at each frequency of observation. The anomalous dust emissivity peaks at $\sim 20$ GHz: it is not consistent with a synchrotron spectrum. Reproduced by permission of the AAS.

Tenerife data were combined with the WMAP data which covered a greater frequency range (de Oliveira-Costa *et al.*, 1999, 2004). Figure 7.11 shows results of this comparison. The low frequency data were inconsistent with a synchrotron spectrum for this intermediate-latitude dust. A similar analysis of COBE, but including 19 GHz data, was made by Banday *et al.* (2003).

Infrared spectroscopy has shown that there is a component of dust that comes from very small interstellar grains that are PAHs. They contain 10 to 100 carbon atoms and are a few Angstroms in radius. In the general interstellar medium PAHs absorb stellar photons which raise them to temperatures of a few 100 K. They then cool by radiating in IR bands between 3 and 17 μm giving the characteristic IR spectrum shown in Figure 7.10. At the same time the PAHs undergo collisions with gas-phase species and as a result they can become charged. Such charged PAHs thereby carry a significant permanent dipole moment of a few tenths of a Debye. As a result of the frequent collisions, the spinning PAHs will produce electric dipole emission from rotational levels.

Draine and Lazarian (1998a) estimated the rotational spectrum in the 1–100 GHz range and found that the spectrum peaked as shown in Figure 7.12. However the emission was found to be sensitive to the physical state of the gas such as its density and temperature. The figure plots the spectrum expected for five environments in the ISM:

- CNM cold neutral medium
- WNM warm neutral medium
- WIM warm ionized medium
- MC molecular clouds
- DC dark clouds

It is seen that both peak frequency and the emissivity change with environment. The peak frequency remains in the range 20 to 30 GHz despite the large differences in environment.

Fullerenes which consist of 60–200 carbon atoms have been suggested as being responsible for the anomalous emission (Iglesias-Groth, 2006). The presence of fullerenes in the ISM has been a matter of debate since their discovery in 1985 (Kroto *et al.*, 1985). They are also potential carriers of the 2175 Å bump and the diffuse interstellar bands in dense

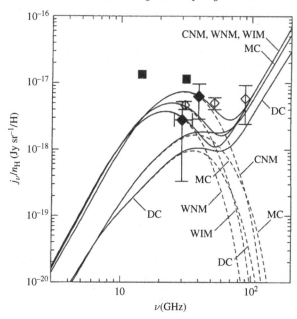

FIGURE 7.12. The spectrum of spinning dust. The radio emissivity is given in terms Jy sr$^{-1}$ per unit hydrogen column density. The dotted curves show the spectra for different environments. The solid lines are the spectra when the corresponding thermal emission is added. Early observational data are included. Taken from Draine & Lazarian (1998b). Reproduced with permission from the AAS.

clouds. The hydrogenated fullerenes would have a diameter and dipole moment that can account for the spectral shape of the anomalous emission as well as the shape of the ultraviolet bump. Following the Draine & Lazarian formulation it can be shown that the most stable fullerene, $C_{60}$, emits at a frequency $\sim 25$ GHz while $C_{240}$ emits at $\sim 3$ GHz. A reasonable mix of the stable fullerenes could fit the observational radio data for the anomalously emitting clouds.

Larger grains of submicron radius may have a significant contribution in the 10 to 100 GHz range. Although these grains stay at low temperatures (10 to 20 K) and have a modified blackbody spectrum, $I \sim \nu^\beta B_\nu(T_d)$, they could have a more complex behavior. There is observational and laboratory evidence that the emissivity index, $\beta$, depends on the grain temperature, $T_d$. This may indicate the presence of two-level systems (TLSs) in silicates (Agladze *et al.*, 1996; Boudet *et al.*, 2005). These systems TLS arise from impurities and/or disorder in the bulk solid. Further laboratory research is required to clarify the situation. It may be relevant that a resonant transition has been found in silicates (Phillips, 1987).

Draine & Lazarian (1999) also proposed that magnetic dipole emission may be emitted by large grains over a broad range between 10 and 100 GHz. In order for the emission to peak at the observed frequency, corresponding to a gyro-frequency of $\sim 6$ GHz, a rather large, but not implausible, spontaneous magnetization is required. Magnetite and metallic iron, which includes most interstellar iron, would be possible. At this time, magnetic dipole emission cannot be ruled out as a possibility. This emission would be a contributor to the anomalous emission and would be important for the emission and polarization of the sky at microwave frequencies. Magnetic susceptibility measurements of likely interstellar material are in progress.

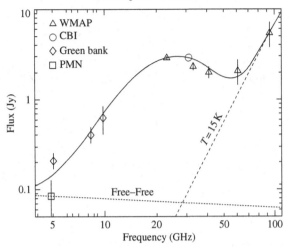

FIGURE 7.13. The spectrum of the compact dark cloud LDN1622 (from Casassus *et al.*, 2006). The free–free (dotted) and thermal dust (dashed) spectral components which fit the data are indicated. The PMN data are from the Parkes–MIT–NRAO survey. The spinning dust component is clearly seen. Reproduced by permission of the AAS.

Observation of individual compact dust clouds provides one of the best opportunities for understanding the physical processes that occur in anomalous dust emission. A morphological study of the Lynds Dark Nebula LDN1622 was made by Casassus *et al.* (2006) operating the CBI with a resolution of 8 arcmin at 31 GHz. This nebula lies in the Orion East molecular cloud at a distance of $\sim 120$ pc. It has a starless core and structure on a scale of $\leq 10$ arcmin within a $20' \times 30'$ region. The integrated free–free emission from the region is $< 0.1$ Jy which is negligible compared with the dust emission of 2.9 Jy at 31 GHz. When combined with WMAP data at 23, 33, 41, 61 and 94 GHz as well as low-frequency data, a well-defined spectrum is produced as shown in Figure 7.13. The 1–100 GHz spectrum is clearly peaked at $\sim 30$ GHz and has the form expected for spinning dust shown in Figure 7.12. The morphological analysis of the 31 GHz and the IRAS data sets showed a close correlation of the 31 GHz emission with all the IRAS bands but was tightest with the 25 μm band. This is what would be expected if the radio emission were associated with very small dust grains responsible for the spinning dust emission.

Watson *et al.* (2005) discovered strong anomalous emission in the Perseus molecular cloud (G 159.6–18.5) with the COSMOSOMAS experiment operating in the frequency range 11–17 GHz. The radio emission is associated with a dust cloud at 260 pc which is seen in IRAS maps as a ring-like structure 1.5° in diameter. A low surface brightness HII region with a flux density of $\sim 7$ Jy underlies the dust emission. Figure 7.14 shows the integrated spectrum of the region based on COSMOSOMAS, WMAP, 0.408 and 1.4 GHz data. A composite Draine & Lazarian model is shown for comparison, indicating an encouraging fit to a spinning dust model. Recent observations with the Super-Extended VSA at 33 GHz show clear emission from features in this ring. The 33 GHz emission shows strongest correlation with the 25 μm IRAS band again suggesting that the anomalous emission is from very small (spinning) grains.

On the Galactic plane strong sources of FIR emission are found mainly associated with HII regions. Dust temperatures here are in the range $20 - 30$ K, significantly higher than in the diffuse ISM. Very small dust grains are not believed to exist in the HII region

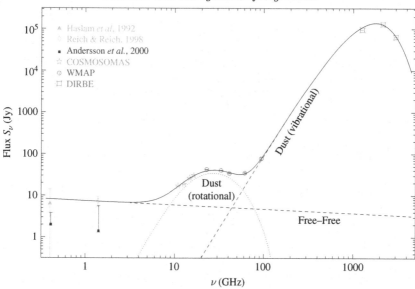

FIGURE 7.14. The spectrum of the Perseus molecular cloud (from Watson *et al.*, 2005). The thermal dust, vibrational dust, and free-free components are indicated. The emission in the 10–100 GHz band can be fitted with a spinning dust spectrum as given by Draine and Lazarian (1998a). Reproduced by permission of the AAS.

environment. It is important to investigate the anomalous emission from dust in HII regions. Dickinson *et al.* (2006) observed the diffuse HII region LPH96 (G201.66+1.64) with the CBI at 31 GHz with a resolution of 6 arcmin and found that the spectrum including existing data at nearby frequencies was mainly free–free. However, 14% of the flux density could be anomalous emission at 31 GHz; the corresponding 31 GHz emissivity relative to 100 μm is somewhat less than in the cooler dust clouds described above. An extensive study has been made by Dickinson *et al.* (2007) of the anomalous emission from dust in the strongest HII regions in the southern sky. When the 31 GHz CBI data were combined with published data at nearby frequencies, the fitted spectral index was found to be close to the theoretical value for optically thin free–free emission, confirming that the majority of the flux at 31 GHz was from ionized gas at an electron temperature of 7000–8000 K. For all six sources in the survey the 31 GHz flux density was slightly higher than expected for a pure free–free spectrum. This excess emission could be either spinning dust or another emission mechanism. Comparisons with the 100 μm data indicated an average dust emissivity of $3.3 \pm 1.7$ K MJy$^{-1}$ sr$^{-1}$ or a 95% confidence upper limit of $< 6.1$ K MJy$^{-1}$ sr$^{-1}$. The most significant detection of anomalous emission was in G284.3–0.3 (RCW49) where the dust emissivity was $13.6 \pm 4.2$ K MJy$^{-1}$ sr$^{-1}$, a value similar that found in higher-latitude dust. The HII-region dust emissivity relative to the 100 μm surface brightness is compared with that in cool dust clouds in Table 7.2. It can be seen that the HII-region dust emissivity is a factor of three to four less than in the high-latitude cooler dust. The data for the intermediate- and high-latitude diffuse dust will be discussed in more detail in Section 7.4.

We have presented here the evidence which suggests that the anomalous emission is from spinning dust grains in terms of the paradigm proposed by Draine & Lazarian. In Section 7.4 we will discuss the results of studies of the more extended cirrus dust where it is necessary to solve simultaneously for all the emission components rather than concentrating on anomalous dust alone.

TABLE 7.2. Dust emissivity at 31 GHz relative to the 100 μm surface brightness in HII regions compared with that in cooler dust clouds (Dickinson *et al.* 2007)

| Source | Dust emissivity μK MJy$^{-1}$ sr$^{-1}$ | Reference |
|---|---|---|
| **HII regions** | | |
| 6 HII regions (mean) | $3.3 \pm 1.7$ | Dickinson *et al.* (2007) |
| LPH96 | $5.8 \pm 2.3$ | Dickinson *et al.* (2006) |
| **Cool dust clouds** | | |
| 15 regions WMAP | $11.2 \pm 1.5$ | Davies *et al.* (2006) |
| All-sky WMAP | $10.9 \pm 1.1$ | Davies *et al.* (2006) |
| LDN1622 | $24.1 \pm 0.7$ | Casassus *et al.* (2006) |
| G159.6–18.5 | $17.8 \pm 0.3$ | Watson *et al.* (2005) |

## 7.3 Extragalactic radio sources and infrared galaxies

### 7.3.1 Radio galaxies

Radio sources are a foreground for CMB studies in two forms. In the first case they can be seen as individual objects and can be studied in their own right but must be removed from the data before the CMB is investigated. The second way in which sources contaminate the CMB is the noise-like foreground they form at a level below that where individual sources can be identified. In each resolution element there will be $N$ sources in a given flux density range with a Poisson uncertainty of $N^{1/2}$. This uncertainty in $N$ translates to a temperature fluctuation and makes a contribution to the power spectrum at the corresponding angular scale of the CMB foreground. In this section we consider the types of radio source which contribute to the foreground power spectrum and show their importance, particularly at small angular scales (large values of circular frequency).

The extragalactic radio sources are synchrotron emitters with a range of spectral indices which are largely a consequence of their level of activity. A typical energetic radio galaxy is Cygnus A which is at a distance of 240 Mpc (redshift $z = 0.056$); it is illustrated in Fig. 7.15. A central black hole energizes two outward moving jets in opposite directions which interact with the intergalactic medium and produce the "woofly" outer lobes. These lobes dominate the integrated spectrum and have a spectral index which steepens from $\alpha = -0.95$ at 1.0 GHz to $\alpha = -1.24$ at 30 GHz. The weaker inner compact object surrounding the black hole has a flat spectrum ($\alpha = 0.0$). The radio quasars in contrast have a dominant central compact component of a higher power than radio galaxies. A typical object is 3C48 at a redshift of 0.367. They have flatter spectra, typically with $\alpha = -0.2$ to $-0.5$, and sometimes with a peaked spectrum reminiscent of optically thick synchrotron emission. The intense quasar activity is short-lived compared with the age of the parent galaxy. Many objects are found intermediate in form between quasars and radio galaxies.

The radio source foreground to the CMB can be derived from the known source counts over a wide range of frequencies. Complete samples taken to faint luminosities are available down to flux density levels which can give a clear picture of the foreground at frequencies $\nu \leq 10$ GHz. Less data are available at higher frequencies in the range

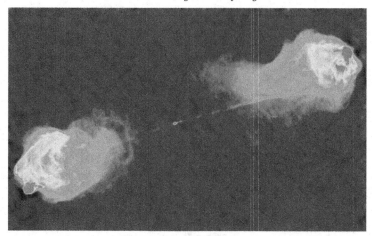

FIGURE 7.15. The radio galaxy Cygnus A. The central object is powered by infall of material into the disk surrounding a central black hole The resultant relativistic jets emerging from the disk interact with the intergalactic medium to produce the outer lobes. (A VLA image at 1.4 GHz.)

10–100GHz where the main thrust of CMB studies is directed as for example in Planck. However this regime is currently under active investigation. Above 100 GHz, dust is the main contributor to the spectrum as we will see below.

The numbers of sources at different flux density levels can be expressed in two ways. The differential source count is given as $N(S)\mathrm{d}S$ sources per steradian with flux densities in the range $S$ to $S + \mathrm{d}S$ at a given frequency. The integral source count is the number of sources per steradian with flux density greater than $S$ and is given by $\int_S^\infty N(S)\mathrm{d}S$; the structure in the source-count function is smeared in this relation and is not generally used. The steep slope of the differential count plots makes the structure less easy to see and so for better visualization counts are multiplied by $S^{\frac{5}{2}}$ as expected for a Euclidean Universe. The plots of $\log[S^{\frac{5}{2}} N(S)\mathrm{d}S]$ versus $\log(S)$ are often normalized to unity at 1 Jy.

Figure 7.16 shows the number counts for 1.4, 5 and 8.4 GHz (Toffolatti *et al.*, 1998) extending from $\sim 10$ Jy down to flux densities as low as 10–100 µJy. The contribution to the number counts of radio sources of different types is indicated. The starburst galaxies like M82 are the principal objects at the lower flux densities. At $S < 1$ mJy the radio-loud Active Galactic Nuclei, which include the flat spectrum radio galaxies, are mostly at substantial redshifts. The steep spectrum ellipticals and quasars dominate the brighter number counts at the lower frequencies.

At the frequencies appropriate to CMB experiments at say $\nu \geq 20$ GHz observational data are only now becoming available for constructing number counts at the important low flux densities. The trend indicated in Fig. 7.16 is for the flat spectrum objects to be in the majority at $S \geq 1$ mJy; this is confirmed by the observations available.

We now consider the contribution of point sources to the foreground sky fluctuations. A Poisson distribution of extragalactic point sources produces a simple white-noise power spectrum with the same power in all multipoles. As a consequence, the point-source fluctuations become increasingly important at small angular scales (high $\ell$), whereas at large angular scales (low $\ell$) the diffuse backgrounds discussed in Sections 7.1 and 7.2 dominate. The point-source fluctuation temperature increases linearly with $\ell$. Figure 7.17 shows the angular power spectra expressed as a fluctuation in temperature

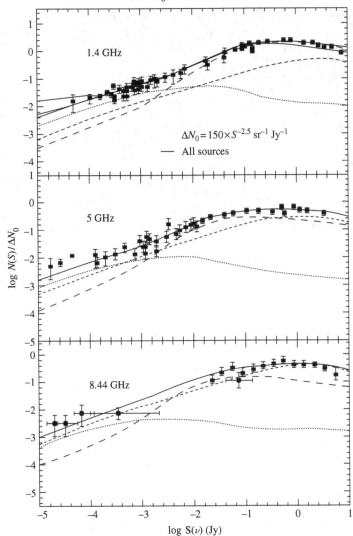

FIGURE 7.16. Comparison between observed and predicted differential source counts at 1.4, 5.0 and 8.44 GHz. The predictions for different classes of radio source are shown: (....) starburst galaxies, (- - -) flat-spectrum ellipticals S0s and QSOs, (- - -) steep spectrum ellipticals, S0s and QSOs. Taken from fig. 1 in Toffolatti *et al.*, 1998, with kind permission from Blackwell Publishing.

of the various foregrounds at frequencies in the range 30 to 100 GHz where radio galaxies are the main source of the point-source fluctuations. The rising diagonal full lines give the contribution from extragalactic sources; the upper line is the situation when sources with $S \geq 1$ Jy are identified and removed while the lower line is the situation for $S \geq 0.1$ Jy. It can be seen that the angular power in the point-source spectrum becomes comparable to the CMB power at $\ell = 500$ at 30 GHz and $\ell = 2000$ at 100 GHz. The point sources exceed the diffuse foregrounds at intermediate Galactic latitudes at $\ell = 100$ and $\ell = 700$ for 30 GHz and 100 GHz respectively. Although the plots in Figure 7.17 were based on earlier data, they show the importance of having an accurate differential source count at each frequency down to mJy flux densities.

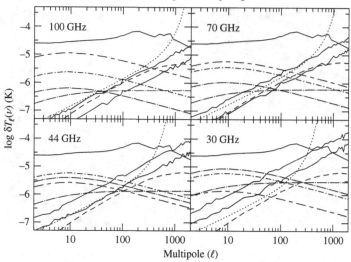

FIGURE 7.17. The angular power spectra of the components contributing to the foreground at frequencies in the range 30 to 100 GHz of the Planck Surveyor. The plots at each frequency are in terms of a temperature fluctuation $\delta T_\ell = [\ell(2\ell+1)C_\ell/4\pi]^{\frac{1}{2}}$ as a function of angular multipole $\ell$. The diagonal solid lines show the contribution of extragalactic sources. The upper of the two lines is where sources brighter than 1 Jy are identified and removed; the lower is for 0.1 Jy. The CMB power spectrum is the heavy line. The heavy dashed lines are the anisotropies due to the Sunyaev–Zeldovich effect. The diffuse Galactic components estimated for Galactic latitudes $|b| > 30°$ are synchrotron (dots + long dashes), free–free (dots + short dashes) and thermal dust (long + short dashes). Reprinted with permission from de Zotti *et al.* (1999). Copyright 1999, American Institute of Physics.

### 7.3.2 Far-infrared galaxies

At frequencies above $\sim 100$ GHz the emission from dust begins to dominate the emission from spiral galaxies. The $\log N/\log S$ plots equivalent to the radio data in Fig. 7.16 can be used to derive the Poisson fluctuations from dusty galaxies at these higher frequencies. The intrinsically weak FIR emitters are the normal galaxies with significant amounts of dust. These are mainly the spirals which have appreciable ongoing star formation which produces and heats the dust. Even early-type galaxies such as ellipticals and S0s can have significant amounts of dust as is clearly demonstrated by the strong optical absorption in the galactic plane of the edge-on galaxy NGC 5866, an S0 galaxy, as seen in Fig. 7.18. The late-type spirals of morphological types Sb to Sd have the greatest amounts of dust amongst the spirals.

More spectacular infrared galaxies are the starburst galaxies. The origin of the intense star-formation phase in these objects is not fully understood. Interaction with a neighbouring object is one process that can fuel the activity. This is almost certainly the situation in M82, the closest and best-studied starburst galaxy. Figure 7.19 shows the gas-rich environment of the tidal field resulting from the tidal interaction between M81 and M82 where HI is spread far beyond the confines of the two galaxies. The spectrum of M82 is illustrated in Fig. 7.20. At frequencies below $\sim 100$ GHz the emission from M82 is synchrotron presumably fuelled by the high supernova rate. High-resolution radio surveys have identified some 30 SNRs in M82. The diffuse synchrotron most likely comes from relativistic electrons generated by earlier supernova explosions. It can be seen in Fig. 7.20 that there is a steepening in the spectral index at a frequency of $\sim 2$ GHz as expected for synchrotron radiation losses.

*Rodney D. Davies*

FIGURE 7.18. NGC 5866 (M102) is an edge-on lenticular galaxy of type $T = -1$. The large amounts of dust, mainly confined to the plane of the galaxy, produce the strong absorption at optical wavelengths. (From the NOT telescope web page, http://www.not.iac.es/.)

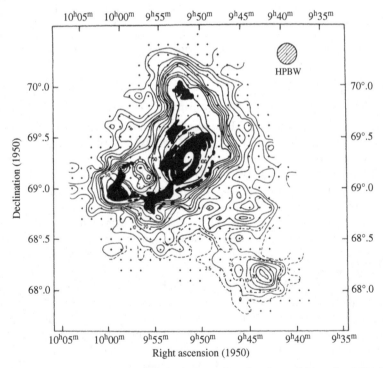

FIGURE 7.19. The $\lambda = 21$ cm neutral hydrogen distribution within the M81/M82 group which lies at a distance of 3.5 Mpc (Appleton *et al.*, 1981). This gas fuels the high level of star formation in the Irregular galaxy M82, the northern object in the group, and the more normal star formation rate in the Sb spiral galaxy M81. Distant counterparts of these two different types of galaxy contribute to the high-$\ell$ power spectrum at radio and millimetre wavelengths. Taken from fig. 4 in Appleton *et al.* (1981), with kind permission of Blackwell Publishing.

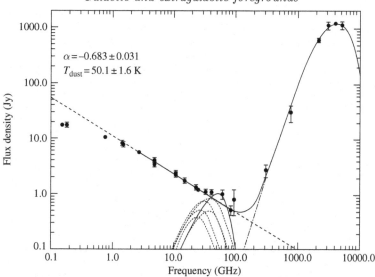

FIGURE 7.20. The integrated spectrum of M82 (Davies and Dickinson, unpublished work). The low-frequency spectrum is synchrotron emission with a characteristic steepening above 2 GHz. The high-frequency spectrum fits thermal dust at a temperature of 50 K. A small amount of free–free and anomalous dust emission may also be present.

The characteristic FIR dust spectrum of M82 is seen at frequencies above $\sim 100$ GHz. Although there is well-documented H$\alpha$ emission from M82, it is not a major contributor to the integrated flux density spectrum. It will be most readily detected in the minimum of the spectrum at frequencies around $\sim 100$ GHz. Similarly there is only weak evidence for anomalous emission.

Another well-known and more powerful starburst galaxy is Arp 220. This ultra-luminous infrared galaxy is at a distance of 75 Mpc ($z = 0.018$), some 25 times further than M82. Objects of this type comprise the main body of bright far-infrared $\log N / \log S$ distribution. The weaker part of the flux density distribution will include both the distant ultra-luminous galaxies as well as the nearby "normal" dusty galaxies.

The dusty galaxies form a point source foreground to the CMB at the higher frequencies of the Planck Surveyor which covers the range 100 GHz to 857 GHz. It is instructive to look at the Poisson contribution of dusty galaxies to the power spectrum in the plots for these frequencies in Fig. 7.21. As is the case for the foreground of radio sources in Fig. 7.17, the dusty sources show a similar linear rise with $\ell$ throughout the $\ell$ range ($\ell = 10$ to 2000). At frequencies of $\sim 300$ GHz the dusty galaxy Poisson contribution is greater than the CMB at $\ell = 1000$. By a frequency of 857 GHz the dusty galaxies exceed the CMB above $\ell = 30$.

The other important feature of Fig. 7.21 is its demonstration of the cosmic far infrared background (CFIRB). At the highest frequencies covered by Planck, the surface density of dusty galaxies has become so high that they form a continuous background of emission. It should also be remembered that the high redshift galaxy dust spectrum is translated to lower frequencies thereby enhancing the flux densities at Planck frequencies. The linear rise with $\ell$ in Figure 7.21 is the Poisson fluctuation in the CFIRB. The two highest frequencies show the relative contributions of the Galactic diffuse cirrus and the CFIRB to the fluctuation power spectrum. At 545 GHz the Galactic emission far exceeds the CMB and the Poisson component over the whole $\ell$ range. On increasing the frequency to 857 GHz the Poisson exceeds both the CMB and the diffuse component.

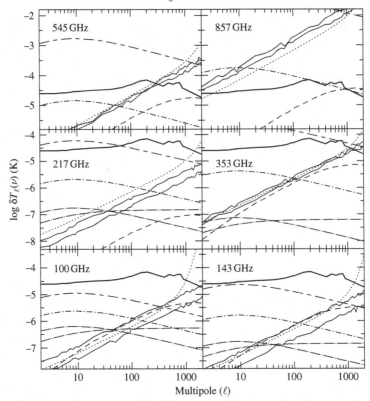

FIGURE 7.21. The fluctuation spectrum of various foreground components of the CMB. The plot format for each of the components is as given in Fig. 7.17. The diffuse dust component (long dash–short dash) is the dominant diffuse foreground over this frequency range. The dusty galaxy Poisson component (linear rising plots) exceeds the CMB for all $\ell$ values at 857 GHz. Reprinted with permission from de Zotti *et al.* (1999). Copyright 1999, American Institute of Physics.

The frequency dependence of the rms fluctuations in each of the foregrounds is compared with the Poisson component from point sources in Fig. 7.22 at Galactic latitude cuts of $|b| > 20°$ and $50°$. The dotted curve shows the fluctuation level produced by discrete extragalactic sources, radio galaxies at the lower frequencies and dusty galaxies at higher frequencies. The minimum in this contribution lies between 100 and 200 GHz. At low frequencies it is the major foreground at $\ell = 2000$ and is comparable with the synchrotron and free–free diffuse components at $\ell < 1000$. Above 100 GHz the fluctuations from the diffuse dust are dominant at all angular scales. Another conclusion that can be drawn from Fig. 7.22 is that the level of foreground contamination to the CMB is only marginally worse for the $|b| > 20°$ cut as compared with the $|b| > 50°$ cut; most of the sky is therefore available for CMB studies.

### 7.3.3 The Sunyaev–Zeldovich effect

Another point-like source contribution to the CMB foregrounds is the Sunyaev–Zeldovich (SZ; 1972) effect produced by galaxy clusters. This has a characteristic spectrum and a distribution with angular size which is different from the galaxy Poisson fluctuations.

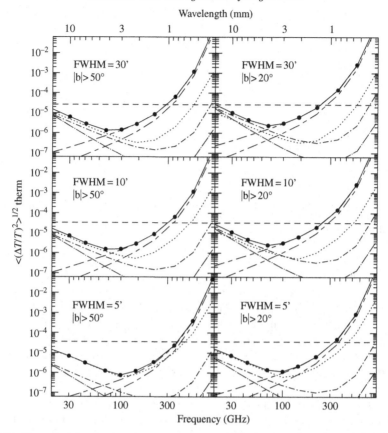

FIGURE 7.22. Temperature fluctuations as a function of frequency for three angular scales and two cuts in Galactic latitude. The angular scales of 30, 10 and 5 arcmin correspond to $\ell = 360$, 1080 and 2000. The horizontal dashed line is the expected level of CMB anisotropies. The dotted curve is the fluctuation level produced by extragalactic radio and dusty galaxies. The solid line is the sum of all the foreground components. There is only a small advantage in making a higher latitude cut (left-hand plots) in reducing the foreground contribution. Reprinted with permission from de Zotti *et al.* (1999). Copyright 1999, American Institute of Physics.

Clusters of galaxies contain a hot electron gas which is responsible for their X-ray emission. The electron temperature is characteristically $\sim 10^8$ K. The inverse Compton scattering of CMB photons against this hot and diffuse electron gas trapped in the potential well of the cluster produces a systematic shift of the photons from the Rayleigh–Jeans side to the Wien side of the CMB spectrum. The spectral form of this "thermal" effect is a negative temperature at low (Rayleigh–Jeans) frequencies and a positive temperature at high (Wien) frequencies. The change from a negative to a positive temperature is at a frequency of 217 GHz. The amplitude of the effect is proportional to $\int n_e dl$. By contrast, the X-ray surface brightness is proportional to $\int n_e^2 dl$ and a comparison of the SZ and the X-ray data gives an independent measure of the cluster distance and hence the Hubble parameter, $H_0$.

Given the $z$ distribution of galaxy clusters, it is possible to derive the power spectrum of the fluctuations in the CMB produced by the SZ effect. Various properties of clusters have to be taken into account such as the dispersion of cluster sizes and electron temperatures. On a cosmological scale, the evolution of clusters as a function of redshift, $z$, must also

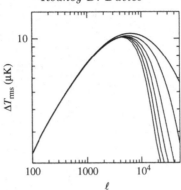

FIGURE 7.23. The power spectrum in terms of a temperature for SZ fluctuations in the CMB in a range of cluster models of preheating (from fig. 5 in Holder *et al.*, 2007, with kind permission of Blackwell Publishing).

be considered. Figure 7.23 plots the power spectrum of cluster SZ effects for a number of models from a recent study by Holder *et al.* (2007). It turns out that the temperature of the decrement of a given cluster is independent of redshift whereas the angular size decreases with redshift. Given that there is a mean redshift of cluster formation, there will be a minimum angular size (maximum $\ell$) for the SZ fluctuations. The peak in the SZ power spectrum is at $\ell = 5000$ in Fig. 7.23 with an rms fluctuation level of 10 μK. The different curves are for various levels of preheating in the cluster due to accretion and infall for example.

The contribution of the SZ power spectrum as shown in Fig. 7.23 is compared with the CMB power spectrum and that of the other foreground components in Figs. 7.17 and 7.21. The first point note is that the SZ contribution (heavy dashed line showing a turn-over at $\ell \sim 2000$) is only important relative to the CMB at $\ell \geq 1000$. Secondly, the characteristic shape of the SZ frequency spectrum gives a minimum in the 217 GHz band of Planck. At lower frequencies the SZ contribution is less than that of the radio point sources while at high frequencies the dusty galaxies give a stronger signal than the SZ effect. In order to identify unambiguously the SZ fluctuation spectrum and separate it from that of the point sources, it is necessary to have sensitive observations in the $\ell$ range 1000 to 5000 and frequencies preferably in the range 40 to 150 GHz. Even then the radio source component is likely to be dominant.

## 7.4 Foregrounds and future CMB ground and space experiments

### 7.4.1 Earth's atmosphere

For Earth observations, the atmosphere constitutes a significant foreground to be dealt with. It is variable in time and position as the atmospheric fluctuations move through the observational field of view. Ground- and balloon-based experiments have been, and will continue to be, major programmes in CMB research as ultra-high sensitivities and/or resolutions are proposed.

The two important constituents of the atmosphere which are of concern at the frequencies relevant to CMB experiments are $H_2O$ and $O_2$. Their spectral features are shown in Fig. 7.24. The emission lines of $H_2O$ are at 22.2, 183.3 and 325.2 GHz while those of $O_2$ are at 60 and 118 GHz. The $H_2O$ line at 22 GHz is pressure broadened in the lower atmosphere; the apparent broadening of the 60 GHz $O_2$ line is in fact due

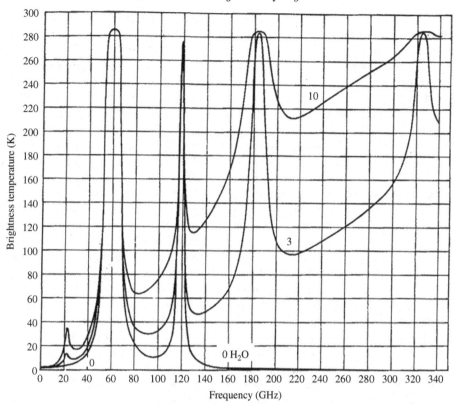

FIGURE 7.24. The water-vapor and $O_2$ emission of the atmosphere as seen in the zenith at sea level. The $O_2$ emission alone is shown as the zero (0) $H_2O$ plot. The plots marked as 3 and 10 g m$^{-3}$ correspond to 6 and 20 mm of precipitable water vapor (pwv) in addition to the $O_2$ emission (adapted from Smith, 1982).

to the presence of some 20 individual lines in this band which are pressure broadened into a smooth band around 60 GHz. In addition to the spectral lines in the atmospheric emission there is an excess of emission which increases with frequency. This underlying continuum is associated with the "wet" atmosphere and is a major contributor to the emission at frequencies lying between the spectral lines; the emission mechanism is not clearly understood (Danese and Partridge, 1989).

The opacity of the atmosphere as shown in Fig. 7.24 has two effects. One is to produce an emitting screen with a temperature of $\tau T_{\mathrm{atm}}$ where $\tau$ is the optical depth of the sum of the $H_2O$ and the $O_2$ opacities and $T_{\mathrm{atm}}$ is the physical temperature of the atmosphere. Figure 7.24 shows the emission spectrum of the atmosphere seen in the zenith from sea level. The other effect is to absorb the incoming radiation, which then requires appropriate correction upwards for the atmospheric opacity.

The structure in the atmosphere can be represented analytically in terms of the Kolmogorov law describing the power of the turbulence (fluctuations) as a function of linear scale. This theoretical law gives a fair match to observations over scales $\leq 5$ km. The corresponding temperature fluctuations, $dT$, increase in proportion to $D^{5/6}$ where $D$ is the scale size of the fluctuation. Accordingly the atmospheric fluctuation noise on temperature maps is less on smaller angular scales. Observations at $\ell = 300$ with the small pencil beams of large antennae or with interferometer arrays are least affected.

The structure in the water vapor is by far the dominant contributor to the fluctuations in radio emissivity of the atmosphere. The $O_2$ is well mixed and only varies slowly with atmospheric pressure.

Figure 7.24 shows clearly that the atmospheric emission at any likely observing site is going to be much greater than the extremely weak signals from the CMB, which are measured in tens of $\mu K$. The best sites will be those with the lowest and most stable water-vapor content. Furthermore the best observing bands will be the minima between the emission lines. The good observing sites are those at high altitude, remembering that the $H_2O$ scale height is 2 km and the $O_2$ scale height is $\sim 7$ km. For example at Mauna Kea in Hawaii at a height of 4.2 km the pressure has fallen from 1013 mbar at sea level to $\sim 600$ mbar so the $O_2$ emission has fallen by a factor of $\sim 2$ and the $H_2O$ by a factor of $\sim 6$. Cosmic microwave background studies at $\sim 30$ GHz at a site such as the Izaña Observatory at an altitude of 2.4 km on Tenerife have shown that the pwv is $< 3$ mm for $\sim 80\%$ of the time, largely because this site is above the atmospheric inversion layer at those times. Another factor contributing to the atmospheric stability at Izaña is its island location, which has also proved to be an excellent site for high-quality solar observations. The emission temperature corresponding to the 3 mm pwv level is 1 K and it is the $\sim 1\%$ fluctuations in this level which produce the background atmospheric noise in the observations. At this site the total ($H_2O$ plus $O_2$) emission is typically 6 K and is a significant contribution to the system noise ($\sim 30$ K). A higher site such as Chajnantor on the Atacama Desert in Chile at 5.1 km should have a factor of $\sim 3$ lower $H_2O$ and $\sim 1.5$ lower $O_2$. The south pole site at 2.8 km has, for its altitude, a low $H_2O$ content because of the low surface temperature at this latitude.

Another component of the Earth's emission that affects measurements of the Galactic H foreground is geocoronal H emission coming from heights of tens of thousands of km. It is variable in time and position. It is comparable in brightness to the Galactic component at intermediate and high latitudes. By using Fabry–Perot interferometry techniques, Haffner *et al.* (2003) have been able to separate the two. Filter imaging cannot make this separation.

### 7.4.2 Zodiacal light emission (ZLE).

The zodiacal dust cloud consists of a population of micron to millimetre size particles distributed between the Sun and the Asteroid Belt at 2 AU. These particles are approximately in equilibrium with the solar radiation and attain temperatures between 240 and 280 K. The near- and far-infrared emission from the ZLE is concentrated towards the ecliptic plane although its geometry is complex with structures down to a scale of $10°$.

The emissivity in the higher Planck bands is expected to fall as $\lambda^{-2}$ as judged from COBE–Far-Infrared Absolute Spectrometer (FIRAS) data. At present this is largely an unexplored regime of the ZLE. Planck should be able to clarify the spectrum and morphology of the ZLE distribution in the Solar System. Figure 7.25 from Maris *et al.* (2006) shows the predicted emission from the ZLE relative to the Galactic emission at 857 GHz as a function of ecliptic longitude over a 14-month period. It should be remembered that at 545 and 857 GHz the Galactic emission outshines the CMB as illustrated in Figs. 7.21 and 7.22. Figure 7.25 shows two slightly different plot shapes as the spacecraft makes two essentially identical coverages of the sky; the differences are due to the different aspects of the zodiacal cloud seen by the spacecraft as it orbits the Sun.

Figure 7.25 shows that the instrumental noise, again plotted relative to the Galactic emission, is negligible at 857 GHz compared with both the Galaxy and the ZLE. At 545 and 353 GHz the ZLE emissivity is 40 and 17% respectively of the value at 857 GHz.

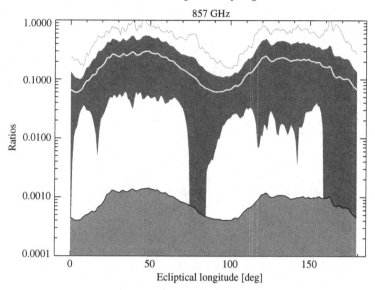

FIGURE 7.25. Zodiacal light emission (ZLE) at 857 GHz expected for the Planck mission (from Maris *et al.*, 2006). The black band is the ratio of the ZLE to the Galactic emission as a function of ecliptic longitude. The ratio for each 2° step in ecliptic longitude is averaged over a circular band 2° wide and 85° in radius drawn around the axis of the given ecliptic longitude. The vertical logarithmic scale runs from $10^{-4}$ to $10^0$. The white line is the mean and the width of the black band is $+/-$ the rms ratio. The dark grey band is the expected instrumental noise contribution as a fraction of the Galactic emission. Throughout the year the ZLE lies between 10 and 30 % of the Galactic emission.

With simulations of the type shown in Fig. 7.25, it appears that the highest frequencies of the Planck satellite should be able to distinguish the Galactic emission from the ZLE. The Planck data are expected to give new information about the size distribution and composition of the zodiacal dust composition as well as the geometrical distribution of the largest grains in the cloud.

### 7.4.3 Analysis techniques using templates

We now turn to the problem of deriving the precision all-sky templates for each of the CMB foregrounds. Two outcomes are required from this study. The first is to obtain a CMB all-sky map that is as free as possible from any contaminating foreground. The second is to use the all-sky observations over a wide range of frequencies to determine the physical properties of each of the foregrounds, namely the synchrotron, the free–free, the anomalous (spinning) dust and the thermal (vibrational) dust. Each of these components has an emissivity that may change with position and frequency in the varying environment of the ISM in different lines of sight through the Galaxy.

One of the serious challenges to performing this task is the CMB itself, which has a power spectrum that falls with increasing $\ell$ and which is not wildly different from that of the components of the ISM. A commonly used approach when using a data set such as WMAP is to construct an internal linear combination (ILC) of the five frequency maps such that the foregrounds cancel out. This is notionally possible if the components change slowly and smoothly with frequency. In fact this is only partially true – the spinning dust component is not well accounted for in this way. Nevertheless, a simple ILC of the form $0.109\,T_{\mathrm{K}} - 0.684\,T_{\mathrm{Ka}} - 0.096\,T_{\mathrm{Q}} + 1.921\,T_{\mathrm{V}} - 0.250\,T_{\mathrm{W}}$ where $T_{\mathrm{K}}$, $T_{\mathrm{Ka}}$, $T_{\mathrm{Q}}$, $T_{\mathrm{V}}$ and $T_{\mathrm{W}}$

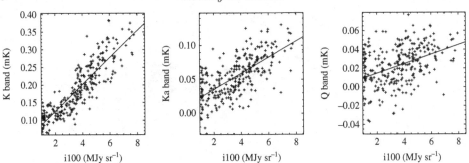

FIGURE 7.26. The $T$–$T$ plots of the dust-correlated emission in the WMAP K, Ka and Q bands for a region around the north celestial pole (adapted from Davies *et al.*, 2006). The dust surface brightness is plotted in terms of MJy sr$^{-1}$ at 100 μm.

are the temperatures at the corresponding frequencies, can be used as a starting point in the ultimate analysis. We discuss this further below.

### *7.4.3.1 The T-T analysis*

The classical T-T approach can provide a detailed look at the local correlation between two data sets. It is particularly useful for identifying the correlated parts of the data when the two sets have different zero levels, which may result from instrumental or other background effects. An example of the use of T-T plots is given in Fig. 7.26, which shows the correlation between the brightness temperatures in the K, Ka and Q bands of WMAP and the 100 μm surface brightness for a $15° \times 15°$ area around the north celestial pole where there is strong dust emission and only weak Hα emission (Davies *et al.*, 2006). The ILC estimate of the CMB brightness has been subtracted from each WMAP band. It is seen that the correlated WMAP emission on a 1° angular scale falls sharply with frequency. At K band (22.8 GHz), the emission is 38 μK per MJy sr$^{-1}$ while at Q band (40.7 GHz) it is 5 μK per MJy sr$^{-1}$. At the resolution of the data in these plots (1°), the rms CMB fluctuations are $\sim 70$ μK, so any uncertainty in the ILC estimate can significantly contribute to the noise on the plots or even the correlation itself. The T-T approach gives a useful first indication of the presence of a correlation, but will be affected by any other foreground in the field that is at least partially correlated with the dust. Free–free emission is the most likely correlated component, which however can be accounted for using the Hα map of the area and the relations in Table 7.1. In the present field the correction for free–free emission reduces the estimated dust emissivity by $\leq 5\%$.

### *7.4.3.2 Cross-correlation analysis in limited-sky areas*

As we have just seen the limitation of the T-T approach is that it deals with only one component at a time. Since maps are available for the three likely foreground components it is important to be able to make simultaneous comparisons between the observed sky and the individual components. The cross-correlation (C-C) method is a least-squares fit of one map to one or more templates. When only one template map is compared with the data, this method is equivalent to the classical T-T method. The advantage of the C-C method is that several components can be fitted simultaneously and that information can be included about the CMB through its signal covariance spectrum rather than having to correct for it.

To compare multiple template components, namely the different foregrounds, to a given data set, the problem becomes a matrix equation. The errors can be determined from the diagonals of the matrices. The power spectrum of the CMB can be taken as that of the WMAP best-fit cold dark matter (CDM) power-law spectrum. Information about the noise covariance can also be included and is determined from the uncorrelated pixel noise in the map data. Gorski *et al.* (1996) describe the method in terms of harmonic space, which is fundamentally no different from working in pixel space.

The C-C analysis has been applied by Davies *et al.* (2006) to regions typically 10° to 20° across, where there is structure on scales of 1° to 3°, which are dominated by one of the three components as identified in the standard templates. Five regions were selected from each template. In each region the WMAP amplitude of each of the three components could be estimated and the level of contamination by the other two components could be assessed. The free–free emission at WMAP frequencies appeared to have an electron temperature of $4000 - 5000$ K, which is less than that expected for a 7000 K electron gas in the solar neighbourhood in the Galaxy; this may have been due to some of the dust-correlated Hα emission being attributed to anomalous emission; however the electron temperatures of the brightest Hα regions were close to the expected values. The synchrotron-correlated emission had a brightness temperature spectral index of $-3.0$ to $-3.2$ as measured between 408 MHz and the WMAP frequencies. The results for the dust-correlated anomalous emission in each of the 15 regions are shown in Fig. 7.27. There is a spread of a factor of 2–3 in the dust emissivity at each WMAP frequency. This variation shows a weak correlation with dust temperature. In the K–Q bands, where the thermal dust emission is small, the anomalous emission has a brightness temperature spectral index of $\beta = -2.4$. It steepens at higher frequencies when corrected for thermal dust as expected on the Draine–Lazarian models.

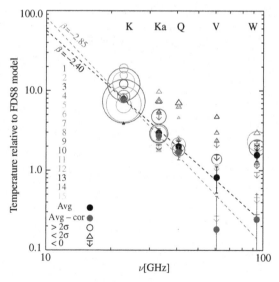

FIGURE 7.27. A summary of the dust emissivities in temperature units relative to model 8 of Finkbeiner *et al.* (1999) in the five WMAP bands for the 15 regions studied. The emissivity in µK per MJy sr$^{-1}$ at 100 µm at each frequency is $\sim 3.5$ times the units plotted. The thermal dust emission dominates in the W band. At the lower frequencies the anomalous emission is dominant. The filled circles are the average of all the clouds and the lower (Avg-cor) circles show the average spectrum corrected for the high-frequency thermal dust component. Adapted from Davies *et al.* (2006).

### 7.4.3.3 Analysis of all-sky foregrounds

In this section we will consider those analyses which have been made on the whole sky rather than individual objects or areas. In most of this work, "full-sky" refers to the sky area away from the Galactic plane where the effects of recent star formation are high and accordingly different approaches are required. A commonly used template is the Kp2 template introduced by the WMAP group; it excludes the regions of high emission around the Galactic plane including the Gould Belt system.

Lagache (2003) used a T-T type method to derive the HI-correlated emission spectrum of the foregrounds. The average emission spectra running from IRAS to WMAP frequencies were estimated in regimes of different HI surface density ranging from $N_H = 3.3$ to $9.9 \times 10^{20}$ atom cm$^{-2}$. An excess emission was found in the WMAP frequency range that was identified with anomalous emission from dust. It was found that the spectral shape of the emission did not change with $N_H$; however the intensity of the emission per unit of $N_H$ fell with increasing $N_H$. Lagache concluded from this latter result that the very small-grain dust responsible for the anomalous emission was preferentially more abundant than the large-grain particles responsible for the 100 μm emission in regions of lower gas density.

An all-sky C-C analysis of the Kp2 sky was included in the analysis of Davies *et al.* (2006) for comparison with the results for the bright regions discussed above in order to see if the dust emissivity over the WMAP bands was different in regions of lower dust brightness (Fig. 7.28). Within the errors there is no difference between the emissivity per unit of dust brightness over the WMAP spectral range between the denser dust clouds at intermediate latitudes than in the general diffuse medium as can be seen in the data at 30 GHz given in Table 7.2. Both these regimes lie in the local spiral arm within 1 kpc of the Sun. The anomalous emission at other Galactocentric distances can only be investigated at lower Galactic latitudes.

A maximum entropy method (MEM) has been used by the WMAP team (Hinshaw *et al.*, 2007) in their most recent assessment of the observations in the Kp2 sky. They used similar templates to those described above for the three classical foreground components, except that the synchrotron spectrum was defined between 408 MHz and the WMAP K

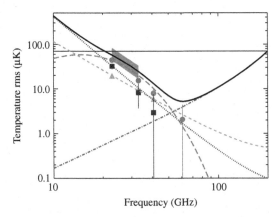

FIGURE 7.28. The rms fluctuation spectrum in thermodynamic temperature units of the foreground components at 1° resolution (adapted from Davies *et al.*, 2006). The analysis is for the Kp2 sky. The plots show the 408 MHz-correlated synchrotron (squares and dotted line with $\beta = -3.1$), the Hα-correlated free–free (triangles and short dashed line with $\beta = -2.14$) and the dust-correlated emission (dots and heavy dashed line with a Draine and Lazarian spectrum). The vibrational dust is shown with $\beta = +1.7$.

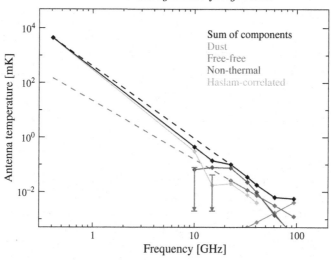

FIGURE 7.29. The WMAP MEM separation of the three foreground components (from Hinshaw *et al.*, 2007). The curves show the mean signal in the Galactic latitude range $|b| = 20°$ to $50°$ computed from the MEM component maps: the dashed black line is thermal dust, the dashed light-grey line is free–free, the dark grey line is the non-thermal signal and the solid black line is the sum of the three. The light-grey solid line is Haslam-correlated. The non-thermal signal is strongly correlated with dust. Reproduced by permission of the AAS.

and Ka bands. No specific solution was made for the anomalous dust as was the case in the C-C analysis. A cosec($b$) function was first subtracted from the data as well as the ILC CMB. The results are shown in Fig. 7.29 where the spectra of the three components are plotted over the WMAP frequency bands and extended to lower frequencies. The "non-thermal" component which is strongly correlated with dust is the dominant component in the frequency range 15 to 40 GHz; it is similar to that ascribed to the dust-correlated emission in Fig. 7.28. The WMAP team interpret this as a new type of synchrotron emission and are cautious about the spinning dust interpretation. The importance of observations at frequencies below those used in WMAP for resolving this problem has already been emphasized and illustrated in Section 7.2 above.

### 7.4.4 Future prospects

As the basic properties of the CMB and the various foregrounds become clearer, new challenges to the observers are emerging. The first of two examples is the requirement for higher sensitivity at high $\ell$ values to clarify the Poissonian point-source contribution to the CMB power spectrum and also the SZ contribution at $\ell \geq 2000$. The second is the emerging strong interest in the $B$-mode of the CMB emission which probes the inflationary epoch; this shows up as a weak linear polarization of the CMB at levels in the range 0.1 to 0.01 µK. In order to be able to detect this, it is vital to know the contribution of the polarized foregrounds to this precision or better and then to remove them.

The major up-coming space mission following the very successful WMAP probe is the Planck satellite to be launched during 2009. It will cover a larger frequency range at higher sensitivity and make substantial contributions to CMB and non-CMB science. Its instrumental characteristics are summarized in Table 7.3. It also has a polarization capability at all the bands between 30 and 353 GHz and has the potential to make important

TABLE 7.3.   Summary of Planck instrument characteristics

| Frequency (GHz) | 30 | 44 | 70 | 100 | 143 | 217 | 353 | 545 | 857 |
|---|---|---|---|---|---|---|---|---|---|
| Beamwidth (arcmin) | 33 | 24 | 14 | 10 | 7.1 | 5.0 | 5.0 | 5.0 | 5.0 |
| $\Delta T/T$ ( $10^{-6}$) | 2.0 | 2.7 | 4.7 | 2.5 | 2.2 | 4.8 | 14.7 | 147 | 6700 |
| Polarization (linear) | P | P | P | P | P | P | P | – | – |

contributions the understanding of cosmic polarization. Launched at the same time is the Herschel telescope which can target and study fields at far-infrared wavelengths; a strong part of its programme will be the follow-up of Planck discoveries.

A major ambition in CMB physics is to understand the origin of its fluctuation spectrum. The inflationary era is widely considered to hold vital clues to this problem, in that the tensor field will produce $B$-mode linear polarization, which will be most easily detectable at low $\ell$ values ($\ell \leq 200$). The associated instrumental challenge is immense. The expected polarized signals are extremely weak, lying in the range $10^{-1}$ to $10^{-2}$ μK. These values may be compared with the scalar $E$-mode signals, which in this $\ell$ range are $\sim 3$ μK and have been detected by WMAP and ground-based experiments. However the greatest obstacle is the Galactic synchrotron foreground emission at lower frequencies and thermal dust emission at $\geq 100$ GHz. At the frequencies where the foreground polarization is a minimum (60 to 150 GHz) at the $\ell$ scales of interest, the rms temperatures of the foregrounds are 3 to 10 μK and need to be accounted for to a precision of a few percent.

A picture of the synchrotron polarization is emerging from sensitive 1.4 GHz all-sky measurements. When taken with data from the lower frequency WMAP maps it is clear that there is significant depolarization at 1.4 GHz at intermediate and low Galactic latitudes. The C-Band All-Sky Survey (CBASS) project aims to cover the whole sky at 5 GHz at a resolution of 1° and at a sensitivity sufficient to measure the polarized signal to a few percent. At 5 GHz the depolarization is small over most of the sky. This project bridges the gap between 1.4 GHz and the frequencies used in WMAP and Planck, thereby establishing the spectrum of synchrotron intensity and polarization which are required for correcting the CMB and for understanding the foregrounds themselves.

Several experiments on the ground are beginning to explore the ultra-sensitive regimes required for detecting the $B$-modes. These include the Clover project which uses a 2 m telescope operating at 97, 150 and 225 GHz and will be located on the Atacama Desert in the Andes. The target sensitivity is 0.1 μK in polarization with integrations several years in duration. Other projects following on ACBAR at the south pole are being planned. Very long duration balloon missions, such as BOOMERanG and others, may contribute in this field. The challenge of the $B$-mode will certainly elicit other experimental initiatives as was the situation three decades ago when the first detections of the CMB fluctuations themselves were beckoning.

REFERENCES

AGLADZE, N. I., SIEVERS, A. J., JONES, S. A. *et al.* (1996). *Astrophys. J.* **462**, 1026.

ALFVEN, H. and HERLOFSON, N. (1950). *Phys Rev.* **78**, 616.

ANDERSSON, B.-G., WANNIES, P. G., MORIARTY-SCHIEVEN, G. H. and BAKKER, E. J. (2000). *Astrophys. J.* **119**, 1325.

APPLETON, P. N., DAVIES, R. D. and STEPHENSON, R. J. (1981). *Mon. Not. R. Astron. Soc.* **195**, 327.

BANDAY, A. J., DICKINSON, C., DAVIES, R. D. *et al.* (2003). *Mon. Not. R. Astron. Soc.* **345**, 897.

BOUDET, N., MUTSCHKE, H., NAVRAL, C. *et al.* (2005). *Astrophys. J.* **633**, 272.

CASASSUS, S., CABRERA, G. F., FÖRSTER, F. *et al.* (2006). *Astrophys. J.* **639**, 951.

DANESE, L. and PARTRIDGE, R. B. (1989). *Astrophys. J.* **342**, 604.

DAVIES, R. D., DICKINSON, C., BANDAY, A. J. *et al.* (2006). *Mon. Not. R. Astron. Soc.* **370**, 1125.

DE OLIVERA-COSTA, A., KOGUT, A., DEVLIN, M. J. *et al*, (1997). *Astrophys. J.* **482**, L7.

DE OLIVEIRA-COSTA, A., TEGMARK, M., GUTIERREZ, C. M. *et al.* (1999). *Astrophys. J.* **527**, L9.

DE OLIVERA-COSTA, A., TEGMARK, M., DAVIES, R. D. *et al*, (2004). *Astrophys. J.* **606**, L89.

DESERT, F. X., BOULANGER, F. and PUGET, J. L. (1990). *Astron. Astrophys.*, **237**, 215.

DE ZOTTI, G., TOFFOLATTI, L., ARGÜESO, F. *et al.* (1999). 3 K cosmology. *Am. Inst. Phys.* **476**, 204.

DICKINSON, C., DAVIES, R. D. and DAVIS, R. J. (2003). *Mon. Not. R. Astron. Soc.* **341**, 369.

DICKINSON, C. CASASSUS, S., PINEDA, J. L. *et al.* (2006). *Astrophys. J.* **643**, L111.

DICKINSON, C., DAVIES, R. D., BRONFMAN, L. *et al.* (2007). *Mon. Not. R. Astron. Soc.* **379**, 297.

DRAINE, B. T. and LAZARIAN, A. (1998a). *Astrophys. J.* **494**, L19 [DL98].

DRAINE, B. T. and LAZARIAN, A. (1998b). *Astrophys. J.* **508**, 157.

DRAINE, B. T. and LAZARIAN, A. (1999). *Astrophys. J.* **512**, 740.

FINKBEINER, D. P., DAVIS, M. and SCHLEGEL, D. J. (1999). *Astrophys. J.* **524**, 867.

GAUSTAD, J. E., MCCULLOUGH, P. R., ROSING, W. and VAN BUREN, D. (2001). *Pub. Astron. Soc. Pacific* **113**, 1326.

GORSKI, K. M., BANDAY, A. J., BENNETT, C. L. *et al.* (1996). *Astrophys. J. Lett.* **464**, L11.

GREEN, D. A. (2004) *Bull. Astron. Soc. India* **32**, 335.

HAFFNER, L. M., REYNOLDS, R. J., TUFTE, S. L. *et al.* (2003). *Astrophys. J. Suppl. Ser.* **149**, 405.

HASLAM, C. G, T., SALTER, C. J., STOFFEL, H. and WILSON, W. (1982). *Astron. Astrophys. Suppl. Ser.* **47**, 1.

HINSHAW, G., NOLTA, M. R., BENNETT, C. L. *et al.* (2007). *Astrophys. J. Suppl. Ser.* **170**, 288.

HOLDER, G. P., MC CARTHY, I. G. and BABUL, A. (2007). *Mon. Not. R. Astron. Soc.* **382**, 1697.

IGLESIAS-GROTH, S. (2006). *Mon. Not. R. Astron. Soc.* **368**, 1925.

KOGUT, A., BANDAY, A. J., BENNETT, C. L. *et al.* (1996). *Astrophys. J.* **464**, L5.

KROTO, H. W., HEATH, J. R., O'BRIEN, S. C. *et al.* (1985). *Nature* **318**, 162.

LAGACHE, G. (2003). *Astron. Astrophys.* **405**, 813.

LEITCH, E. M., READHEAD, A. C. S., PEARSON, T. J. and MYERS, S. T. (1997). *Astrophys. J.* **486**, L23.

MARIS, M., BURIGANA, C. and FOGLIANI, S. (2006). *Mem. Soc. Astron. It. Suppl.* **11**, 83.

MILNE, D. K. (1970). *Aust. J. Phys.* **23**, 425.

PALADINI, R., DAVIES, R.D. and DE ZOTTI, G. (2004). *Mon. Not. R. Astron. Soc.* **347**, 237.

PHILLIPS, W.A. (1987). *Rep. Prog. Phys.* **50**, 1657.

REICH, P. and REICH, W. (1988). *Astron. Astrophys. Suppl. Ser.* **74**, 7.

RYBICKI, G. B. and LIGHTMAN, A. R. (1979). *Radiative Processes in Astrophysics.* Wiley.

SCHLEGEL, D. J., FINKBEINER, D. P. and DAVIS, M. (1998). *Astrophys. J.* **500**, 525.

SMITH, E. K. (1982). *Radio Sci.* **17**, 1455.

SUNYAEV, R. A. and ZELDOVICH, Y. B. (1972). *Comm. Astrophys. Space. Phys.* **4**, 173.

TOFFOLATTI, L., ARGÜESO GÓMEZ, F., DE ZOTTI, G. *et al.* (1998). *Mon. Not. R. Astron. Soc.* **297**, 117.

WATSON, R. A., REBOLO, R., RUBIÑO-MARTÍN, J. A. *et al.* (2005). *Astrophys. J.* **624**, L89.

WOOD, K. and MATHIS, J. S. (2004). *Mon. Not. R. Astron. Soc.* **353**, 1126.

# 8. Probes of fundamental cosmology

## MALCOLM LONGAIR

## Abstract

The chapter discusses many aspect of cosmology which are independent of studies of the cosmic microwave background (CMB) radiation. The specific topics that are addressed include the basic framework of cosmological models, tests of the standard cosmological models, the evolution of perturbations in the pre- and post-recombination Universe and the physics of the very early Universe. The exposition concentrates on how the key results can be understood in terms of elementary physical concepts and on some of the trickier aspects of the analyses which need special care. Many the cosmological tests besides those involving the CMB radiation are discussed. An elementary treatment is given of the origin of fluctuations during the inflationary era which can reproduce the Harrison–Zeldovich power spectrum.

## 8.1 Introduction

My task in these four lectures is to summarise the results of the recent explosion of information on observational cosmology and their interpretation, emphasising the parts which are *not the cosmic microwave background radiation*. Fortunately, the second edition of my book *Galaxy Formation* was published by Springer-Verlag in January 2008 (Longair, 2008). Many more details of the observations and calculations are given there. In my approach to the subject, the emphasis is upon the basic physics involved in astrophysical cosmology, trying to keep it as simple, but rigorous, as possible.

If you are interested in the history of how our present understanding of cosmology came about, you may find my book, *The Cosmic Century: A History of Astrophysics and Cosmology* useful. It covers the history of cosmology up to 2005. Part of the value of this approach is that it gives some understanding of the problems which faced the pioneers of cosmology and the numerous wrong turns which were taken. It also provides help in understanding the various prejudices and reasons for the positions adopted by different cosmologists.

During the course of the lectures at this school, you have been learning many complex and difficult things. I want to go right back to basics and look hard at fundamental issues in the construction of cosmological models. This is important because we are now living in the era of *precision cosmology*. The general opinion is that the basic parameters of the cosmological models are now known to better than 10%. The next step is to advance the observational limits to 1% precision and so we need to be sure that the physics we use is known to better than 1%. We need to look hard at many of the assumptions we usually take for granted.

The plan for the four lectures is therefore as follows:
- The basic framework for cosmology
- The tests of cosmological models
- Perturbations – the post-recombination Universe
- Reflections on the very early Universe

## 8.2 The basic structure of cosmological models

First, we examine the fundamentals of the cosmological models used in modern cosmology. These turn out to be remarkably successful, but we need to ask how secure these foundations are. In particular, we need to review:

- the basic observations on which the models are based;
- the basic assumptions made in the construction of cosmological models;
- how the models really work.

When we come to study specific examples of the models, we will normally assume for illustrative purposes the preferred concordance values of the cosmological parameters which can be taken to be

$$\Omega_0 = 0.3, \quad \Omega_\Lambda = 0.7, \quad h = 0.7. \tag{8.1}$$

We will add in other cosmological parameters later.

### 8.2.1 Basic observations

The starting points for modern cosmological studies are the observations of the cosmic microwave background (CMB) radiation by the COBE satellite. The following fundamental facts for cosmology were established:

- The spectrum is very precisely that of a perfect blackbody at radiation temperature $T = 2.726$ K (Fixsen *et al.*, 1996).
- A perfect dipole component is detected, corresponding to the motion of the Earth through the frame in which the radiation would be perfectly isotropic, at a speed of about 600 km s$^{-1}$.
- Away from the Galactic plane, the radiation is isotropic to better than one part in $10^5$. Significant temperature fluctuations $\Delta T/T \approx 10^{-5}$ of cosmological origin were detected on scales $\theta \geq 10°$ at high Galactic latitudes (Fig. 8.1a).

It is intriguing that the same large-scale features seen in the COBE map are also present in the WMAP image of the sky (Fig. 8.1b). The WMAP experiment had much higher angular resolution than COBE, $\theta = 0.3°$ compared with $\theta = 7°$ and yet the same overall features can be observed. In the WMAP image, the Galactic foreground emission has been subtracted. As explained in detail by the other lecturers, the cosmic microwave background radiation originated on the last scattering surface at a redshift $z \sim 1000$, or scale factor $a \sim 10^{-3}$.

The COBE and WMAP observations established the isotropy of the Universe on a large scale, but we also need to know about the *homogeneity* of the distribution of matter in the Universe. Evidence has been provided by large surveys of galaxies, starting with the local distribution as determined by Geller and Huchra (1989; Fig. 8.2a). This diagram shows a slice through the local Universe in which the redshifts of 15 000 galaxies were observed. Our Galaxy is located at the centre of the diagram and the bounding circle corresponds to a redshift $z = 0.05$.

These 'cone diagrams' have been extended to much larger redshifts by the AAO 2dF survey (Figure 8.2b) and by the Sloan Digital Sky Survey, by factors of 4 and 5 respectively. Both surveys included more than 200 000 galaxies. The large-scale distribution of galaxies is irregular with giant walls and holes on scales very much greater than those of clusters of galaxies. The distributions display the same degree of inhomogeneity as we observe to larger distances in the Universe. This is quantified by the two-point correlation functions for galaxies to different distances,

$$n(r) = n_0[1 + \xi(r)], \tag{8.2}$$

FIGURE 8.1. (a) The map of the whole sky in a Hammer–Aitoff projection as observed by the Cosmic Background Explorer (COBE) with angular resolution $\theta = 7°$ (Bennett *et al.*, 1996). (b) The Wilkinson Microwave Anisotropy Probe (WMAP) map of the whole sky with angular resolution $\theta = 0.3°$. In the WMAP image, the Galactic foreground emission has been subtracted (Bennett *et al.*, 2003). Images can be found at the WMAP page at http://wmap.gsfc.nasa.gov/.

which scale exactly as expected if the large-scale clustering is statistically stationary throughout the Universe at the present epoch (Fig. 8.3). There is also the consequence that we are no longer observing the distant Universe through a non-distorting screen. The images of distant objects are necessarily observed through an irregular gravitationally lensing screen.

The second result we need is the redshift–distance relation for galaxies. All classes of galaxy follow the same Hubble's law, $v = H_0 r$ where $H_0$ is *Hubble's constant*. The combination of the large-scale isotropy and homogeneity of the Universe with Hubble's law means that the Universe as a whole is expanding uniformly.

## 8.3 Basic assumptions and the Robertson–Walker metric

The standard cosmological models are based upon two assumptions and a theory of relativistic gravity:[1]

- The *cosmological principle* asserts that we are not located at any special location in the Universe. Combined with the observations that the Universe is isotropic, homogeneous and expanding uniformly on a large scale, this leads to the *Robertson–Walker*

---

[1] Many more details of the arguments in this section are given in Longair, 2008, ch. 5.

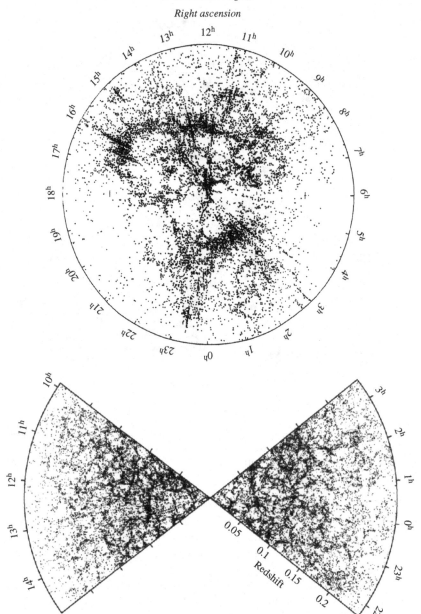

FIGURE 8.2. (a) The distribution of galaxies in the nearby Universe as derived from the Harvard–Smithsonian Center for Astrophysics survey of galaxies. The map contains over 14 000 galaxies. The bounding circle corresponds to $z = 0.05$. Rich clusters of galaxies appear as 'fingers' pointing radially towards our Galaxy at the centre of the diagram. The distribution of galaxies is highly irregular with huge holes, filaments and clusters of galaxies throughout the local Universe (Geller and Huchra, 1989). (b) The distribution of galaxies to four times greater distance than the Geller–Huchra survey as observed from the 2dF galaxy redshift survey (Colless *et al.*, 2001). Image from the 2dF web page (http://www2.aao.gov.au/2dFGRS/).

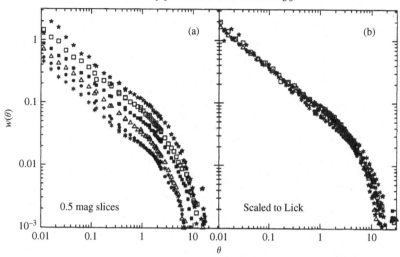

FIGURE 8.3. The two-point correlation function for galaxies over a wide range of angular scales. (a) The scaling test for the homogeneity of the distribution of galaxies can be performed using the correlation functions for galaxies derived from the APM surveys at increasing limiting apparent magnitudes in the range $17.5 < m < 20.5$. The correlation functions are displayed in intervals of 0.5 magnitudes. (b) The two-point correlation functions scaled to the correlation function derived from the Lick counts of galaxies. Both figures are taken from fig. 2 in Maddox *et al.* (1990), with kind permission from Blackwell Publishing.

*metric.* Only special relativity and the postulates of isotropy and homogeneity are needed to derive this metric.

- *Weyl's postulate* is the statement that the world lines of particles meet at a singular point in the finite or infinite past. This solves the clock synchronisation problem and means that there is a unique world line passing through every point in space-time. The fluid moves along streamlines in the expansion and behaves like a perfect fluid with energy–momentum tensor $T^{\alpha\beta}$.
- *General relativity* enables the energy–momentum tensor of matter and radiation to be related to the geometrical properties of space-time.

To derive the Robertson–Walker metric, it is simplest to begin with the Minkowski metric for any isotropic curved space, beginning with the simplest case of an isotropic 2-space, the surface of a sphere (Fig. 8.4). The distance $\rho$ round the arc of a great circle from the point O to P is $\rho = \theta R_c$ and so the increment of distance on the surface of the 2-sphere is

$$\mathrm{d}l^2 = \mathrm{d}\rho^2 + R_c^2 \sin^2\left(\frac{\rho}{R_c}\right)\,\mathrm{d}\phi^2. \tag{8.3}$$

The distance $\rho$ is the shortest distance between O and P on the surface of the sphere since it is part of a great circle and is therefore the *geodesic distance* between O and P in the isotropic curved space. Geodesics play the role of straight lines in curved space.

We can write the metric in an alternative form if we introduce the distance measure

$$x = R_c \sin\left(\frac{\rho}{R_c}\right). \tag{8.4}$$

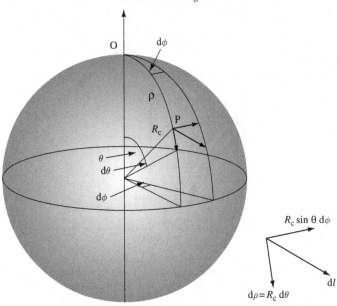

FIGURE 8.4. Illustrating the geometry of an isotropic 2-space as the surface of a sphere. Taken from fig. 5.4 in Longair (2008), with kind permission from Springer Science and Business Media.

Then,

$$dl^2 = \frac{dx^2}{1 - \kappa x^2} + x^2 d\phi^2 \tag{8.5}$$

The form of (8.5) indicates that $x$ is an *angular diameter distance*. Since any section through an isotropic 3-space must be an isotropic 2-space, (8.5) can be extended to three-dimensional polar coordinates and the *Minkowski metric* for any isotropic 3-space written as

$$ds^2 = dt^2 - \frac{1}{c^2} dl^2; \tag{8.6}$$

$dl$ is given by either

$$dl^2 = d\rho^2 + R_c^2 \sin^2\left(\frac{\rho}{R_c}\right) [d\theta^2 + \sin^2\theta \, d\phi^2], \tag{8.7}$$

or by

$$dl^2 = \frac{dx^2}{1 - \kappa x^2} + x^2 [d\theta^2 + \sin^2\theta \, d\phi^2]. \tag{8.8}$$

Notice that $x$ and $\rho$ are equivalent but physically quite distinct distance measures. We can now proceed to derive from this metric the *Robertson–Walker metric*.

There is a problem in applying the metric (8.6) to the expanding Universe as is illustrated by a space-time diagram for the expanding Universe (Fig. 8.5). Since light travels at a finite speed, all astronomical objects are observed along our *past light cone*, which is centred on the Earth at the present epoch $t_0$. Therefore, when we observe distant objects, we do not observe them at the present epoch, but rather at an earlier epoch $t_1$ when the distances between galaxies were smaller and the spatial curvature different. The problem

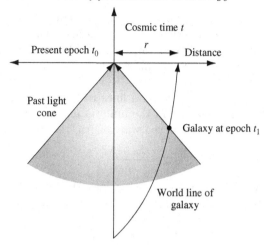

FIGURE 8.5. A simple space-time diagram illustrating the definition of the comoving radial coordinate distance. Taken from fig. 5.5 in Longair (2008), with kind permission from Springer Science and Business Media.

is that the metric (8.6) can only be applied to an isotropic curved space defined *at a single epoch*.

To determine a proper distance which can be included in the metric (8.6), we line up a set of observers between the Earth and the galaxy whose distance we wish to measure. The observers measure the distance $d\rho$ to the next fundamental observer at a particular cosmic time $t$. By adding together all the $d\rho$s, we find a proper distance $\rho$ which is measured *at a single epoch* and which can be used in the metric (8.6). The distance $\rho$ is a *fictitious distance* since we do not know how to project the distances of galaxies to the present epoch until we know the kinematics of the expanding Universe. Thus, *the distance measure $\rho$ depends upon the choice of cosmological model.* Referring comoving distances *to the present epoch*, $\rho = a(t)r$ where $r$ is the *comoving radial distance coordinate* and $a(t)$ is the *scale factor* which describes how the distance between galaxies partaking in the uniform expansion of the Universe changes with cosmic time $t$, normalised to unity at the present epoch.

Because of the isotropy of the expansion, it is simple to show that the *Robertson–Walker metric* can be written in the following form:

$$ds^2 = dt^2 - \frac{a^2(t)}{c^2}[dr^2 + \Re^2 \sin^2(r/\Re)(d\theta^2 + \sin^2\theta \, d\phi^2)]. \quad (8.9)$$

The metric contains one unknown function $a(t)$, the variation of the *scale factor* with cosmic time and the constant $\Re = R_c(t_0)$, the radius of curvature of the geometry of the Universe at the present epoch. The terms of the metric have the following meanings:

- $t$ is *cosmic time*, the time measured by a clock carried by an observer who partakes in the uniform expansion of the Universe;
- $r$ is the *comoving radial distance coordinate* which is fixed to a galaxy for all time;
- $a(t) \, dr$ is the element of proper (or geodesic) distance in the radial direction at the epoch $t$;
- $a(t)[\Re \sin(r/\Re)] \, d\theta$ is the element of proper distance perpendicular to the radial direction subtended by the angle $d\theta$ at the origin;
- similarly, $a(t)[\Re \sin(r/\Re)] \sin\theta \, d\phi$ is the element of proper distance in the $\phi$-direction.

Note the following features of the Robertson–Walker metric. The physics of the expansion of the Universe is built into the function $a(t)$. The radius of curvature of space $\Re$ changes with scale factor as $\Re(t) = \Re a$. The curvature $\kappa = 1/\Re^2$ and can be positive, negative or zero so that $\Re$ can be real, zero or imaginary corresponding to spherical, flat and hyperbolic geometries.

By redshift, we mean the shift of spectral lines to longer wavelengths because of the recession of galaxies from our Galaxy in the uniform expansion. If $\lambda_e$ is the wavelength of the line as emitted and $\lambda_0$ the observed wavelength, the redshift $z$ is defined to be

$$z = \frac{\lambda_0 - \lambda_e}{\lambda_e}. \tag{8.10}$$

It follows directly from the Robertson–Walker metric that the redshift is directly related to the scale factor $a$ through the relation

$$\boxed{a(t) = \frac{1}{1+z}.} \tag{8.11}$$

This is the real meaning of redshift in cosmology – the redshift determines the scale factor of the Universe when the light from the distant source was emitted – it has nothing to do with velocities.

A key test of the Robertson–Walker metric is that the same formula which describes the redshift of spectral lines should also apply to time intervals in the emitted and received reference frames. This test can be carried out using type 1a supernovae, which have remarkably similar light curves (Fig. 8.6a). The lower figure (Fig. 8.6b) shows the width of the light curves of Type 1a supernovae as a function of redshift. A clear time dilation effect is observed exactly proportional to $(1 + z)$, as predicted by the Robertson–Walker metric. The lower diagram of Fig. 8.6b shows the width of the light curves divided by $(1 + z)$ and can be seen to be consistent with the time dilation result.

## 8.4 Radiation-dominated universes

For a gas of photons, massless particles or a relativistic gas in the ultra-relativistic limit $E \gg mc^2$, the pressure $p$ is related to the energy density $\varepsilon$ by $p = \frac{1}{3}\varepsilon$ and the inertial mass density of the radiation $\rho_r$ to its energy density $\varepsilon$ by $\varepsilon = \rho_r c^2$. If $N(\nu)$ is the number density of photons of energy $h\nu$, the energy density of radiation is found by summing over all frequencies

$$\varepsilon = \sum_\nu h\nu N(\nu). \tag{8.12}$$

If the number of photons is conserved, their number density varies as $N = N_0 a^{-3} = N_0(1 + z)^3$ and the energy of each photon changes with redshift by the usual redshift factor $\nu = \nu_0(1 + z)$. Therefore, the variation of the energy density of radiation with cosmic epoch is

$$\varepsilon = \sum_{\nu_0} h\nu_0 N_0(\nu_0)(1 + z)^4 = \varepsilon_0(1 + z)^4 = \varepsilon_0 a^{-4}. \tag{8.13}$$

In the case of blackbody radiation, the energy density of the radiation is given by the Stefan–Boltzmann law $\varepsilon = aT^4$ and its spectral energy density, that is, its energy density per unit frequency range, by the Planck distribution

$$\varepsilon(\nu)\,d\nu = \frac{8\pi h\nu^3}{c^3}\frac{1}{e^{h\nu/kT} - 1}\,d\nu. \tag{8.14}$$

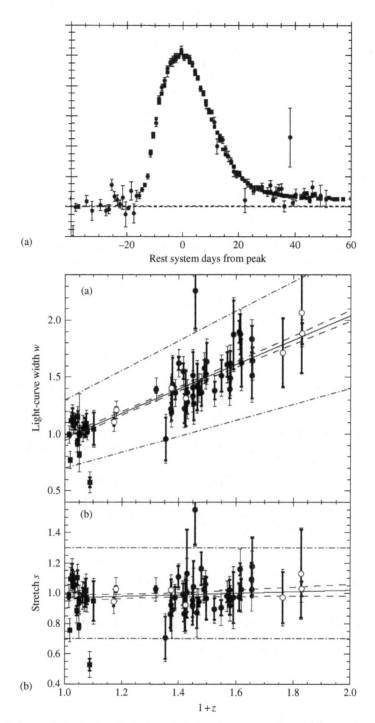

(a)

Rest system days from peak

(a)

Light-curve width $w$

(b)

Stretch $s$

(b)

$1 + z$

FIGURE 8.6. (a) The average time variation of the brightness of a type 1a supernovae from a large sample of supernovae observed in the Calan-Tololo and Supernova Cosmology Program projects. The light curves have been corrected for the effects of time dilation and the luminosity–width correlation. (b) The observed width $w$ of the light curves of type 1a supernovae plotted against $(1 + z)$. In the lower diagram, the stretch $s$ plotted against $(1 + z)$, stretch being defined as the observed light curve width $w$ divided by $(1 + z)$ for each supernova (adapted from Goldhaber et al., 2001). Figures reproduced by permission from the AAS.

TABLE 8.1. Estimates of the radiation temperature of the cosmic microwave background radiation from observations of fine-structure splittings of the ground state of neutral carbon atoms CI

| Author | Quasar | Redshift | Predicted | Observed |
|---|---|---|---|---|
| Songaila *et al.*, 1994 | Q1331+170 | $z_{abs} = 1.776$ | 7.58 K | $7.4 \pm 0.8$ K |
| Ge *et al.*, 1997 | QSO 0013-004 | $z_{abs} = 1.9731$ | 8.105 K | $7.9 \pm 1.0$ K |
| Ledoux *et al.*, 2006 | PSS J1443+2724 | $z_{abs} = 4.224$ | 14.2 K | Consistency |

It immediately follows that, for blackbody radiation, the radiation temperature $T_r$ varies with redshift as $T_r = T_0(1 + z)$ and the spectrum of the radiation changes as

$$
\begin{aligned}
\varepsilon(\nu_1)\,d\nu_1 &= \frac{8\pi h \nu_1^3}{c^3}[(e^{h\nu_1/kT_1} - 1)]^{-1}\,d\nu_1 \\
&= \frac{8\pi h \nu_0^3}{c^3}[e^{h\nu_0/kT_0} - 1)^{-1}](1 + z)^4\,d\nu_0 \\
&= (1 + z)^4\,\varepsilon(\nu_0)\,d\nu_0.
\end{aligned}
\tag{8.15}
$$

This provides another key test of the change of the time dilation formula since the background radiation temperature inferred from observations of fine-structure lines in the spectra of distant quasars should increase as $(1 + z)$.

The fine-structure splittings of the ground state of neutral carbon atoms CI enable this test to be carried out. The photons of the background radiation excite the fine-structure levels of the ground state of the neutral carbon atoms and the relative strengths of the absorption lines originating from the ground and first excited states are determined by the energy density and temperature of the background radiation. Some examples of the results of this test are given in Table 8.1, which shows that the expected increase in radiation temperature with redshift is indeed observed.

## 8.5 Space-time diagrams for the standard world models

### 8.5.1 Times and distances

Let us summarise the various times and distances used in cosmological analyses.
*Comoving radial distance coordinate.* In terms of cosmic time and scale factor, the comoving radial distance coordinate $r$ is defined to be

$$
r = \int_t^{t_0} \frac{c\,dt}{a} = \int_a^1 \frac{c\,da}{a\dot{a}}.
\tag{8.16}
$$

*Proper radial distance coordinate.* We encounter the same problems as in defining the comoving radial distance coordinate, since it only makes sense to define distances at a particular cosmic epoch $t$. Therefore, we *define* the proper radial distance $r_{prop}$ to be the comoving radial distance coordinate projected back to the epoch $t$, that is

$$
r_{prop} = a \int_t^{t_0} \frac{c\,dt}{a} = a \int_a^1 \frac{c\,da}{a\dot{a}}.
\tag{8.17}
$$

*Particle horizon.* The particle horizon $r_H$ is defined as the maximum proper distance over which there could have been causal communication from $t = 0$ to the epoch $t$

$$r_H = a \int_0^t \frac{c\,dt}{a} = a \int_0^a \frac{c\,da}{a\dot{a}}. \tag{8.18}$$

*Event horizon.* The event horizon $r_E$ is defined as the greatest proper radial distance an object could have if it is ever to be observable by an observer at cosmic time $t_1$.

$$r_E = a \int_{t_1}^{t_{max}} \frac{c\,dt}{a(t)} = a \int_{a_1}^{a_{max}} \frac{c\,da}{a\dot{a}}. \tag{8.19}$$

*Cosmic time.* Cosmic time $t$ is defined as time measured by a observer who partakes in the uniform expansion of the Universe.

$$t = \int_0^t dt = \int_0^a \frac{da}{\dot{a}}. \tag{8.20}$$

*Conformal time.* The *conformal time* $\tau$ is similar to the definition of comoving radial distance coordinate. Time intervals are projected forward to present epoch

$$dt_{conf} = d\tau = \frac{dt}{a}. \tag{8.21}$$

At any epoch, the conformal time has value

$$\tau = \int_0^t \frac{dt}{a} = \int_0^a \frac{da}{a\dot{a}}. \tag{8.22}$$

Because of the effects of time dilation, the interval of conformal time $d\tau$ is the time which would be measured by an observer at the present epoch.

### 8.5.2 The past light cone

This topic requires a little care. First, because of the assumptions of isotropy and homogeneity, Hubble's linear relation $v = H_0 r$ applies at the present epoch *to recessions speeds which exceed the speed of light*, where $r$ is the radial comoving distance coordinate. Recall how $r$ is defined. The observers measured increments of distance $\Delta r$ at the present epoch $t_0$. If we consider fundamental observers who are far enough apart, this speed can exceed the speed of light. There is nothing in this assertion that contradicts the special theory of relativity – it is a geometric result because of the requirements of isotropy and homogeneity of the world model.

Consider the surface of an expanding spherical balloon. As the balloon inflates, a linear velocity–distance relation is found on the surface of the sphere, not only about any point on the sphere, but also at arbitrarily large distances on that surface. At very large distances on the surface, the speed of separation can exceed the speed of light, but there is no causal connection between these points – they are simply partaking in the uniform expansion of the underlying space-time geometry of the Universe.

The proper distance between two observers at some epoch $t$ is $r_{prop} = a(t)r$, where $r$ is comoving radial distance. Differentiating with respect to cosmic time,

$$\frac{dr_{prop}}{dt} = \dot{a}r + a\frac{dr}{dt}. \tag{8.23}$$

The first term on the right-hand side represents the motion of the substratum and, at the present epoch, is $H_0 r$. Consider, for example, the case of a very distant object in the critical world model, $\Omega_0 = 1, \Omega_\Lambda = 0$. As $a$ tends to zero, the comoving radial distance coordinates tends to $r = 2c/H_0$. Therefore, the local rest frame of objects at these large distances moves at twice the speed of light relative to our local frame of reference *at the present epoch*. At the epoch at which the light signal was emitted along our past light cone, the recessional velocity of the local rest frame $v_{\text{rec}} = \dot{a}r$ was greater than this value, because $\dot{a} \propto a^{-1/2}$.

The second term on the right-hand side of (8.23) corresponds to the velocity of peculiar motions in the local rest frame at $r$, since it corresponds to changes of the comoving radial distance coordinate. The element of proper radial distance is $a\,dr$ and so, for a light wave travelling along our past light cone towards the observer at the origin, we find

$$\boxed{v_{\text{tot}} = \dot{a}r - c.}$$
(8.24)

This result defines the propagation of light from the source to the observer in space-time diagrams for the expanding Universe.

We can now plot the trajectories of light rays from their source to the observer at $t_0$. The proper distance from the observer at $r = 0$ to the past light cone $r_{\text{PLC}}$ is

$$r_{\text{PLC}} = \int_0^t v_{\text{tot}}\,\mathrm{d}t = \int_0^a \frac{v_{\text{tot}}\,\mathrm{d}a}{\dot{a}}.$$
(8.25)

Notice that, initially, the light rays from distant objects are propagating away from the observer – this is because the local isotropic cosmological rest frame is moving away from the observer at $r = 0$ at a speed greater than that of light. The light waves are propagated to the observer at the present epoch through local inertial frames which expand with progressively smaller velocities until they cross the *Hubble sphere* at which the recession velocity of the local frame of reference is the speed of light. The definition of the radius of the Hubble sphere $r_{\text{HS}}$ at epoch $t$ is thus given by

$$c = H(t)\,r_{\text{HS}} = \frac{\dot{a}}{a}r_{\text{HS}} \qquad \text{or} \qquad r_{\text{HS}} = \frac{ac}{\dot{a}}.$$
(8.26)

Note that $r_{\text{HS}}$ is a proper radial distance. The Hubble sphere will prove to be a key quantity in the analysis of the evolution of small perturbations. From this epoch onwards, propagation is towards the observer until, as $t \to t_0$, the speed of propagation towards the observer is the speed of light.

### 8.5.3 The Davis and Lineweaver diagrams

It is simplest to illustrate how the various scales change with time for specific examples of standard cosmological models. These diagrams were elegantly presented by Davis and Lineweaver (2004) in a paper which merits close study.

We consider first the critical world model (Fig. 8.7). Different versions of the space-time diagram for the critical world model are shown in Fig. 8.7. In all three presentations, the world lines of galaxies having redshifts 0.5, 1, 2 and 3 are shown. When plotted against comoving radial distance coordinate in Figs. 8.7b and c, these are vertical lines. The Hubble sphere and particle horizon, as well as the past light cone, are shown in all three diagrams. There is no event horizon in this model. These diagrams illustrate a number of interesting features.

- Fig. 8.7a is the most intuitive diagram and illustrates many of the points discussed above. For example, the Hubble sphere intersects the past light cone at the point where the $v_{\text{tot}} = 0$ and the tangent to the past light cone at that point is vertical.
- In Fig. 8.7b and c, the initial singularity at $t = 0$ has been stretched out to become a singular line.

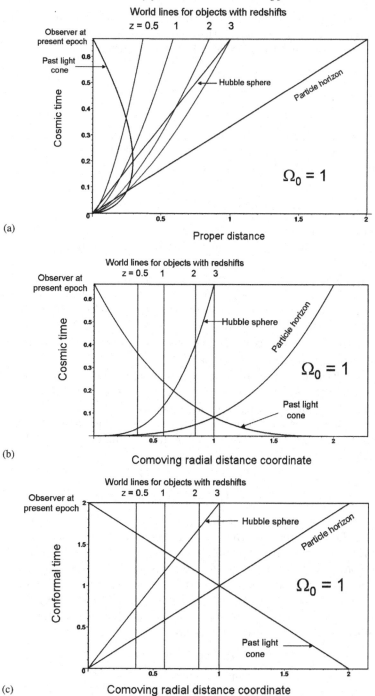

FIGURE 8.7. Space-time diagrams for the Einstein–de Sitter world model, $\Omega_0 = 1; \Omega_\Lambda = 0$. (a) The plot of cosmic time against proper distance; (b) The plot of cosmic time against comoving distance coordinate; (c) The plot of conformal time against comoving distance coordinate. In all three cases, times and distances are measured in units of $H_0^{-1}$ and $c/H_0$ respectively. Taken from fig. 12.1a-c in Longair (2008), with kind permission from Springer Science and Business Media.

- Fig. 8.7c is the simplest diagram in which cosmic time has been replaced by conformal time. In the critical model, the relations are particularly simple, the particle horizon, the past light cone and the Hubble sphere being given by

$$r_H(\text{comoving}) = \tau,$$
$$r_{PLC}(\text{comoving}) = 2 - \tau,$$
$$r_{HS}(\text{comoving}) = \tau/2.$$

The similarities and differences with the standard concordance model can now be appreciated. Adopting $\Omega_0 = 0.3$ and $\Omega_\Lambda = 0.7$, the rate of change of the scale factor with cosmic time in units in which $c = 1$ and $H_0 = 1$ is

$$\dot{a} = \left[ \frac{0.3}{a} + 0.7(a^2 - 1) \right]^{1/2}. \tag{8.27}$$

The diagrams shown in Fig. 8.8 have many of the same general features as Fig. 8.7, but there are significant differences, the most important of these being associated with the dominance of the dark energy term at late epochs.

- First, note that the cosmic timescale is stretched out relative to the critical model.
- The world lines of galaxies begin to diverge at the present epoch as the repulsive effect of the dark energy dominates over the attractive force of gravity.
- The Hubble sphere converges to a proper distance of 8.12 in units of $c/H_0$. The reason for this is that the expansion becomes exponential in the future and Hubble's constant tends to a constant value of $\Omega_\Lambda^{1/2}$.
- Unlike the critical model, there is an event horizon in the reference model. The reason is that, although the geometry is flat, the exponential expansion drives galaxies beyond distances at which there could be causal communication with an observer at epoch $t$. It can be seen from Fig 8.8a that the event horizon tends towards the same asymptotic value of 8.12 in proper distance units as the Hubble sphere. To demonstrate this, we evaluate the integral

$$r_E = a \int_a^\infty \frac{da}{[0.3a + 0.7(a^4 - a^2)]^{\frac{1}{2}}}. \tag{8.28}$$

For large values of $a$, terms other than that in $a^4$ under the square root in the denominator can be neglected and the integral becomes $1/0.7^{1/2} = 1.12$, as found above for the Hubble sphere. In Fig. 8.8b and c, the comoving distance coordinates for the Hubble sphere and the event horizon tend to zero as $t \to \infty$ because, for example, (8.26) has to be divided by $a$ to convert it to a comoving distance and $a \to \infty$.

- Just as in the case of the critical model, the simplest diagram is that in which conformal time is plotted against comoving radial distance coordinate. It is a useful exercise to show that the various lines are:

$$r_H(\text{comoving}) = \tau,$$
$$r_{PLC}(\text{comoving}) = \tau_0 - \tau, \tag{8.29}$$
$$r_E(\text{comoving}) = r_0 - \tau,$$

where $\tau_0 = 3.305$ and $r_0 = 4.446$ for our reference cosmological model. These forms of the relations in terms of comoving distance coordinate and conformal time are true for all models.

The remarkable Appendix B in the paper by Davis and Lineweaver indicates how even some of the most distinguished cosmologists and astrophysicists can lead the newcomer to the subject astray.

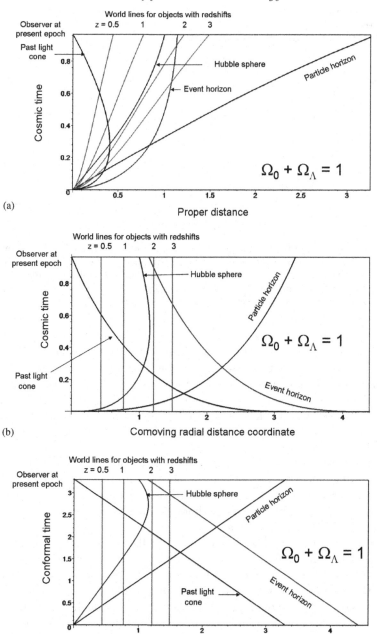

FIGURE 8.8. Space-time diagrams for the concordance model, $\Omega_0 = 0.3; \Omega_\Lambda = 0.7$. (a) The plot of cosmic time against proper distance; (b) The plot of cosmic time against comoving distance coordinate; (c) The plot of conformal time against comoving distance coordinate. In all three cases, times and distances are measured in units of $H_0^{-1}$ and $c/H_0$ respectively. Taken from fig. 12.2a-c in Longair (2008), with kind permission from Springer Science and Business Media.

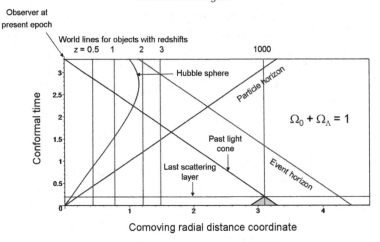

FIGURE 8.9. The plot of conformal time against comoving distance coordinate for the concordance model with the epoch of recombination and the past light cone from that epoch back to the initial singularity which is represented by the singular line along the abscissa. Times and distances are measured in units of $H_0^{-1}$ and $c/H_0$ respectively. Taken from Fig. 20.2a in Longair (2008), with kind permission from Springer Science and Business Media.

Let us illustrate the origin of the *horizon problem* in these cosmological models. This can be stated: 'Why is the Universe so isotropic?' At earlier cosmological epochs, the particle horizon $r \sim ct$ encompassed less and less mass and so the scale over which particles could be causally connected became smaller and smaller. On the last scattering surface at $z \approx 1500$, the particle horizon corresponds to an angle $\theta \approx 2°$ on the sky. How did opposite sides of the sky know they had to have the same properties within one part in $10^5$?

The problem can be pleasantly illustrated in a version of the conformal diagram in which we have included the epoch of recombination and the past light cone from that epoch back to the initial singularity (Fig. 8.9). It can be seen that the past light from the last scattering surface to the initial (line) singularity spans only a very small range of comoving radial distance coordinates.

The inflationary picture solves the horizon and flatness problems by assuming there was a period of exponential growth of the scale factor in the very early Universe. In the extended conformal time diagram (see Fig. 8.10), the time coordinate is set to zero at the end of the inflationary era at, say, $10^{-32}$ s and evolution of the Hubble sphere and the past light cone at recombination extrapolated back through the inflationary era.

From the perspective of smoothing out irregularities in the distribution of matter and radiation, the radius of the Hubble sphere is important parameter since it corresponds to the the distance of causal contact *at a particular epoch*, $c = Hr$. Notice that the past light cones from opposite directions on the sky intersect in the early inflationary era and so isotropisation is perfectly feasible. Note also that the point at which the Hubble sphere crosses the comoving radial distance coordinate of the last scattering surface, exactly corresponds to the time when the past light cones from opposite directions on the sky touch at conformal time $-3$.

Because any object preserves its comoving radial distance coordinate for all time, in the early Universe, objects lie within the Hubble sphere, but during the inflationary expansion, they pass through it and remain outside it for the rest of the inflationary expansion. Only when the Universe transforms back into the standard Friedman model

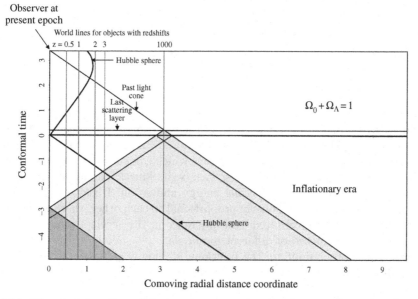

FIGURE 8.10. The plot of conformal time against comoving distance coordinate for the extended concordance model in which the inflationary epoch has been included from $10^{-34}$ to $10^{-32}$ seconds. Times and distances are measured in units of $H_0^{-1}$ and $c/H_0$ respectively. Taken from fig. 20.2b in Longair (2008), with kind permission from Springer Science and Business Media.

does the Hubble sphere begin to expand again and objects can then 'reenter the horizon'. This behavior occurs for all scales and masses of interest in understanding the origin of structure in the present Universe. Since causal connection is no longer possible on scales greater than the Hubble sphere, objects 'freeze out' when they pass through the Hubble sphere during the inflationary era, but they come back in again and regain causal contact when they recross the Hubble sphere.

The inflationary expansion also drives the geometry towards a flat Euclidean geometry since $R_c = a\Re$ and so $\Re \to \infty$ if there is a huge exponential expansion, $a \propto \exp(Ht)$.

## 8.6 Testing Einstein's theory of general relativity

### 8.6.1 Einstein's equivalence principle

The statement that, locally, inertial and gravitational mass are the same is known as the *weak equivalence principle*. Einstein's version of the principle is much stronger:

All local, freely falling, non-rotating laboratories are fully equivalent for the performance of all physical experiments.

Clifford Will (2006) gives a more transparent statement of the principle.

The *Einstein equivalence principle (EEP)* is a more powerful and far-reaching concept; it states that:
 (i) The weak equivalence principle is valid.
 (ii) The outcome of any local non-gravitational experiment is independent of the velocity of the freely-falling reference frame in which it is performed – *local Lorentz invariance (LLI)*.
 (iii) The outcome of any local non-gravitational experiment is independent of where and when in the universe it is performed – *local position invariance (LPI)*.

How good is Einstein's equivalence principle? Following Will's exposition, the deviations from linearity can be written

$$m_{\rm g} = m_{\rm I} + \sum_{\rm A} \frac{\eta^{\rm A} E^{\rm A}}{c^2}, \tag{8.30}$$

where $E^{\rm A}$ is the internal energy of the body generated by interaction A, $\eta$ is a dimensionless parameter that measures the strength of the violation of the linearity of the relation between $m_{\rm g}$ and $m_{\rm I}$ induced by that interaction, and $c$ is the speed of light. The internal energy terms include all the mass-energy terms which can contribute to the inertial mass of the body, for example, the body's rest energy, its kinetic energy, its electromagnetic energy, weak-interaction energy, binding energy and so on. If the inertial and gravitational masses were not exactly linearly proportional to each other, there would be a finite value $\eta^{\rm A}$, which would be exhibited as a difference in the accelerations of bodies of the same inertial mass composed of different materials.

A measurement of, or limit to, the fractional difference in accelerations between two bodies yields the quantity known as the "Eötvös ratio,"

$$\eta = 2 \frac{|a_1 - a_2|}{a_1 + a_2} = \sum_{\rm A} \eta^{\rm A} \left( \frac{E_1^{\rm A}}{m_1 c^2} - \frac{E_2^{\rm A}}{m_2 c^2} \right). \tag{8.31}$$

Will has provided an excellent diagram showing the present limits to the value of the Eötvös ratio (Fig. 8.11). These range from the original measurement carried out by Eötvös to recent experiments involving lunar laser ranging and experiments which searched for evidence of a 'fifth force', as a by-product resulting in strong constraints on the Eötvös ratio. The present limits are at the level of about 1 part in $10^{13}$.

### 8.6.2 Parameterised post-Newtonian models of relativistic gravity

The objectives of the tests of general relativity can be expressed in terms of how well the standard metric of general relativity can account for the results of very precise Solar System and other experiments. In other words, how well does Einstein's equation

$$R_{\mu\nu} - \tfrac{1}{2} g_{\mu\nu} R + \Lambda g_{\mu\nu} = -\frac{8\pi G}{c^2} T_{\mu\nu}$$

account for the experiments.

A convenient way of expressing the comparison of theory and experiment is to adopt parameterised post-Newtonian (PPN) models of relativistic gravity. As expressed by Will,

The comparison of metric theories of gravity with each other and with experiment becomes particularly simple when one takes the slow-motion, weak-field limit. This approximation, known as the post-Newtonian limit, is sufficiently accurate to encompass most solar-system tests that can be performed in the foreseeable future. It turns out that, in this limit, the spacetime metric predicted by nearly every metric theory of gravity has the same structure. It can be written as an expansion about the Minkowski metric in terms of dimensionless gravitational potentials of varying degrees of smallness.

To give some flavour of this approach, Table 8.2 gives a list of possible modifications to general relativity and the corresponding PPN parameters. The metric tensor $g_{\mu\nu}$ is then modified as follows:

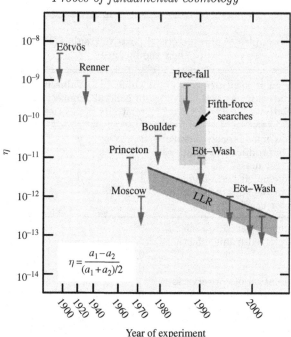

FIGURE 8.11. Selected tests of the weak equivalence principle, showing bounds on $\eta$, which measures the fractional differences in accelerations of different materials or bodies. The free-fall and Eöt–Wash experiments were originally performed to search for a fifth force (light grey region, representing many experiments). The dark grey band shows bounds on $\eta$ for gravitating bodies from lunar laser ranging (LLR). From Will (2006), with permission.

$$
\begin{aligned}
g_{00} =& -1 + 2U - 2\beta U^2 - 2\xi\Phi_W + (2\gamma + 2 + \alpha_3 + \zeta_1 - 2\xi)\Phi_1 \\
&+ 2(3\gamma - 2\beta + 1 + \zeta_2 + \xi)\Phi_2 + 2(1 + \zeta_3)\Phi_3 + 2(3\gamma + 3\zeta_4 - 2\xi)\Phi_4 \\
&- (\zeta_1 - 2\xi)\mathcal{A} - (\alpha_1 - \alpha_2 - \alpha_3)w^2 U - \alpha_2 w^i w^j U_{ij} \\
&+ (2\alpha_3 - \alpha_1)w^i V_i + \mathcal{O}(\epsilon^3) \\
g_{0i} =& -\frac{1}{2}(4\gamma + 3 + \alpha_1 - \alpha_2 + \zeta_1 - 2\xi)V_i - \frac{1}{2}(1 + \alpha_2 - \zeta_1 + 2\xi)W_i \\
&- \frac{1}{2}(\alpha_1 - 2\alpha_2)w^i U - \alpha_2 w^j U_{ij} + \mathcal{O}(\epsilon^{5/2}) \\
g_{ij} =& (1 + 2\gamma U)\delta_{ij} + \mathcal{O}(\epsilon^2)
\end{aligned}
$$

The quantities $U$, $U_{ij}$, $\Phi_W$, $\Phi_1$, $\Phi_2$, $\Phi_3$, $\Phi_4$, $\mathcal{A}$, $V_i$, $W_i$ are various metric potentials which can be interpreted in terms of Newtonian gravity. For example, $U$ is just the Newtonian gravitational potential. Some of these post-Newtonian corrections have quite obvious meanings. For example, inspection of the first three terms of the expression for $g_{00}$ shows that, for a point mass, the first two are just the familiar metric coefficient $(1 + \phi/c^2)$ and the third is a non-linear term in the square of the potential $(\phi/c^2)^2$.

These expressions look rather forbidding at first sight, but they provide a very convenient means of testing theories in which the Einstein equivalence principle is relaxed and provide further constraints upon acceptable theories. In some extensions of the standard model of particle physics, even some of our most cherished theories, such as Lorentz

TABLE 8.2. Examples of parameterised post-Newtonian (PPN) parameters and their significance

| Parameter | What it measures relative to general relativity | Value in general relativity | Value in semi-conservative theories | Value in fully conservative theories |
|---|---|---|---|---|
| $\gamma$ | How much space-curvature is produced by unit rest mass? | 1 | $\gamma$ | $\gamma$ |
| $\beta$ | How much "non-linearity" in the superposition law for gravity? | 1 | $\beta$ | $\beta$ |
| $\xi$ | Preferred-location effects? | 0 | $\xi$ | $\xi$ |
| $\alpha_1$ | Preferred-frame effects? | 0 | $\alpha_1$ | 0 |
| $\alpha_2$ | | 0 | $\alpha_2$ | 0 |
| $\alpha_3$ | | 0 | 0 | 0 |
| $\zeta_1$ | Violation of conservation of total momentum? | 0 | 0 | 0 |
| $\zeta_2$ | | 0 | 0 | 0 |
| $\zeta_3$ | | 0 | 0 | 0 |
| $\zeta_4$ | | 0 | 0 | 0 |

invariance, might have to be sacrificed. Another example involves scalar-tensor modifications of general relativity which appear in unification schemes such as string theory, and in cosmological model building.

### *8.6.3 Recent results*

Traditionally, there are four tests of general relativity

- The *gravitational redshift* of electromagnetic waves in a gravitational field. Hydrogen masers in rocket payloads confirm the prediction at the level of about 5 parts in $10^5$. This is essentially a test of the conservation of energy in a gravitational field.
- The *advance of the perihelion of Mercury*. Continued observations of Mercury by radar ranging have established the advance of the perihelion of its orbit to about 0.1% precision with the result $\dot{\omega} = 42.98(1 \pm 0.001)$ arcsec per century. General relativity predicts a value of $\dot{\omega} = 42.98$ arcsec per century. In terms of the PPN parameters, $(2\gamma - \beta - 1) < 3 \times 10^{-3}$, taking account of upper limit to the Sun's quadrupole moment.
- The *gravitational deflection of light by the Sun* has been measured by VLBI and the values found are $(1 + \gamma)/2 = 0.99992 \pm 0.00023$. The historical development of the precision of these measurements from the original results of the eclipse expeditions by Eddington and his colleagues to the present day is shown in Fig. 8.12.
- The *time delay of electromagnetic waves* propagating through a varying gravitational potential was proposed by Irwin Shapiro (1964). While en route to Saturn, the Cassini spacecraft found a time delay corresponding to $(\gamma - 1) = (2.1 \pm 2.3) \times 10^{-5}$. Hence the coefficient $\frac{1}{2}(1 + \gamma)$ must be within at most 0.0012 percent of unity. The historical development of the precision of the time-delay measurements is also shown in Fig. 8.12.

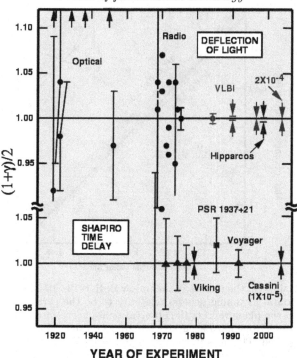

FIGURE 8.12. Measurements of the quantity $(1 + \gamma)/2$ from light deflection and time delay experiments. The value of $\gamma$ according to general relativity is unity. The arrows at the top of the diagram denote anomalously large values from early eclipse expeditions. The time-delay measurements from the Cassini spacecraft yielded agreement with general relativity at the level of $10^{-3}$ percent. Very long baseline interferometry (VLBI) radio deflection measurements have reached 0.02 percent accuracy. The *Hipparcos* limits were derived from precise measurements of the positions of stars over the whole sky and resulted in a precision of 0.1 percent in the measurement of $\gamma$. From Will (2006), with permission.

Some of the best tests of general relativity have been made using pulsars in binary neutron star systems. In these cases, the pulsar is one of a pair of neutron stars in binary orbits about their common centre of mass. There is a displacement between the axis of the magnetic dipole and the rotation axis of the neutron star. The radio pulses are assumed to be caused by beams of radio emission from the poles of the magnetic field distribution. Many parameters of the binary orbit and the masses of the neutron stars can be measured with very high precision by accurate timing measurements.

An example of the precision available for the pulsar PSR 1913+16, the first known example of a pulsar in a close binary neutron star system, is shown in Fig. 8.13 which shows the determination of the masses of the two neutron stars in the $M_1$–$M_2$ plane. The width of each strip reflects the observational uncertainties in the timing measurements, shown as a percentage. The inset shows the same three most accurate constraints on the full mass plane; the intersection region has been magnified 400 times in the full figure. If general relativity were not the correct theory of gravity, the lines would not intersect at one point. It is evident that general relativity passes this test rather successfully.

One of the most spectacular discoveries of this remarkable story is the fact that the binary pulsar emits gravitational radiation which leads to a speeding up of the stars in their binary orbit. Figure 8.14 shows the change of orbital phase as a function of time for the binary neutron star system PSR 1913+16 compared with the expected changes due

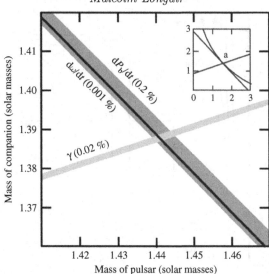

FIGURE 8.13. Constraints on the masses of the pulsar PSR 1913+16 and its invisible companion from precise timing data, assuming general relativity to be the correct theory of gravity. The width of each strip in the plane reflects the observational uncertainties, shown as a percentage. The inset shows the same three most accurate constraints on the full mass plane; the intersection region has been magnified 400 times in the full figure. From Will (2006), with permission.

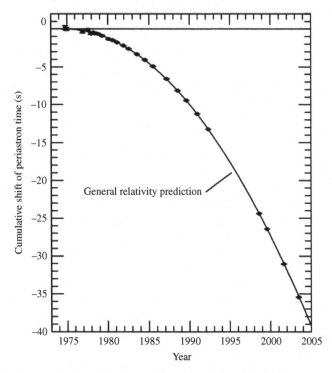

FIGURE 8.14. The change of orbital phase as a function of time for the binary neutron star system PSR 1913+16 compared with the expected changes due to gravitational radiation energy loss by the binary system (Taylor, 1992). Taken from Will (2006), with permission.

TABLE 8.3. The measured properties of the binary neutron star system
J0737-3039 compared with the predictions of general relativity

| PK parameter | Observed | GR expectation | Ratio |
|---|---|---|---|
| $\dot{P}_{\rm b}$ | 1.252(17) | 1.24787(13) | 1.003(14) |
| $\gamma$ (ms) | 0.3856(26) | 0.38418(22) | 1.0036(68) |
| $s$ | 0.99974(−39,+16) | 0.99987(−48,+13) | 0.99987(50) |
| $r(\mu s)$ | 6.21(33) | 6.153(26) | 1.009(55) |

TABLE 8.4. Upper limits to the rate of change of the
gravitational constant $G$ with cosmic epoch (Will, 2006)

| Method | $(\dot{G}/G)/10^{-13}$ year$^{-1}$ |
|---|---|
| Lunar laser ranging | $4 \pm 9$ |
| Binary pulsar PSR 1913+16 | $40 \pm 50$ |
| Helioseismology | $0 \pm 16$ |
| Big Bang nucleosynthesis | $0 \pm 4$ |

to gravitational radiation energy loss by the binary system. These observations enable
many alternative theories of gravity to be excluded.

In 2004, the most extreme relativistic binary system J0737-3039 was discovered with
an orbital period of 2.45 hours and a remarkably high value of its periastron advance,
$d\omega/dt = 16.9°$ year$^{-1}$ (Table 8.3). In this case, both neutron stars are observed as pulsars
and so their orbits can be determined with extraordinary precision (Lyne *et al.*, 2004). As
a result of energy loss by gravitational waves, the common orbit shrinks by 7 mm per day.
The two neutron stars will coalesce in about 85 million years. In 2.5 years, the precision
with which general relativity can be tested has approached that of PSR 1913+16.

### 8.6.4 Variations of the constants of physics with cosmic epoch

In Table 8.4, various Solar System, astrophysical and cosmological tests are listed,
which provide limits to the rate at which the gravitational constant could have varied
over cosmological timescales. For the binary pulsar, the bounds are dependent upon the
theory of gravity in the strong-field regime and on the neutron star equation of state.
Big Bang nucleosynthesis bounds assume a specific form for time dependence of $G$. It
is apparent that there can have been little variation in the value of the gravitational
constant over the last $10^{10}$ years.

There has been some debate about whether or not the fine-structure constant $\alpha$ has
changed very slightly with cosmic epoch from observations of fine-structure lines in large
redshift absorption line systems in quasars. In 2001, Webb and his colleagues reported a
significant decrease in the value of the fine structure constant at large redshifts, shown

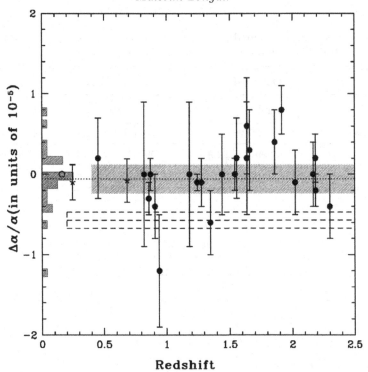

FIGURE 8.15. Measurements of the variation of the fine structure constant with cosmic epoch. The dashed box shows the range of values found by Webb and his colleagues (2001). The data points are from the results of Chand and his colleagues (2004). Figure adapted from Chand *et al.* (2004).

by the open rectangle in Fig. 8.15 (Webb *et al.*, 2001). Chand and his colleagues found little evidence for such a change (Chand *et al.*, 2004).

## 8.7 Inhomogeneous universes

One issue that will undoubtedly become important for precision cosmology is the importance of inhomogeneities in the large-scale distribution of matter. To illustrate the magnitude of the effects, we consider the simplest case of an inhomogeneous world model within the background of the critical Einstein–de Sitter world model, $\Omega_0 = 1, \Omega_\Lambda = 0$, for which the spatial geometry is flat, $\kappa = 0, \Re = \infty$.

Suppose the Universe were so inhomogeneous that all the matter was condensed into point-like objects. The following argument is due to Zeldovich (1964). Then, there is only a small probability that there will be any matter within the light cone subtended by a distant object of small angular size. Because of the long-range nature of gravitational forces, the background metric remains the standard flat Einstein–de Sitter metric

$$ds^2 = dt^2 - \frac{a^2(t)}{c^2} \left[ dr^2 + r^2(d\theta^2 + \sin^2\theta \, d\phi^2) \right]$$

$$= dt^2 - \frac{a^2(t)}{c^2} \left[ dx^2 + dy^2 + dz^2 \right],$$

where $a(t) = (t/t_0)^{\frac{2}{3}}$ and $t_0 = \frac{2}{3} H_0$ is the present age of the Universe; $r$, $x$, $y$ and $z$ are comoving coordinates referred to the present epoch $t_0$.

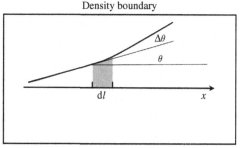

FIGURE 8.16. Illustrating the divergence of a light ray because of the "negative" mass, indicated by the grey shaded area, due to the absence of a disk of material in the interval d*l* within the light cone. Taken from fig. 7.12b in Longair (2008), with kind permission from Springer Science and Business Media.

In the model of an inhomogeneous Universe, we consider the propagation of the light rays in this background metric, but include in addition the effect of the absence of matter within the light cone subtended by the source at the observer. The angular deflection $\Delta\theta$ of a light ray by a point mass, or by an axially symmetric distribution of mass at the same distance, is

$$\Delta\theta = \frac{4GM(<p)}{pc^2}, \tag{8.32}$$

where $M(<p)$ is the mass within 'collision parameter' $p$, that is, the distance of closest approach of the light ray to the point mass.

Because of the principle of superposition, the effect of the "missing mass" within the light cone may be precisely found by supposing that the distribution of mass has negative density $-\rho(t)$ within the light cone. The deviations of the light cones from the homogeneous result, $\Delta\theta = dy/dx = $ constant, are due to the influence of the "negative mass" within the light cone. As a result, the light rays bend *outwards* rather than inwards, as in the usual picture (Fig. 8.16). The resulting deflection is

$$L = \frac{2c\Theta}{5H_0}\left[1 - (1+z)^{-5/2}\right]. \tag{8.33}$$

Corresponding results have been obtained for Friedman models with $\Omega_0 \neq 1$ by Dashevsky and Zeldovich (1964) and by Dyer and Roeder (1972, 1973). In these cases, if $\Omega_0 > 1$,

$$L = \frac{3c\Omega_0^2\Theta}{4H_0(\Omega_0-1)^{\frac{5}{2}}}\left[\sin^{-1}\left(\frac{\Omega_0-1}{\Omega_0}\right)^{\frac{1}{2}} - \sin^{-1}\left(\frac{\Omega_0-1}{\Omega_0(1+z)}\right)^{\frac{1}{2}}\right]$$
$$- \frac{3c\Omega_0\Theta}{4H_0(\Omega_0-1)^2}\left[1 - \frac{(1+\Omega_0z)^{\frac{1}{2}}}{(1+z)}\right] + \frac{1}{2(\Omega_0-1)}\left[1 - \frac{(1+\Omega_0z)^{\frac{1}{2}}}{(1+z)^2}\right].$$

If $\Omega_0 < 1$, the inverse trigonometric functions are replaced by inverse hyperbolic functions according to the rule $\sin^{-1}\mathrm{i}x = \mathrm{i}\sinh^{-1}x$.

Dyer and Roeder have presented the analytic results for intermediate cases in which a certain fraction $\alpha$ of the total mass density is uniformly distributed within the light cone. For the Einstein–de Sitter model, they find the simple result:

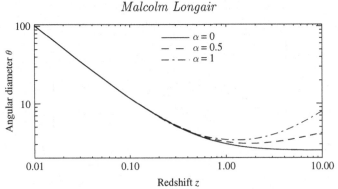

FIGURE 8.17. Comparison between the angular diameter–redshift relation in the homogeneous, uniform Einstein–de Sitter world model ($\alpha = 1$), the same background model in which there is no mass within the light cone subtended by the source ($\alpha = 0$) and the case in which half of the total mass is uniformly distributed and the rest is contained in point masses ($\alpha = 0.5$).

$$L = \Theta D_{\mathrm{A}} = \Theta \frac{2}{\beta}(1+z)^{(\beta-5)/4}[1-(1+z)^{-\beta/2}], \qquad (8.34)$$

where $\beta = (25 - 24\alpha)^{\frac{1}{2}}$. The minimum in the standard $\theta-z$ relation disappears in the maximally inhomogeneous model (Fig. 8.17). Thus, if no minimum is observed in the $\theta-z$ relation for a class of standard rods, it does not necessarily mean that the Universe must have $\Omega_0 \approx 0$. It might just mean that the Universe is of high density and is highly inhomogeneous.

If a minimum *is* observed in the $\theta-z$ relation, there must be matter within the light cone and limits can be set to the inhomogeneity of the matter distribution in the Universe. The effects upon the observed intensities of sources may be evaluated using the usual procedures, that is, the $\theta-z$ relation may be used to work out the fraction of the total luminosity of the source incident upon the observer's telescope using the reciprocity theorem. The end results are not so very different from those of the standard models, but remember, we are in the era of precision cosmology.

## 8.8 Determination of cosmological parameters

First of all, let us list the traditional set of cosmological parameters.
- *Hubble's constant*, $H_0$ – the present rate of expansion of the Universe,

$$H_0 = \left(\frac{\dot{a}}{a}\right)_{t_0} = \dot{a}(t_0). \qquad (8.35)$$

- The *deceleration parameter*, $q_0$ – the present dimensionless deceleration of the Universe.

$$q_0 = -\left(\frac{\ddot{a}a}{\dot{a}^2}\right)_{t_0} = -\frac{\ddot{a}(t_0)}{H_0^2}. \qquad (8.36)$$

- The *density parameter*, $\Omega_0$ – the ratio of the present mass-energy density of the Universe $\rho_0$ to the critical density $\rho_{\mathrm{c}} = 3H_0^2/8\pi G$,

$$\Omega_0 = \frac{\rho_0}{\rho_{\mathrm{c}}} = \frac{8\pi G \rho_0}{3H_0^2}. \qquad (8.37)$$

For many aspects of astrophysical cosmology, it is important to determine separately the density parameter in baryonic matter $\Omega_b$ and the overall density parameter $\Omega_0$, which includes all forms of baryonic and non-baryonic dark matter.

- The density parameter of the vacuum fields, or the dark energy,

$$\Omega_\Lambda = 8\pi G \rho_v / 3H_0^2 = \Lambda/3H_0^2, \tag{8.38}$$

where $\Lambda$ is the cosmological constant.
- The *curvature of space,*

$$\kappa = c^2/\Re^2. \tag{8.39}$$

- The *age of the Universe, $T_0$,*

$$T_0 = \int_0^1 \frac{da}{\dot{a}}. \tag{8.40}$$

Within the context of the Friedman world models, these are not independent parameters. Specifically, from Einstein's field equations for a uniformly expanding Universe,

$$\ddot{a} = -\frac{4\pi G}{3} a \left(\rho + \frac{3p}{c^2}\right) + \tfrac{1}{3}\Lambda a \; ; \tag{8.41}$$

$$\dot{a}^2 = \frac{8\pi G \rho}{3} a^2 - \frac{c^2}{\Re^2} + \tfrac{1}{3}\Lambda a^2 \; , \tag{8.42}$$

it is straightforward to show that

$$\kappa \left(\frac{c}{H_0}\right)^2 = (\Omega_0 + \Omega_\Lambda) - 1, \tag{8.43}$$

and

$$q_0 = \frac{\Omega_0}{2} - \Omega_\Lambda. \tag{8.44}$$

We should attempt to measure all the quantities independently and find out if these indeed describe our Universe.

At small redshifts the differences between the world models depend only upon the deceleration parameter and *not* upon the density parameter and the curvature of space. In order to relate observables to intrinsic properties, we need to know how the distance measure $D = \Re \sin(r/\Re)$ depends upon redshift and this involves two steps. First, we work out the dependence of the comoving radial distance coordinate $r$ upon redshift $z$ and then form the distance measure $D$.

Let us first carry out this calculation in terms of the *kinematics* of a world model decelerating with deceleration parameter $q_0$. We can write the variation of the scale factor $a$ with cosmic epoch in terms of a Taylor series:

$$a = a(t_0) + \dot{a}(t_0)\,\Delta t + \tfrac{1}{2}\ddot{a}(t_0)(\Delta t)^2 + \dots$$
$$= 1 - H_0\tau - \tfrac{1}{2}q_0 H_0^2 \tau^2 + \dots,$$

where we have introduced $H_0$, $q_0$ and the look-back time $\tau = t_0 - t = -\Delta t$; $t_0$ is the present epoch and $t$ is some earlier epoch.

It is straightforward to show that

$$\frac{1}{1+z} = 1 - x - \frac{q_0}{2}x^2 + \dots; \qquad z = x + \left(1 + \frac{q_0}{2}\right)x^2 + \dots. \tag{8.45}$$

The dependence upon the curvature only appears in third order in $z$ and so, to second order, we find the kinematic result

$$D = \left(\frac{c}{H_0}\right)\left[z - \frac{z^2}{2}(1 + q_0)\right]. \tag{8.46}$$

The distance measure $D$ is the quantity which appears in the relation between observables and intrinsic properties, $L = 4\pi D^2(1 + z)^2$ for bolometric luminosities.

From the full solution of the dynamical field equations, we obtain a similar result (see Longair, 2008, ch. 8 for details). Preserving quantities to third order in $z$, we find

$$D = \frac{c}{H_0}\left[z - \frac{z^2}{2}(1 + q_0) + \frac{z^3}{6}\left(3 + 6q_0 + 3q_0^2 - 3\Omega_0\right)\right]. \tag{8.47}$$

The identity of (8.46) and (8.47) to second order in redshift is apparent.

The list of cosmological parameters to be estimated is greater than the traditional list given above. In confronting theories of the formation of large-scale structure with the observations of the power spectrum of fluctuations in the CMB radiation and its polarisation, the extended list of parameters shown in Table 8.5 has been used. The revolution of the last 10 years has been that there are now numerous convincing independent approaches to estimating basic cosmological parameters. These include:

- Fluctuation spectrum and polarisation of the CMB radiation.
- The value of Hubble's constant from the HST Key Project and other methods.
- The $m-z$ relation for supernovae of type 1a.
- Mass density of the Universe from the infall velocities of galaxies into large-scale structures.
- The formation of the light elements by primordial nucleosynthesis.
- Cosmic timescale from the theory of stellar evolution and nucleocosmochronology.
- The power spectrum of galaxies from the Sloan Digital Sky Survey and the AAO 2dF galaxy survey.

### 8.8.1 Temperature fluctuations and polarization of the CMB radiation

There is no need to describe these observations and their interpretation at this Winter School. You should know by now what the "answers" are and their uncertainties. For completeness, I include as Figure 8.18, the remarkable three-year power spectrum and its polarisation from the WMAP experiment. These are really extraordinary observations.

### 8.8.2 Hubble's constant

The controversies of the 1970s and 1980s have been resolved thanks to a very large effort by many observers to improve knowledge of the distances to nearby galaxies. This approach can be thought of as the ultimate in the traditional 'distance indicator' approach to the determination of cosmological distances. The final result of the *Hubble space telescope* HST key project, led by Wendy Freedman was (Freedman *et al.*, 2001):

- $H_0 = 72 \pm 7$ (1-$\sigma$) km s$^{-1}$ Mpc$^{-1}$.

In addition to the traditional route to the determination of Hubble's constant, physical methods of estimating physical distances at large distance, which are independent of redshift, are now producing excellent results. An example is the use of the time delays in the *gravitational lensing* of distant variable quasars. A time delay of 418 days has been measured for the two components of the double quasar 0957+561 (Kundic *et al.*, 1997).

TABLE 8.5. List of cosmological parameters to be determined
from observations of the power spectrum and polarisation
of the CMB radiation

| Parameter | Definition |
|---|---|
| $\omega_{\rm b} = \Omega_{\rm b}h^2$ | Baryon density parameter |
| $\omega_{\rm cdm} = \Omega_{\rm cdm}h^2$ | Cold dark matter density parameter |
| $h$ | Hubble's constant |
| $\Omega_\Lambda$ | Dark energy density parameter |
| $n_{\rm s}$ | Scalar spectral index |
| $\tau$ | Reionisation optical depth |
| $\sigma_8$ | Density variance in 8 Mpc spheres |
| $w$ | Dark energy equation of state |
| $\Omega_k$ | Curvature density parameter |
| $f_\nu = \Omega_\nu/\Omega_{\rm cdm}$ | Massive neutrino fraction |
| $N_\nu$ | Number of relativistic neutrino species |
| $\Delta_{\mathcal{R}}^2$ | Amplitude of curvature perturbations |
| $r$ | Tensor-scalar ratio |
| $A_{\rm s}$ | Amplitude of scalar power-spectrum |
| $\alpha = {\rm d}\ln n_{\rm s}/{\rm d}\ln k$ | Running of scalar spectral index |
| $A_{\rm SZ}$ | SZ marginalization factor |
| $b$ | Bias factor |
| $z_{\rm s}$ | Weak lensing source redshift |

This observation enables physical scales at the lensing galaxy to be determined, the main uncertainty resulting from the modelling of the mass distribution in the lensing galaxy. Kundic and his colleagues derived a value of Hubble's constant of

- $H_0 = 64 \pm 13$ km s$^{-1}$ Mpc$^{-1}$

at the 95% confidence level.

A statistical analysis of 16 multiply imaged quasars by Oguri (2007) found the following value of Hubble's constant,

- $H_0 = 68 \pm 6\,({\rm stat}) \pm 8\,({\rm syst})$ km s$^{-1}$ Mpc$^{-1}$.

Another very promising approach is through the use of the *Sunyaev–Zeldovich effect* in conjunction with the X-ray emission from hot intracluster gas to determine the dimensions of the gas cloud, independent of its redshift. Bonamente and his colleagues studied 38 clusters of galaxies in the redshift interval $0.14 \le z \le 0.89$ using X-ray data from the Chandra X-ray Observatory and measurements of the corresponding Sunyaev–Zeldovich decrements from the Owens Valley Radio Observatory and the Berkeley–Illinois–Maryland Association interferometric arrays (Bonamente *et al.*, 2006). Their best estimate of Hubble's constant was:

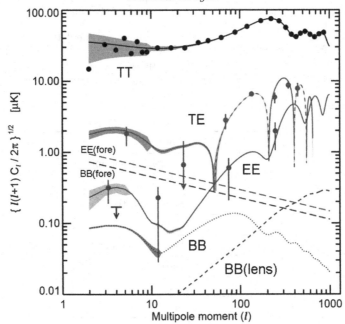

FIGURE 8.18. The measured power spectrum of fluctuations in the intensity and polarisation of the CMB radiation. Plots for the total intensity, the polarized intensity and the cross correlation between the total intensity and the polarized intensity are labelled TT, EE and TE respectively. The best-fitting model is shown by the corresponding lines. The dashed sections of the TE curve indicate multipoles in which the polarization signal is anticorrelated with the total intensity. The model predictions are binned in $\ell$ in the same way as the data. The binned EE polarization data are divided into bins of $2 \leq \ell \leq 5$, $6 \leq \ell \leq 49$, $50 \leq \ell \leq 199$, and $200 \leq \ell \leq 799$. The dotted line labelled BB shows the expected power spectrum of $B$-mode gravitational waves if the primordial ratio of tensor to scalar perturbations was $r = \Delta_t^2/\Delta_s^2 = 0.3$. The WMAP experiment found only upper limits to this signal, the $1\sigma$ upper limit corresponding to $0.17$ $\mu$K for the weighted average of multipoles $\ell = 2$–$10$. The $B$-mode signal due to gravitational lensing of the $E$-modes is shown as a dashed line labelled BB(lens). The upturn in the polarized signal at $\ell \leq 10$ is associated with polarization originating during the reionisation era. The foreground model for Galactic synchrotron radiation plus dust emission is shown as straight dashed lines labelled EE(fore) and BB(fore) for the scalar and tensor modes respectively, both being evaluated at $\nu = 65$ GHz. From Page *et al.* (2007). NASA and the WMAP science team. Reproduced with permission from the AAS.

- $H_0 = 76.9\,^{+3.9}_{-3.4}$ (stat) $^{+10.0}_{-8.0}$ (syst) km s$^{-1}$ Mpc$^{-1}$,

assuming $\Omega_0 = 0.3$ and $\Omega_\Lambda = 0.7$.

### 8.8.3 The redshift–magnitude relation

Traditionally, it was hoped that the redshift–magnitude relation for distant galaxies would provide a route to estimating cosmological parameters. However, this route has proved to be challenging because of the Galactic evolution and other astrophysical phenomena, which mask the effects due to the space-time geometry of the Universe. The example I know best is the work carried out by Simon Lilly and me in the early 1980s using the 3CR radio galaxies in the near-infrared wavebands when these were the only samples of galaxies which extended to redshifts greater than 1 (Lilly and Longair, 1984). We had to take account of the passive evolution of the stellar populations and that seemed to be quite a stable process (Fig. 8.19).

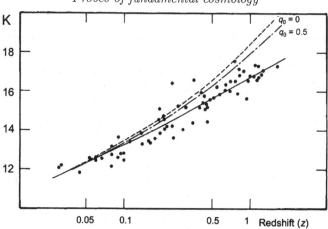

FIGURE 8.19. The K magnitude–redshift relation for a complete sample of narrow-line radio galaxies from the 3CR catalog. The infrared apparent magnitudes were measured at a wavelength of 2.2 µm. The dashed lines show the expectations of world models with $q_0 = 0$ and $\frac{1}{2}$. The solid line is a best-fitting line for standard world models with $\Omega_\Lambda = 0$ and includes the effects of stellar evolution of the old stellar population of the galaxies. From fig. 8 in Lilly and Longair (1984), with kind permission from Blackwell Publishing.

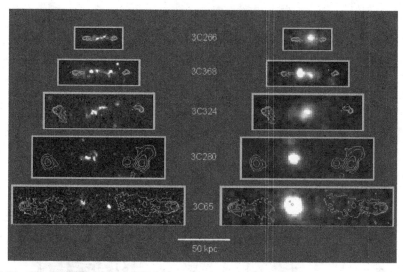

FIGURE 8.20. The HST (left) and UK Infrared telescope (right) images of the radio galaxies 3C 266, 368, 324, 280 and 65 with the VLA radio contours superimposed (Best *et al.*, 1996). The images are drawn on the same physical scale. The angular resolution of the HST images is 0.1 arcsec while that of the ground-based infrared images is about 1 arcsec. From fig. 1 in Best *et al.* (1996), with kind permission from Blackwell Publishing.

This approach was complicated by the discovery that the optical images of the galaxies at redshifts $z \sim 1$ are aligned with the radio structures, implying that the radio structures are influencing the total luminosities of the galaxies (Fig. 8.20). This effect is illustrated by our HST observations of the 3CR radio galaxies at redshifts $z \sim 1$ (Best *et al.*, 1996). In fact, in our most recent work, we found that, once account was taken of the alignment effect, the radio galaxies at $z \sim 1$–2 are in fact more luminous than those nearby (Inskip *et al.*, 2002).

Fortunately, the *type 1a supernovae* have turned out to be remarkably good standard candles for estimating cosmological distances. The type 1a supernovae are associated with the explosions of white dwarfs in binary systems. The explosion mechanism is probably a nuclear deflagration associated with mass transfer from the companion onto the surface of the white dwarf. The accretion process results in strong heating of the surface layers, which eventually become hot enough to initiate nuclear burning. This process is expected to result in a uniform class of explosions.

Type 1a supernovae have dominated methods for extending the redshift–distance relation to large redshifts. They are very luminous explosions and the standard luminosities become even more impressive when account is taken of the correlation between luminosity and the width of light curve. The results of the combined ESSENCE and SNLP projects are shown in Fig. 8.21 (Wood-Vasey *et al.*, 2007). The ESSENCE project has the objective of measuring the redshifts and distances of about 200 supernovae while the Supernova Legacy Project (SNLP) aims to obtain distances for about 500 supernovae. The observations are consistent with a finite and positive value of the cosmological constant. In Fig. 8.21, the luminosity distance modulus versus redshift diagram is plotted

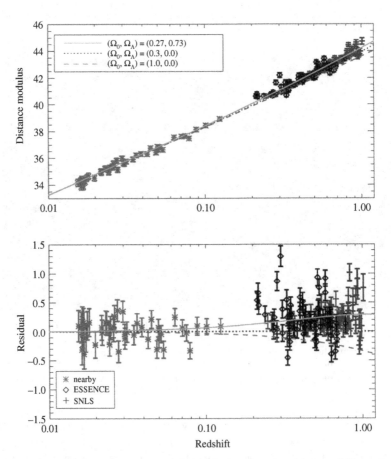

FIGURE 8.21. The luminosity distance–redshift relation for supernovae of Type 1a from the combined ESSENCE and Supernova Legacy Survey data. For comparison the overplotted solid line and residuals are for a $\Lambda$CDM model with $w = -1$, $\Omega_0 = 0.27$ and $\Omega_\Lambda = 0.73$. The dotted and dashed lines are for models with $\Omega_\Lambda = 0$, as indicated in the figure legend. Figure from Wood-Vasey *et al.* (2007). Reproduced by permission from the AAS.

for the ESSENCE, SNLS, and nearby SNe Ia. For comparison, the overplotted line and residuals are shown for a $\Lambda$CDM model with $w = -1$, $\Omega_0 = 0.27$ and $\Omega_\Lambda = 0.73$. The dashed line shows the expectation of the model with $\Omega_0 = 1$ and $\Omega_\Lambda = 0$ and the dotted line the model with $\Omega_0 = 0.3$ and $\Omega_\Lambda = 0$.

### 8.8.4 Redshift distortions due to infall into large-scale density perturbations

Galaxies in the vicinity of a supercluster are accelerated towards it, thus providing a measure of the mean over-density of gravitating matter within the system. The velocities induced by large-scale density perturbations depend upon the *density contrast* $\delta\rho/\rho$ between the system studied and the mean background density. A typical formula for the infall velocity $u$ of test particles into a density perturbation is (Gunn, 1978)

$$u \propto H_0 r \Omega_0^{0.6} \left( \frac{\delta\rho}{\rho} \right)_0 . \tag{8.48}$$

In the case of small spherical perturbations, a result correct to second order in the density perturbation was presented by Lightman and Schechter (1990),

$$\frac{\delta v}{v} = -\frac{1}{3} \Omega_0^{4/7} \left( \frac{\delta\rho}{\rho} \right)_0 + \frac{4}{63} \Omega_0^{13/21} \left( \frac{\delta\rho}{\rho} \right)_0^2 . \tag{8.49}$$

In Gunn's analysis, values of $\Omega_0 \sim 0.2$–$0.3$ were found, similar to those from studies of clusters of galaxies by Bahcall (2000) who used their virial masses.

The best approach to determining the mean density of matter in the Universe over large scales is to use the two-dimensional correlation function for galaxies in three dimensions. This has been carried out using 2dF Galaxy Redshift Survey by Peacock and his colleagues (2001). In Fig. 8.22, the flattening of the dimensions of the "sphere" in the vertical direction is due to the infall of galaxies into large-scale density perturbations. The elongations along the central vertical axis are associated with the velocity dispersion in clusters of galaxies. The inferred overall density parameter is $\Omega_0 = 0.25$, if $h = 0.7$, consistent with Bahcall's estimate.

An important issue is whether or not the visible light of galaxies traces the dark matter. Dekel and his colleagues devised numerical procedures for deriving the distribution of mass in the local Universe entirely from the measured velocities and distances of complete samples of nearby galaxies (Hudson *et al.*, 1995). Applying Poisson's equation, the mass distribution responsible for the observed peculiar velocity distribution can be reconstructed numerically (Fig. 8.23). Despite using only the velocities and distances, and *not* their number densities, many of the familiar features of our local Universe are recovered – the Virgo supercluster and the "Great Attractor" can be seen as well as voids in the mean mass distribution.

One of the most promising approaches to understanding the relation between the visible and dark matter is through the use of weak gravitational lensing for very large samples of galaxies by foreground galaxies. This provides one of the very few ways of determining directly by observation the extent of the dark matter haloes about galaxies (Schneider *et al.*, 2006). Indeed, one of the most exciting prospects for future observations is the use of weak gravitational lensing to map the large-scale distribution of dark matter in three dimensions, which can provide information both about the growth of dark matter structures and about the large-scale geometry of the Universe (see also chapter 4 by Matthias Bartelmann in this volume).

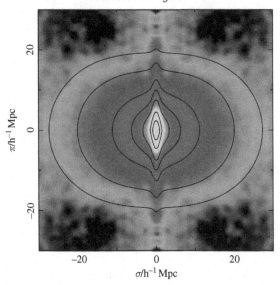

FIGURE 8.22. The two-dimensional correlation function for galaxies selected from the 2dF Galaxy Redshift Survey, $\xi(\sigma, \pi)$, plotted as a function of the inferred transverse ($\sigma$) and radial ($\pi$) pair separation. To illustrate deviations from circular symmetry, the data from the first quadrant have been repeated with reflection in both axes. This plot shows clearly the redshift distortions associated (a) with the "fingers of God" elongations along the central vertical axis and (b) the coherent Kaiser flattening of the correlation function in the radial direction at large radii (Peacock *et al.*, 2001). Image from the 2dF web page (http://www2.aao.gov.au/2dFGRS/).

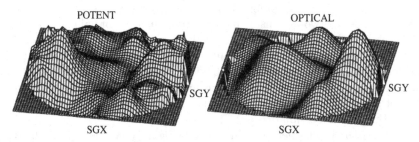

FIGURE 8.23. Surface density plots of the density field in the local Supergalactic (SG) plane. The left-hand panel shows the mass distribution reconstructed from the peculiar velocity and distance information for the galaxies in this region using the POTENT numerical procedure. The right-hand panel shows the density field of optical galaxies, both images smoothed with a Gaussian filter of radius 1200 km s$^{-1}$. The density contrast is proportional to the height of the surface above (or below) the plane of the plot. Taken from fig. 2 in Hudson *et al.* (1995), with kind permission from Blackwell Publishing.

### 8.8.5 Formation of the light elements by primordial nucleosynthesis

The predicted primordial abundances of the light elements according to the standard concordance model is a sensitive probe of the primordial baryon density parameter (Steigman *et al.*, 2004, 2007). The predicted abundances depend upon the present baryon-to-photon ratio which can be written in the form $\eta = 10^{10} n_{\rm b}/n_\gamma = 274\,\Omega_{\rm b}h^2$. The predictions are shown as a function of $\eta$ in Fig. 8.24. The widths of the bands reflect the theoretical uncertainties in the predictions.

Various approaches have been taken to determining the primordial abundances of the light elements and these are summarised graphically in Fig. 8.25 and in the figure caption.

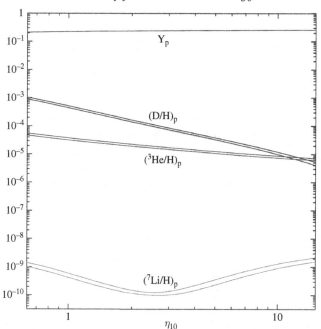

FIGURE 8.24. The predicted primordial abundances of the light elements as a function of the present baryon-to-photon ratio in the form $\eta = 10^{10} n_{\rm b}/n_\gamma = 274\,\Omega_{\rm b} h^2$. The parameter $Y_{\rm p}$ is the abundance of helium by mass, whereas the abundances for D, $^3$He and $^7$Li are plotted as ratios by number relative to hydrogen. The widths of the bands reflect the theoretical uncertainties in the predictions. The computations were carried out using Big Bang nucleosynthesis codes developed at Ohio State University (Steigman *et al.*, 2004, 2007).

Steigman finds a best-fitting value $\Omega_{\rm b} h^2 = 0.022^{+0.003}_{-0.002}$ (Steigman *et al.*, 2004, 2007). This can be compared with the independent estimate from the power spectrum of fluctuations of the CMB radiation $\Omega_{\rm b} h^2 = 0.0224 \pm 0.0009$ (Spergel *et al.*, 2007). This is one of the most remarkable convergences of two totally different approaches to the determination of density parameters in cosmology.

### 8.8.6 Cosmic timescales

The classic approach to the determination of cosmological timescales is to estimate the ages of the oldest globular clusters through the procedure of matching the observed Hertzsprung–Russell diagrams to theoretical isochrones. An example of the application of this method to the globular cluster 47 Tucanae is shown in Fig. 8.26 (Hesser *et al.*, 1987).

Similar analysis have been carried out by Bolte (1997) and Chaboyer (1998) who found the following values:

$$\text{Bolte}: T_0 = 15 \pm 2.4 \text{ (stat)} {}^{+4}_{-1} \text{ (syst) Gy}, \tag{8.50}$$

$$\text{Chaboyer}: T_0 = (11.5 \pm 1.3) \text{ Gy}. \tag{8.51}$$

Chaboyer also showed that these values are consistent with the cooling ages of white dwarfs.

Another approach to determining cosmic timescales is to use *nucleocosmochronology*. In this approach, the ages of old stars are determined from the abundances of radioactive

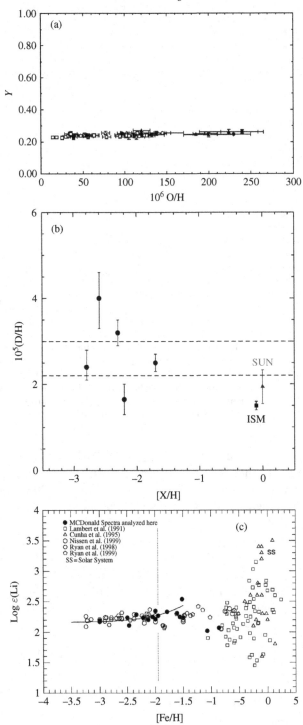

FIGURE 8.25

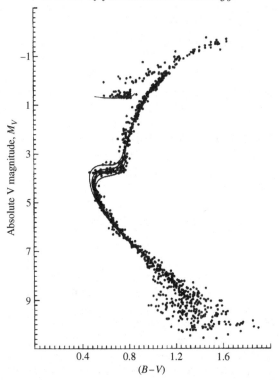

FIGURE 8.26. The Hertzsprung–Russell diagram for the globular cluster 47 Tucanae. The solid lines show fits to the data using theoretical models of the evolution of stars of different masses from the main sequence to the giant branch due to VandenBerg. The isochrones shown have ages of 10, 12, 14 and 16 $\times 10^9$ years, the best-fitting values lying in the range $(12-14) \times 10^9$ years. The cluster is metal-rich relative to other globular clusters, the metal abundance corresponding to about 20% of the solar value. Taken from Hesser *et al.* (1987).

species with lifetimes almost as long as the age of the Universe. The remarkable star CS 22892-052 has iron abundance 1000 times less than the solar value, and so is one of the oldest stars known (Sneden *et al.*, 1992). A number of species never previously observed in such metal-poor stars were detected, as well a single line of the radioactive element thorium. A lower limit to the age of the star was found to be

$$(15.2 \pm 3.7) \times 10^9 \text{ years.}$$

Using standard nuclear dating procedures, Schramm (1997) found a lower limit to the age of the Galaxy of $9.6 \times 10^9$ years, his best estimates of the age of the Galaxy

Caption for FIGURE 8.25. (a) The helium mass fraction $Y$ as a function of oxygen abundance for a large sample of low-metallicity extragalactic HII regions. (b) The cosmic deuterium abundance (D/H) as a function of the silicon abundance from observations of quasar absorption line systems (filled circles). Also shown are the D abundances for the local interstellar medium (ISM, filled square) and the Solar System (Sun, filled triangle). (c) A compilation of data on the cosmic lithium abundance from observations of metal-poor and metal-rich stars (For the references to the data points shown the reader is referred to the original paper). $\varepsilon$(Li) is defined to be $10^{12}$(Li/H) and [Fe/H] is the logarithmic metallicity relative to the standard solar value. The "Spite plateau" in (Li/H) is observed at metallicities less than 100 times the solar value, [Fe/H] $\leq -2$. Figure from Steigman *et al.* (2004).

being somewhat model-dependent, but typically ages of about $(12-14) \times 10^9$ years were inferred.

### 8.8.7 The Statistics of gravitational lensing

The statistics of gravitational lenses is an important approach to determining cosmological parameters because the numbers observed depend upon the stretching of distance measures due to the presence of the cosmological constant in Einstein's field equations. Figure 8.27, due to Carroll and his colleagues, shows the probability of strong gravitational lensing relative to the case of the Einstein–de Sitter model for different combinations of $\Omega_0$ and $\Omega_\Lambda$ (Carroll *et al.*, 1992). The probability for any other model becomes

$$p(z_{\mathrm{S}}) = \frac{15H_0^2}{4c^2} \left[ 1 - \frac{1}{(1+z_{\mathrm{S}})} \right]^{-3} \int_0^{z_{\mathrm{S}}} \left( \frac{D_{\mathrm{L}} D_{\mathrm{LS}}}{D_{\mathrm{S}}} \right)^2$$
$$\times \frac{(1+z)^2 \, \mathrm{d}z}{[(1+z)^2(\Omega_0 z + 1) - \Omega_\Lambda z(z+2)]^{\frac{1}{2}}}. \tag{8.52}$$

It can be seen that the numbers observed depends strongly on $\Omega_\Lambda$.

The largest survey to date designed specifically to address this problem has been the Cosmic Lens All Sky Survey (CLASS) in which a very large sample of flat-spectrum radio sources were imaged by the Very Large Array (VLA), Very Long Baseline Array (VLBA) and the MERLIN long-baseline interferometer. The CLASS collaboration has reported the point-source lensing rate to be one per $690 \pm 190$ targets (Mitchell *et al.*, 2005). The CLASS collaboration found that the observed fraction of multiply-lensed sources was consistent with flat world models, $\Omega_0 + \Omega_\Lambda = 1$, in which the density parameter in the matter $\Omega_0$ was

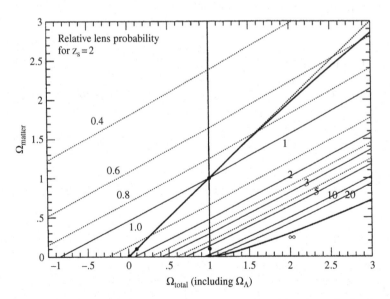

FIGURE 8.27. The probability of observing strong gravitational lensing relative to that of the critical Einstein–de Sitter model, $\Omega_0 = 1, \Omega_\Lambda = 0$ for a quasar at redshift $z_{\mathrm{S}} = 2$ (Carroll *et al.*, 1992). The contours show the relative probabilities derived from the integral (8.52). Reprinted, with permission, from the *Annual Review of Astronomy and Astrophysics*, Volume 30 (c) 1992 by Annual Reviews (www.annualreviews.org).

$$\Omega_0 = 0.31 \, {}^{+0.27}_{-0.14} (68\%) \, {}^{+0.12}_{-0.10} (\text{syst}). \tag{8.53}$$

Alternatively, for a flat universe with an equation of state for the dark energy of the form $p = w\rho c^2$, they found an upper limit to $w$,

$$w < -0.55 \, {}^{+0.18}_{-0.11} (68\%), \tag{8.54}$$

consistent with the standard value for the cosmological constant $w = -1$ .

### 8.8.8 Acoustic peaks in the galaxy power spectra

One of the key predictions of the standard concordance cosmological model is that the acoustic peaks seen in the power spectrum of the CMB radiation should have their counterpart in the correlation function for galaxies. The detection of the *baryon acoustic oscillations* requires huge samples of galaxies with known redshifts and this has now been achieved by the massive surveys of galaxies undertaken by the AAO 2dF survey (Cole *et al.*, 2005) and the Sloan Digital Sky Survey (Eisenstein *et al.*, 2005), the results of which are shown in Figs. 8.28 and 8.29.

The importance of these observations for the determination of cosmological parameters can be appreciated from the model fitting shown in Fig. 8.29. The predicted spectrum for a model with no baryonic matter contains no acoustic peak while models including baryonic matter have a clear baryonic signature. The distinction is that the baryons oscillated in the potential wells defined by the dark matter perturbations up to the epoch when the temperature perturbations were imprinted on the CMB radiation, while the dark matter perturbations grew linearly on all scales (see Section 8.12.1).

It is interesting that on smaller angular scales, the models which fit the data best include the results of semi-analytic modelling of the formation of galaxies within the dark matter gravitational potential wells, as is illustrated by the simulations shown in Fig. 8.30 (Springel *et al.*, 2005).

### 8.8.9 Properties of the concordance model

Given this wealth of data on the determination of cosmological parameters, it is quite remarkable that they are all consistent with a single *concordance cosmological model*. Adopting $H_0 = 70.1$ km s$^{-1}$ Mpc$^{-1}$, the self-consistent set of parameters listed in Table 8.6 can account for all the observations described above within the uncertainties of each determination (Komatsu *et al.*, 2008). The amazing result is that, granted how different all the techniques are, the cosmological parameters are now known to better than 10% accuracy. This is an extraordinary revolution – it has never happened before in the history of astrophysical cosmology. We certainly live in the era of precision cosmology. But this is also a huge challenge – we need to understand the physics to better than 10% to determine the cosmological parameters with improved precision.

There are many ways of combining the different data sets in order to understand how the data are constrained by different types of cosmological data. These calculations are available on the WMAP website[2] for the five-year WMAP data. It is well worth the effort to study these results in some detail to understand how well the concordance values are constrained by different sets of data.

---

[2] http://lambda.gsfc.nasa.gov/product/map/current/parameters.cfm

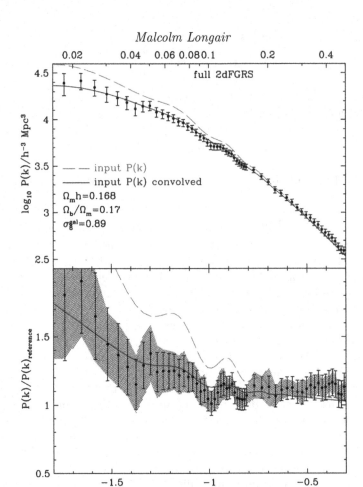

Figure 8.28. The power spectrum of the three-dimensional distribution of galaxies in the 2dF Galaxy Redshift Survey. The points with error bars are the best estimates of the observed power spectrum once the biases and corrections for incompleteness are taken into account. In the lower panel, the data from the upper panel have been divided by a reference cold dark matter model with $\Omega_{cdm} = 0.2$, $\Omega_\Lambda = 0$ and $\Omega_b = 0$ which has a smooth power spectrum. The grey dashed line is a best-fitting model before convolution with the window function for the survey. The solid line shows the best fit once the model is convolved with the window function. From fig. 12 in Cole *et al.*, 2005, with kind permission from Blackwell Publishing.

It is useful to illustrate the dynamics of the preferred cosmological model. The scale-factor–cosmic-time relation for the $\Omega_0 + \Omega_\Lambda = 1$ model is

$$ t = \int_{t_1}^{t_0} \mathrm{d}t = -\frac{1}{H_0} \int_{\infty}^{z} \frac{\mathrm{d}z}{(1+z)[\Omega_0(1+z)^3 + \Omega_\Lambda]^{\frac{1}{2}}} \, . \tag{8.55} $$

The results of carrying out this integration for a range of flat cosmological models is shown in Fig. 8.31. The model which is very close to the concordance model given in Table 8.6 is that corresponding to $\Omega_0 = 0.3$. It can be seen that, although the model was decelerating in the past, it is now in an accelerating phase and is tending towards exponential acceleration under the influence of the dark energy.

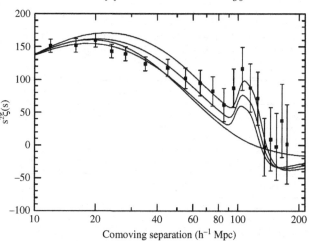

FIGURE 8.29. The large-scale redshift–space correlation function of the Sloan Digital Sky Survey Luminous Red Galaxy sample plotted as the correlation function times $s^2$. This presentation was chosen to show the curvature of the power spectrum at small physical scales. The models have $\Omega_0 h^2 = 0.12$ (top), 0.13 (middle), and 0.14 (bottom), all with $\Omega_b h^2 = 0.024$ and $n = 0.98$. The smooth line through the data with no acoustic peak is a pure cold dark matter model with $\Omega_0 h^2 = 0.105$. Taken from Eisenstein *et al.*, 2005. Reproduced by permission from the AAS.

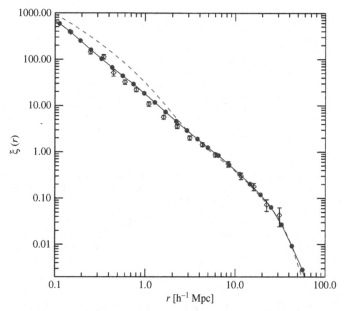

FIGURE 8.30. The two-point correlation function for galaxies at the present epoch (Springel *et al.*, 2005). Filled circles show estimates of the function for model galaxies brighter than $M_K = -23$ selected from the Millennium Simulation. The observed relation for the 2dF Galaxy Redshift Survey is shown as diamonds with error bars. Both the observational data and the simulated galaxies have correlation functions which are very close to power laws on scales $r \leq 20\,\mathrm{h}^{-1}$ Mpc. The correlation function for the dark matter alone, shown above by a dashed line, deviates significantly from a power law. Reprinted by permission from Macmillan Publishers Ltd: Nature (Springel *et al.*, 2005), copyright 2005.

TABLE 8.6. A concordance set of parameters for the origin of structure in the Universe. The values are for a flat $\Lambda$CDM model with power-law initial perturbation spectrum (Komatsu *et al.*, 2008).

| | |
|---|---|
| Hubble's constant | $H_0 = 70.1 \text{ km s}^{-1} \text{ Mpc}^{-1}$ |
| Curvature of space | $\Omega_\Lambda + \Omega_0 = 1$ |
| Baryonic density parameter | $\Omega_b = 0.046$ |
| Cold dark matter density parameter | $\Omega_{cdm} = 0.233$ |
| Total matter density parameter | $\Omega_0 = 0.279$ |
| Density parameter in vacuum fields | $\Omega_\Lambda = 0.721$ |
| Optical depth for Thomson scattering on reheating | $\tau = 0.084$ |
| Scalar spectral index | $n_s = 0.960$ |

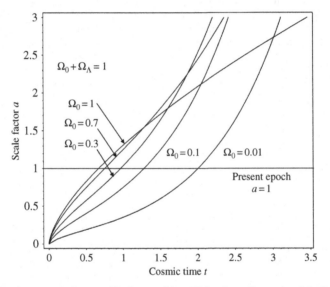

FIGURE 8.31. The dynamics of spatially flat world models, $\Omega_0 + \Omega_\Lambda = 1$, with different combinations of $\Omega_0$ and $\Omega_\Lambda$. The abscissa is plotted in units of $H_0^{-1}$. The concordance model corresponds closely to the model with $\Omega_0 = 0.3$. (Taken from fig. 7.5 in Longair (2008), with kind permission from Springer Science and Business Media).

For the world model with $\Omega_0 = 0.3$ and $\Omega_\Lambda = 0.7$, the age of the Universe is $0.964 H_0^{-1}$, which is remarkably close to $H_0^{-1}$. Thus, $T_0 \approx H_0^{-1}$, which is an interesting result. Is this a coincidence? According to the standard concordance model, the answer is "Yes."

## 8.9 Perturbations – linear and non-linear

The next challenge is to understand the physics of the development of perturbations in the standard picture in the linear and non-linear regimes. The aim of the cosmologist is to explain how large-scale structures formed in the expanding Universe in the sense that, if $\delta\rho$ is the enhancement in density of some region over the average background density $\rho$, the *density contrast* $\Delta = \delta\rho/\rho$ reached amplitude 1 from initial conditions which must have been remarkably isotropic and homogeneous. Once the initial perturbations have grown in amplitude to $\Delta = \delta\rho/\rho \approx 1$, their growth becomes non-linear and they rapidly evolve towards bound structures in which star formation and other astrophysical processes lead to the formation of galaxies and clusters of galaxies as we know them.

The density contrasts $\Delta = \delta\rho/\rho$ for galaxies, clusters of galaxies and superclusters at the present day are about $\sim 10^6$, 1000 and a few, respectively. Since the average density of matter in the Universe $\rho$ changes as $(1 + z)^3$, it follows that typical galaxies must have had $\Delta = \delta\rho/\rho \approx 1$ at a redshift $z \approx 100$. The same argument applied to clusters and superclusters suggests that they could not have separated out from the expanding background at redshifts greater than $z \sim 10$ and 1 respectively.

### 8.9.1 The wave equation for the growth of small density perturbations

The standard equations of gas dynamics for a fluid in a gravitational field consist of three partial differential equations which describe (i) the conservation of mass, or the equation of continuity; (ii) the equation of motion for an element of the fluid, Euler's equation; and (iii) the equation for the gravitational potential, Poisson's equation.

$$\text{Equation of continuity}: \frac{\partial \rho}{\partial t} + \nabla \cdot (\rho \boldsymbol{v}) = 0 \; ; \tag{8.56}$$

$$\text{Equation of motion}: \frac{\partial \boldsymbol{v}}{\partial t} + (\boldsymbol{v} \cdot \nabla)\boldsymbol{v} = -\frac{1}{\rho}\nabla p - \nabla\phi \; ; \tag{8.57}$$

$$\text{Gravitational potential}: \nabla^2\phi = 4\pi G\rho. \tag{8.58}$$

These equations describe the dynamics of a fluid of density $\rho$ and pressure $p$ in which the velocity distribution is $\boldsymbol{v}$. The gravitational potential $\phi$ at any point is given by Poisson's equation in terms of the density distribution $\rho$. The partial derivatives describe the variations of these quantities *at a fixed point in space*. This coordinate system is often referred to as *Eulerian coordinates*.

We perturb the system about the uniform expansion $\boldsymbol{v}_0 = H_0\boldsymbol{r}$:

$$\boldsymbol{v} = \boldsymbol{v}_0 + \delta\boldsymbol{v}, \quad \rho = \rho_0 + \delta\rho, \quad p = p_0 + \delta p, \quad \phi = \phi_0 + \delta\phi. \tag{8.59}$$

After a bit of algebra, we find the following equation for the peculiar velocity induced by the growth of the perturbation:

$$\frac{d}{dt}\left(\frac{\delta\rho}{\rho_0}\right) = \frac{d\Delta}{dt} = -\nabla \cdot \delta\boldsymbol{v} \; , \tag{8.60}$$

where $\Delta = \delta\rho/\rho_0$ is the density contrast. After a bit more algebra, we obtain the wave equation for $\Delta$

$$\frac{d^2\Delta}{dt^2} + 2\left(\frac{\dot{a}}{a}\right)\frac{d\Delta}{dt} = \Delta(4\pi G\rho_0 - k^2 c_\mathrm{s}^2), \tag{8.61}$$

where the adiabatic sound speed $c_\mathrm{s}^2$ is given by $\partial p/\partial\rho = c_\mathrm{s}^2$ and $k$ is the proper wavevector.

### 8.9.2 The Jeans' instability

The differential equation for gravitational instability in a static medium is obtained by setting $\dot{a} = 0$. For waves of the form $\Delta = \Delta_0 \exp \mathrm{i}(\boldsymbol{k} \cdot \boldsymbol{r} - \omega t)$, the dispersion relation,

$$\omega^2 = c_{\mathrm{s}}^2 k^2 - 4\pi G \rho_0, \tag{8.62}$$

is obtained.

- If $c_{\mathrm{s}}^2 k^2 > 4\pi G \rho_0$, the right-hand side is positive and the perturbations are oscillatory, that is, they are sound waves in which the pressure gradient is sufficient to provide support for the region. Writing the inequality in terms of wavelength, stable oscillations are found for wavelengths less than the critical *Jeans' wavelength* $\lambda_{\mathrm{J}}$

$$\lambda_{\mathrm{J}} = \frac{2\pi}{k_{\mathrm{J}}} = c_{\mathrm{s}} \left( \frac{\pi}{G\rho} \right)^{\frac{1}{2}}. \tag{8.63}$$

- If $c_{\mathrm{s}}^2 k^2 < 4\pi G \rho_0$, the right-hand side of the dispersion relation is negative, corresponding to unstable modes. The solutions can be written

$$\Delta = \Delta_0 \exp(\Gamma t + i\boldsymbol{k} \cdot \boldsymbol{r}), \tag{8.64}$$

where

$$\Gamma = \pm \left[ 4\pi G \rho_0 \left( 1 - \frac{\lambda_{\mathrm{J}}^2}{\lambda^2} \right) \right]^{\frac{1}{2}}. \tag{8.65}$$

The positive solution corresponds to exponentially growing modes. For wavelengths much greater than the Jeans' wavelength, $\lambda \gg \lambda_{\mathrm{J}}$, the characteristic growth time for the instability is

$$\tau = \Gamma^{-1} = (4\pi G \rho_0)^{-\frac{1}{2}} \sim (G\rho_0)^{-\frac{1}{2}}. \tag{8.66}$$

This is the famous *Jeans' instability* and the timescale $\tau$ is the typical collapse time for a region of density $\rho_0$.

### 8.9.3 *The Jeans' instability in an expanding medium*

We return to the differential equation for $\Delta$ in an expanding Universe,

$$\frac{\mathrm{d}^2 \Delta}{\mathrm{d}t^2} + 2 \left( \frac{\dot{a}}{a} \right) \frac{\mathrm{d}\Delta}{\mathrm{d}t} = \Delta (4\pi G \rho - k^2 c_{\mathrm{s}}^2). \tag{8.67}$$

The second term $2(\dot{a}/a)(\mathrm{d}\Delta/\mathrm{d}t)$ modifies the classical Jeans' analysis in crucial ways. It is apparent from the right-hand side of (8.67) that the Jeans' instability criterion applies in this case also but the growth rate is significantly modified. Let us work out the growth rate of the instability in the long-wavelength limit $\lambda \gg \lambda_{\mathrm{J}}$, in which case we can neglect the pressure term $c_{\mathrm{s}}^2 k^2$. We therefore have to solve the equation

$$\frac{\mathrm{d}^2 \Delta}{\mathrm{d}t^2} + 2 \left( \frac{\dot{a}}{a} \right) \frac{\mathrm{d}\Delta}{\mathrm{d}t} = 4\pi G \rho_0 \Delta. \tag{8.68}$$

Let us first consider the special cases $\Omega_0 = 1$ and $\Omega_0 = 0$ for which the scale-factor–cosmic-time relations are

$$a = (\tfrac{3}{2} H_0 t)^{2/3} \quad \text{and} \quad a = H_0 t, \tag{8.69}$$

respectively.

*The Einstein–de Sitter critical model* $\Omega_0 = 1$. In this case,

$$4\pi G \rho = \frac{2}{3t^2} \quad \text{and} \quad \frac{\dot{a}}{a} = \frac{2}{3t}. \tag{8.70}$$

We immediately find the key result

$$\boxed{\Delta = \frac{\delta\rho}{\rho} \propto a = (1+z)^{-1}.}$$ (8.71)

In contrast to the *exponential* growth found in the static case, the growth of the perturbation in the case of the critical Einstein–de Sitter universe is *algebraic*.

*The empty, Milne model* $\Omega_0 = 0$ In this case,

$$\rho = 0 \quad \text{and} \quad \frac{\dot{a}}{a} = \frac{1}{t},$$ (8.72)

and $\Delta = \text{constant}$.

In the early stages of the matter-dominated phase, the dynamics of the world models tend to $a \propto t^{\frac{2}{3}}$, and the density contrast grows linearly with $a$. At redshifts $\Omega_0 z \ll 1$, the amplitudes of the perturbations grow very slowly and tend to the empty Milne model.

### 8.9.4 Perturbing the Friedman solutions

Let us derive the same results from the dynamics of the Friedman models. The development of a spherical perturbation in the expanding Universe can be modelled by embedding a spherical region of density $\rho + \delta\rho$ in an otherwise uniform Universe of density $\rho$. The parametric solutions for the dynamics of the world models, or the spherical region, can be written

$$a = A(1 - \cos\theta) \qquad t = B(\theta - \sin\theta) \;;$$

$$A = \frac{\Omega_0}{2(\Omega_0 - 1)} \qquad B = \frac{\Omega_0}{2H_0(\Omega_0 - 1)^{\frac{3}{2}}}.$$

We now compare the dynamics of the region of slightly greater density with that of the background model. We expand the expressions for $a$ and $t$ to fifth order in $\theta$ with the result

$$a = \Omega_0^{1/3} \left(\frac{3H_0 t}{2}\right)^{\frac{2}{3}} \left[1 - \frac{1}{20}\left(\frac{6t}{B}\right)^{\frac{2}{3}}\right].$$ (8.73)

We can now write down an expression for the change of density of the spherical perturbation with cosmic epoch

$$\rho(a) = \rho_0 a^{-3} \left[1 + \frac{3}{5}\frac{(\Omega_0 - 1)}{\Omega_0} a\right].$$ (8.74)

Notice that, if $\Omega_0 = 1$, there is no growth of the perturbation. The density perturbation may be considered to be a mini-Universe of slightly higher density than $\Omega_0 = 1$ embedded in an $\Omega_0 = 1$ model, as illustrated in Fig. 8.32. Therefore, the density contrast changes with scale factor as

$$\Delta = \frac{\delta\rho}{\rho} = \frac{\rho(a) - \rho_0(a)}{\rho_0(a)} = \frac{3}{5}\frac{(\Omega_0 - 1)}{\Omega_0}a.$$ (8.75)

This result indicates why density perturbations grow only linearly with cosmic epoch. The instability corresponds to the slow divergence between the variation of the scale factors with cosmic epoch of the model with $\Omega_0 = 1$ and one with slightly greater density. This is the essence of the argument developed by Tolman and Lemaître in the 1930s and developed more generally by Lifshitz in 1946 to the effect that, because the instability develops only algebraically, galaxies could not have formed by gravitational collapse.

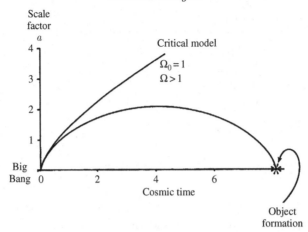

FIGURE 8.32. Illustrating the growth of a spherical perturbation in the expanding Universe as the divergence between two Friedman models with slightly different densities. (Taken from fig. 11.2 in Longair 2008, with kind permission from Springer Science and Business Media).

### 8.9.5 The general solutions

A general solution of (8.68) for the growth of the density contrast with scale factor for all pressure-free Friedman world models can be rewritten in terms of the density parameter $\Omega_0$ as follows:

$$\frac{\mathrm{d}^2\Delta}{\mathrm{d}t^2} + 2\left(\frac{\dot{a}}{a}\right)\frac{\mathrm{d}\Delta}{\mathrm{d}t} = \frac{3\Omega_0 H_0^2}{2}a^{-3}\Delta, \tag{8.76}$$

where, in general,

$$\dot{a} = H_0\left[\Omega_0\left(\frac{1}{a} - 1\right) + \Omega_\Lambda(a^2 - 1) + 1\right]^{\frac{1}{2}}. \tag{8.77}$$

The solution for the growing mode can be written as follows (Heath, 1977):

$$\Delta(a) = \frac{5\Omega_0}{2}\left(\frac{1}{a}\frac{\mathrm{d}a}{\mathrm{d}t}\right)\int_0^a \frac{\mathrm{d}a'}{(\mathrm{d}a'/\mathrm{d}t)^3}, \tag{8.78}$$

where the constants have been chosen so that the density contrast for the standard critical world model with $\Omega_0 = 1$ and $\Omega_\Lambda = 0$ has unit amplitude at the present epoch, $a = 1$. With this scaling, the density contrasts for all the examples we will consider correspond to $\Delta = 10^{-3}$ at $a = 10^{-3}$. It is simplest to carry out the calculations numerically for a representative sample of world models.

*Models with $\Omega_\Lambda = 0$.* The development of density fluctuations from a scale factor $a = 10^{-3}$ to $a = 1$ is shown for a range of world models with $\Omega_\Lambda = 0$ in Fig. 8.33. These results are consistent with the calculations carried out above, in which it was argued that the amplitudes of the density perturbations vary as $\Delta \propto a$ so long as $\Omega_0 z \gg 1$, but the growth essentially stops at smaller redshifts. For example, the density contrast in the model with $\Omega = 0.01$ only grows by a factor of 24 between $a = 10^{-3}$ and the present epoch $a = 1$.

*Models with finite $\Omega_\Lambda$.* The models of greatest interest are the flat models for which $(\Omega_0 + \Omega_\Lambda) = 1$, in all cases, the fluctuations having amplitude $\Delta = 10^{-3}$ at $a = 10^{-3}$. The growth of the density contrast is somewhat greater in the cases $\Omega_0 = 0.1$ and $0.3$ as compared with the corresponding cases with $\Omega_\Lambda = 0$ (Fig. 8.34). The fluctuations

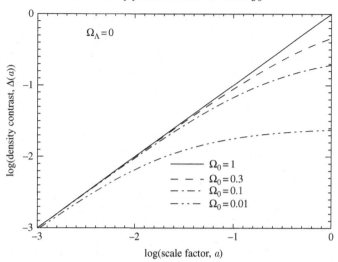

FIGURE 8.33. The growth of density perturbations over the range of scale factors $a = 10^{-3}$ to 1 for world models with $\Omega_\Lambda = 0$ and density parameters $\Omega_0 = 0.01$, 0.1, 0.3 and 1.

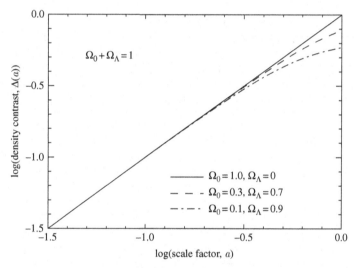

FIGURE 8.34. The growth of density perturbations over the range of scale factors $a = 1/30$ to 1 for world models with $\Omega_0 + \Omega_\Lambda = 1$ and density parameters $\Omega_0 = 0.1$, 0.3 and 1.

continue to grow to greater values of the scale factor $a$, corresponding to smaller redshifts, as compared with the models with $\Omega_\Lambda = 0$.

The reason for these differences is that, if $\Omega_\Lambda = 0$, the condition $\Omega_0 z = 1$ corresponds to the change from flat to hyperbolic geometry and so neighbouring geodesics diverge and the strength of the gravitational force is weakened relative to the flat case. In the case $\Omega_0 + \Omega_\Lambda = 1$, the geometry is forced to be Euclidean and so the growth continues until the repulsive effect of the $\Lambda$ term overwhelms the attractive force of gravity at small redshifts. The changeover takes place at much smaller redshifts, $(1 + z) \approx \Omega_0^{-1/3}$ if $\Omega_0 \ll 1$. As we will see, this is good news if we need to suppress the observed amplitude of fluctuations in the CMB radiation.

### 8.9.6 Peculiar velocities in the expanding universe

The development of velocity perturbations in the expanding Universe can be investigated for the case in which we neglect pressure gradients so that the velocity perturbations are only driven by the potential gradient $\delta\phi$.

$$\frac{d\boldsymbol{u}}{dt} + 2\left(\frac{\dot{a}}{a}\right)\boldsymbol{u} = -\frac{1}{a^2}\nabla_c\delta\phi. \tag{8.79}$$

In (8.79), $\boldsymbol{u}$ is the perturbed *comoving* velocity and $\nabla_c$ is the gradient in comoving coordinates. We split the velocity vectors into components parallel and perpendicular to the gravitational potential gradient, $\boldsymbol{u} = \boldsymbol{u}_\| + \boldsymbol{u}_\perp$, where $\boldsymbol{u}_\|$ is parallel to $\nabla_c\delta\phi$. The velocity associated with $\boldsymbol{u}_\|$ is often referred to as *potential motion* since it is driven by the potential gradient. On the other hand, the perpendicular velocity component $\boldsymbol{u}_\perp$ is not driven by potential gradients and corresponds to *vortex* or *rotational motions*.

*Rotational velocities.* Consider first the rotational component $\boldsymbol{u}_\perp$. The equation for the peculiar velocity reduces to

$$\frac{d\boldsymbol{u}_\perp}{dt} + 2\left(\frac{\dot{a}}{a}\right)\boldsymbol{u}_\perp = 0. \tag{8.80}$$

The solution of this equation is straightforward $\boldsymbol{u}_\perp \propto a^{-2}$. Since $\boldsymbol{u}_\perp$ is a comoving perturbed velocity, the proper velocity is $\delta\boldsymbol{v}_\perp = a\boldsymbol{u}_\perp \propto a^{-1}$. Thus, the rotational velocities decay as the Universe expands. This is no more than the conservation of angular momentum in an expanding medium, $mvr = \text{constant}$. This poses a grave problem for models of galaxy formation involving primordial turbulence. Rotational turbulent velocities decay and there must be sources of turbulent energy, if the rotational velocities are to be maintained.

*Potential motions.* The development of potential motions can be directly derived from the equation

$$\frac{d\Delta}{dt} = -\nabla \cdot \delta\boldsymbol{v}, \tag{8.81}$$

that is, the divergence of the peculiar velocity is proportional to minus the rate of growth of the density contrast. For the case $\Omega_0 = 1$,

$$|\delta v_\||| = |au| = \frac{H_0 a^{\frac{1}{2}}}{k}\left(\frac{\delta\rho}{\rho}\right)_0 = \frac{H_0}{k}\left(\frac{\delta\rho}{\rho}\right)_0 (1+z)^{\frac{-1}{2}}, \tag{8.82}$$

where $(\delta\rho/\rho)_0$ is the density contrast at the present epoch. Thus, $\delta v_\| \propto t^{1/3}$.

The peculiar velocities are driven both by the amplitude of the perturbation and its scale. Equation (8.82) shows that, if $\delta\rho/\rho$ is the same on all scales, the peculiar velocities are driven by the smallest values of $k$, that is, by the perturbations on the largest physical scales. This is an important result for understanding the origin of the peculiar motion of the Galaxy with respect to the frame of reference in which the microwave background radiation is 100% isotropic and of large-scale streaming velocities.

### 8.9.7 The relativistic case

In the radiation-dominated phase of the Big Bang, the primordial perturbations are in a radiation-dominated plasma, for which the relativistic equation of state $p = \frac{1}{3}\varepsilon$ is appropriate. The equation of energy conservation becomes

$$\frac{\partial \rho}{\partial t} = -\nabla \cdot \left( \rho + \frac{p}{c^2} \right) \boldsymbol{v};$$ (8.83)

$$\frac{\partial}{\partial t} \left( \rho + \frac{p}{c^2} \right) = \frac{\dot{p}}{c^2} - \left( \rho + \frac{p}{c^2} \right) (\nabla \cdot \boldsymbol{v}).$$ (8.84)

Substituting $p = \frac{1}{3}\rho c^2$ into (8.83) and (8.84), the relativistic continuity equation is obtained:

$$\frac{\mathrm{d}\rho}{\mathrm{d}t} = -\tfrac{4}{3}\rho(\nabla \cdot \boldsymbol{v}).$$ (8.85)

Euler's equation for the acceleration of an element of the fluid in the gravitational potential $\phi$ becomes

$$\left( \rho + \frac{p}{c^2} \right) \left[ \frac{\partial \boldsymbol{v}}{\partial t} + (\boldsymbol{v} \cdot \nabla)\boldsymbol{v} \right] = -\nabla p - \left( \rho + \frac{p}{c^2} \right) \nabla \phi.$$ (8.86)

If we neglect the pressure gradient term, (8.86) reduces to the familiar equation

$$\frac{\mathrm{d}\boldsymbol{v}}{\mathrm{d}t} = -\nabla \phi.$$ (8.87)

Finally, the differential equation for the gravitational potential $\phi$ becomes

$$\nabla^2 \phi = 4\pi G \left( \rho + \frac{3p}{c^2} \right).$$ (8.88)

For a fully relativistic gas, $p = \frac{1}{3}\rho c^2$ and so

$$\nabla^2 \phi = 8\pi G \rho.$$ (8.89)

The net result is that the equations for the evolution of the perturbations in a relativistic gas are of similar mathematical form to the non-relativistic case. The same type of analysis that was carried out above leads to the following equation

$$\boxed{\frac{\mathrm{d}^2 \Delta}{\mathrm{d}t^2} + 2 \left( \frac{\dot{a}}{a} \right) \frac{\mathrm{d}\Delta}{\mathrm{d}t} = \Delta \left( \frac{32\pi G \rho}{3} - k^2 c_{\mathrm{s}}^2 \right).}$$ (8.90)

The relativistic expression for the Jeans' length is found by setting the right-hand side equal to zero,

$$\lambda_{\mathrm{J}} = \frac{2\pi}{k_{\mathrm{J}}} = c_{\mathrm{s}} \left( \frac{3\pi}{8G\rho} \right)^{\frac{1}{2}},$$ (8.91)

where $c_{\mathrm{s}} = c/\sqrt{3}$ is the relativistic sound speed. The result is similar to the standard expression for the Jeans' length.

Neglecting the pressure-gradient terms in (8.90), the following differential equation for the growth of the instability is obtained

$$\frac{\mathrm{d}^2 \Delta}{\mathrm{d}t^2} + 2 \left( \frac{\dot{a}}{a} \right) \frac{\mathrm{d}\Delta}{\mathrm{d}t} - \frac{32\pi G \rho}{3} \Delta = 0.$$ (8.92)

We again seek solutions of the form $\Delta = at^n$, recalling that in the radiation-dominated phases, the scale-factor–cosmic-time relation is given by $a \propto t^{1/2}$. We find solutions $n = \pm 1$. Hence, for wavelengths $\lambda \gg \lambda_{\mathrm{J}}$, the growing solution corresponds to

$$\Delta \propto t \propto a^2 \propto (1+z)^{-2}.$$ (8.93)

Thus, once again, the unstable mode grows algebraically with cosmic time.

*8.9.8 The basic problem of structure formation*

Let us summarise the implications of the key results derived above. Throughout the matter-dominated era, the growth rate of perturbations on physical scales much greater than the Jeans' length is

$$\Delta = \frac{\delta\rho}{\rho} \propto a = \frac{1}{1+z}. \tag{8.94}$$

Since galaxies and astronomers certainly exist at the present day, $z = 0$, it follows that $\Delta \geq 1$ at $z = 0$ and so, at the last scattering surface, $z \sim 1000$, fluctuations must have been present with amplitude at least $\Delta = \delta\rho/\rho \geq 10^{-3}$. This result contains both bad and good news.

- The slow growth of density perturbations is the source of a fundamental problem in understanding the origin of galaxies – large-scale structures did not condense out of the primordial plasma by the exponential growth of infinitesimal statistical perturbations.
- Because of the slow development of the density perturbations, we have the opportunity of studying the formation of structures on the last scattering surface at a redshift $z \sim 1000$.

## 8.10 The thermal history of the Universe

Let us deal very briefly with the thermal history of the Universe and key physical effects which play a role in the overall picture. Figure 8.35 summarizes a number of well-known epochs in the thermal history of the Universe. We recall that, overall, the temperature history of the CMB radiation is adiabatic $T_{\mathrm{r}} \propto a^{-1}$ except at epochs when there is injection of energy into the pre- or intergalactic gas. Of special importance are *the epoch of recombination* at $z \approx 1000$ and *the epoch of equality of matter and radiation,* including neutrinos at $z \approx 3530$. These important topics have been discussed in other lectures during this course.

*8.10.1 The sound speed as a function of cosmic epoch*

The speed of sound $c_{\mathrm{s}}$ is given by

$$c_{\mathrm{s}}^2 = \left(\frac{\partial p}{\partial \rho}\right)_S, \tag{8.95}$$

where the subscript $S$ means 'at constant entropy', that is, adiabatic sound waves. The dominant contributors to $p$ and $\rho$ change dramatically as the Universe changes from being radiation- to matter-dominated. The sound speed can then be written

$$c_{\mathrm{s}}^2 = \frac{(\partial p/\partial T)_{\mathrm{r}}}{(\partial \rho/\partial T)_{\mathrm{r}} + (\partial \rho/\partial T)_{\mathrm{m}}}, \tag{8.96}$$

where the partial derivatives are taken at constant entropy. This reduces to the following expression:

$$c_{\mathrm{s}}^2 = \frac{c^2}{3} \frac{4\rho_{\mathrm{r}}}{4\rho_{\mathrm{r}} + 3\rho_{\mathrm{m}}}. \tag{8.97}$$

Thus, in the radiation-dominated era, the speed of sound tends to the relativistic sound speed, $c_{\mathrm{s}} = c/\sqrt{3}$.

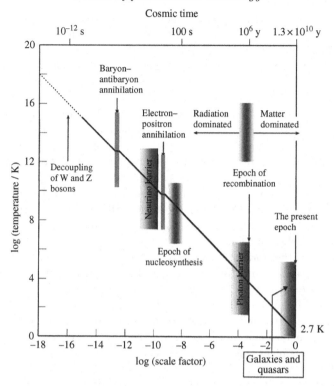

FIGURE 8.35. The thermal history of the CMB radiation according to the standard Big Bang picture. The radiation temperature decreases as $T_r \propto a^{-1}$ except for small discontinuities as different particle–antiparticle pairs annihilate at $kT \approx mc^2$. Various important epochs in the standard picture are indicated, including the neutrino and photon barriers. In the standard model, the Universe is optically thick to neutrinos and photons prior to these epochs. An approximate timescale is indicated along the top of the diagram. Taken from fig. 9.3 in Longair 2008, with kind permission from Springer Science and Business Media.

### 8.10.2 The damping of sound waves

Although the matter and radiation are closely coupled throughout the pre-recombination era, the coupling is not perfect and radiation can diffuse out of the density perturbations. Since the radiation provides the restoring force for support for the perturbation, the perturbation is damped out if the radiation has time to diffuse out of it. This process is often referred to as *Silk damping*.

At any epoch, the mean free path for scattering of photons by electrons is $\lambda = (N_e \sigma_T)^{-1}$, where $\sigma_T = 6.665 \times 10^{-29}$ m$^2$ is the Thomson cross section. The distance that the photons can diffuse is

$$r_D \approx (Dt)^{1/2} = \left(\tfrac{1}{3}\lambda c t\right)^{1/2}, \qquad (8.98)$$

where $t$ is cosmic time. The baryonic mass within this radius, $M_D = (4\pi/3)r_D^3 \rho_B$, can therefore be evaluated for the pre-recombination era.

## 8.11 Particle horizons, the Hubble sphere and superhorizon scales

At early times during the matter-dominated era, all the Friedman models tend toward the dynamics of the critical model and the particle horizon is $r_H(t) = 3ct$. One might have expected the typical distance that light could have travelled by the epoch $t$ to be of order $ct$. The factor 3 takes account of the fact that the world lines of galaxies were closer together at early epochs and so distances greater than $ct$ could be causally connected. A similar calculation for the radiation-dominated era gives the result $r_H(t) = 2ct$.

Equally important for astrophysical cosmology is the *Hubble sphere*. This is the distance at which $v = c$ according to Hubble's law $v = Hr$ at any epoch, where $H = \dot{a}/a$. This is the distance over which causal phenomena can take place *at a particular epoch*. The distinction between the particle horizon and the Hubble sphere is important – for causal phenomena at a particular epoch, such as baryon oscillations, the Hubble sphere is the maximum scale, whereas the maximum scale over which causal connection could have taken place to solve the isotropy problem is determined by the particle horizon.

The particle horizon shrinks to vanishingly small values as cosmic, or conformal, time tends to zero. At early enough epochs the horizon becomes smaller than the scales of galaxies, clusters of galaxies and other large-scale structures. We cannot avoid tackling the problem of what happens to perturbations on scales greater than the particle horizon, which are called *superhorizon scales*.

On scales less than the horizon scale there was always an "unperturbed background," which acts as a reference frame for the growth of small perturbations. We are also able to synchronise the clocks of all fundamental observers within the particle horizon. If, however, the scale of the perturbation exceeds the horizon scale, what do we mean by the "unperturbed background"? Each perturbation then carries its own clock and the whole issue of the synchronization of clocks and the selection of the appropriate reference frame provide real technical challenges. This leads to the problem of the *choice of gauge in general relativity*.

To cut a long story short, one can work in a number of different gauges and, provided one does the sums correctly, one obtains the same result in any gauge. For example, in the *conformal Newtonian gauge* or *longitudinal gauge*, the perturbations are described by the single scalar function $\phi$ which is just the Newtonian gravitational potential. The metric then has the form

$$ds^2 = a^2(\tau)[(1 + 2\phi)d\tau^2 - (1 - 2\phi)(dx^2 + dy^2 + dz^2)], \qquad (8.99)$$

where $\tau$ is conformal time. The Newtonian gravitational potential $\phi$ provides an accurate description of cosmological perturbations on scales greater than the particle horizon.

An alternative gauge is the *synchronous gauge* in which the term $(1 + 2\phi)$ multiplying $d\tau^2$ is not present, with consequent changes of the other components of the metric. The metric in the synchronous gauge can be written

$$ds^2 = a^2(\tau)\{d\tau^2 - [(1 + 2D)\delta_{ij} + 2E_{ij}]\,dx^i dx^j\}, \qquad (8.100)$$

This different slicing through space-time illustrates the point that the appearance of the metric depends upon the choice of gauge, although they all contain the same physics in the end.

An example of the difference the choice of gauge makes can be seen in the instructive diagrams presented by Ma and Bertschinger (1995) which shows the development of the same set of perturbations in the conformal Newtonian and synchronous gauges (Fig. 8.36). These diagrams reinforce the point that the development of density perturbations can appear very different in the two gauges on superhorizon scales, because of the different slicings through space-time. Once the perturbations came through their particle horizons, however, the same evolution is found for all five matter and radiation components.

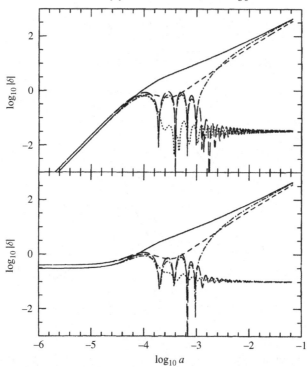

FIGURE 8.36. The evolution of the perturbations in the conformal Newtonian (lower panel) and synchronous (upper panel) gauges. These are mixed hot and cold dark matter models with $\Omega_\nu = 0.2$ and flat geometry. The different lines show the development of perturbations for cold dark matter (solid line), baryons (dot–dashed line), photons (long dashed line), massless neutrinos (dotted line) and massive neutrinos (short dashed lines). For $k \geq 0.1$ Mpc$^{-1}$, the perturbations came through the horizon before the epoch of equality. Taken from Ma and Bertschinger (1995). Reproduced with permission from the AAS.

## 8.12 The cold dark matter scenario of structure formation

Let us build up the necessary components of what has become the standard *cold dark matter* (CDM) picture of structure formation. We begin with the simple baryonic model, which includes many of the features that will reappear in the standard $\Lambda$CDM picture. Figure 8.37 shows how the horizon mass $M_{\rm H}$, the Jeans mass $M_{\rm J}$ and the Silk mass $M_{\rm D}$ change with scale factor $a$ in a purely baryonic picture.

When the perturbations come through the horizon, they almost immediately become sound waves with the pressure support provided by the radiation, which is strongly coupled to the baryonic plasma by Compton scattering. Sunyaev and Zeldovich realised that these sound waves would leave their imprint of the last scattering surface at the epoch of recombination. Their "stability diagram" shown in Fig. 8.38a illustrates the evolution of perturbations on different mass scales (Sunyaev and Zeldovich, 1970a). Because the perturbations have different phase lengths between coming through the horizon and arriving at the epoch of recombination, the phenomenon of *baryon acoustic oscillations* is expected to be imprinted on the last scattering surface at $z \approx 1000$ (Fig. 8.38b).

The problem with this model was that the temperature fluctuations on the last scattering surface were expected to be at least $\Delta T / T \sim 10^{-3}$, far in excess of the observed limits. The solution to this problem came with the realisation that the dark matter is the dominant contribution to $\Omega_0$. To understand how the dark matter helps resolve this

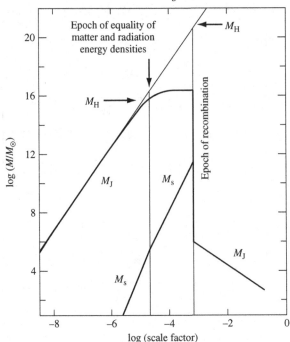

FIGURE 8.37. The evolution of the baryonic Jeans' mass $M_J$ and the baryonic mass within the particle horizon $M_H$ with scale factor. Also shown is the evolution of the Silk mass $M_S$, which was damped by photon diffusion, or Silk damping. (Taken from fig. 12.4 in Longair, 2008, with kind permission from Springer Science and Business Media).

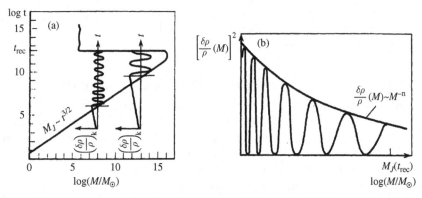

FIGURE 8.38. The "stability diagram" of Sunyaev and Zeldovich. (a) The region of instability is to the right of the solid line. The two additional graphs illustrate the evolution of density perturbations of different masses as they come through the horizon up to the epoch of recombination. (b) Perturbations corresponding to different masses arrive at the epoch of recombination with different phases, resulting in a periodic dependence of the amplitude of the perturbations upon mass. Taken from Sunyaev and Zeldovich (1970a), with kind permission from Springer Science and Business Media.

problem, we need to understand the development of the instabilities in the presence of dark matter.

Neglecting the internal pressure of the fluctuations, the expressions for the density contrasts in the baryons and the dark matter, $\Delta_b$ and $\Delta_{cdm}$ respectively, can be written as a pair of coupled equations

$$\ddot{\Delta}_b + 2\left(\frac{\dot{a}}{a}\right)\dot{\Delta}_b = A\rho_b\Delta_b + A\rho_{cdm}\Delta_{cdm}, \tag{8.101}$$

$$\ddot{\Delta}_{cdm} + 2\left(\frac{\dot{a}}{a}\right)\dot{\Delta}_{cdm} = A\rho_b\Delta_b + A\rho_{cdm}\Delta_{cdm}. \tag{8.102}$$

Let us find the solution for the case in which the dark matter has $\Omega_0 = 1$ and the baryon density is negligible compared with that of the dark matter. Then (8.102) reduces to the equation for which we have already found the solution $\Delta_{cdm} = Ba$ where $B$ is a constant. Therefore, the equation for the evolution of the baryon perturbations becomes

$$\ddot{\Delta}_b + 2\left(\frac{\dot{a}}{a}\right)\dot{\Delta}_b = 4\pi G\rho_{cdm}Ba. \tag{8.103}$$

Since the background model is the critical model, (8.103) simplifies to

$$a^{3/2}\frac{d}{da}\left(a^{-1/2}\frac{d\Delta}{da}\right) + 2\frac{d\Delta}{da} = \tfrac{3}{2}B. \tag{8.104}$$

The solution, $\Delta = B(a - a_0)$, satisfies (8.104). Thus, suppose that, at some redshift $z_0$, the amplitude of the baryon fluctuations is very small, very much less than that of the perturbations in the dark matter. The above result shows how the amplitude of the baryon perturbation develops in the presence of the larger-amplitude perturbation in the dark matter. In terms of redshift we can write

$$\Delta_b = \Delta_{cdm}\left(1 - \frac{z}{z_0}\right). \tag{8.105}$$

Thus, the amplitude of the perturbations in the baryons grows rapidly to the same amplitude as that of the dark matter perturbations. The baryons fall into the dark matter perturbations and rapidly attain amplitudes the same as those of the dark matter.

The resulting evolution of the density perturbations in cold dark matter, the baryonic matter and the radiation are illustrated in Fig. 8.39, which is taken from the presentation by Coles and Lucchin (1995). This diagram shows the evolution of density perturbations

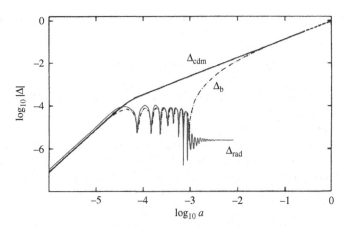

FIGURE 8.39. Illustrating the evolution of density perturbations in the baryonic matter $\Delta_b$, the radiation $\Delta_{rad}$ and the dark matter $\Delta_{cdm}$ according to the cold dark matter scenario. The mass of the perturbation is $M \sim 10^{15}\,M_\odot$. This diagram shows how structure develops in a cold dark matter dominated Universe. The amplitudes of the baryonic perturbations were very much smaller than those in the cold dark matter at the epoch of recombination. From Coles and Lucchin (1995).

of mass $M \sim 10^{15} M_\odot$. The dark matter perturbations on this scale continue to grow as soon as they come through the horizon, but the perturbations in the photon-dominated plasma oscillate as sound waves until the matter and radiation decouple at the epoch of recombination. The baryonic perturbations then grow to the same amplitude as the dark matter perturbations soon after recombination. Note, however, the key point that at the epoch of recombination the coupled baryon–photon perturbations are about an order of magnitude less than those in the dark matter. This resolves the problem of the excessive amplitude of the predicted temperature perturbations according to the baryonic theory of structure formation.

## 8.13 The initial power spectrum

The next task is to work out the initial power spectrum of the primordial perturbations. The smoothness of the two-point correlation function for galaxies suggests that the spectrum of initial fluctuations must have been very broad with no preferred scales and it is therefore natural to begin with a power spectrum of power-law form

$$P(k) = |\Delta_k|^2 \propto k^n. \tag{8.106}$$

The correlation function $\xi(r)$ should then have the form

$$\xi(r) \propto \int \frac{\sin kr}{kr} k^{(n+2)} \, dk. \tag{8.107}$$

Because the function $\sin(kr)/kr$ has value unity for $kr \ll 1$ and decreases rapidly to zero when $kr \gg 1$, we can integrate $k$ from 0 to $k_{\max} \approx 1/r$ to estimate the dependence of the amplitude of the correlation function on the scale $r$.

$$\xi(r) \propto r^{-(n+3)}. \tag{8.108}$$

Since the mass of the fluctuation is proportional to $r^3$, this result can also be written in terms of the mass within the fluctuations on the scale $r$, $M \sim \rho r^3$.

$$\xi(M) \propto M^{-(n+3)/3}. \tag{8.109}$$

Finally, to relate $\xi$ to the root-mean-square density fluctuation on the mass scale $M$, $\Delta(M)$, we take the square root of $\xi$, that is,

$$\Delta(M) = \frac{\delta\rho}{\rho}(M) = \langle\Delta^2\rangle^{1/2} \propto M^{-(n+3)/6}. \tag{8.110}$$

This spectrum has the important property that the density contrast $\Delta(M)$ had the same amplitude on all scales when the perturbations came through their particle horizons, *provided* $n = 1$. Let us illustrate how this comes about.

Before the perturbations came through their particle horizons and before the epoch of equality of matter and radiation energy densities, the density perturbations grew as $\Delta(M) \propto a^2$, although the perturbation to the gravitational potential was frozen-in. Therefore, the development of the spectrum of density perturbations can be written

$$\Delta(M) \propto a^2 M^{-(n+3)/6}. \tag{8.111}$$

A perturbation of scale $r$ came through the horizon when $r \approx ct$, and so the mass of dark matter within it was $M_{\rm cdm} \approx \rho_{\rm cdm}(ct)^3$. During the radiation-dominated phases, $a \propto t^{1/2}$ and the number density of dark matter particles, which will eventually form bound structures at $z \sim 0$, varied as $N_{\rm cdm} \propto a^{-3}$. Therefore, the horizon dark matter mass

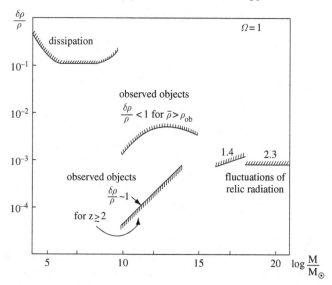

FIGURE 8.40. The diagram presented by Sunyaev and Zeldovich showing the constraints on the amplitude of primordial perturbations when they came through the horizon in order to create the large scale structures observed in the Universe today. Taken from Sunyaev and Zeldovich (1970b), with kind permission from Springer Science and Business Media.

increased as $M_{\mathrm{H}} \propto a^3$, or, $a \propto M_{\mathrm{H}}^{1/3}$. The mass spectrum $\Delta(M)_{\mathrm{H}}$ when the fluctuations came through the horizon at different cosmic epochs was

$$\Delta(M)_{\mathrm{H}} \propto M^{2/3} \, M^{-(n+3)/6} = M^{-(n-1)/6}. \tag{8.112}$$

Thus, if $n = 1$, the density perturbations $\Delta(M) = \delta\rho/\rho(M)$ all had the same amplitude when they came though their particle horizons during the radiation-dominated era.

In the late 1960s, Sunyaev and Zeldovich used a variety of constraints to derive the form of the initial power spectrum of density perturbations as they came through the horizon (Fig. 8.40). They found that a scale-invariant spectrum with $n = 1$ and amplitude $\delta\rho/\rho = 10^{-4}$ on mass scales from $10^5$ to $10^{20} \, M_\odot$ was consistent with all the available observations at that time (Sunyaev and Zeldovich, 1970a,b). Harrison (1970) studied the form the primordial spectrum must have in order to prevent the overproduction of excessively large-amplitude perturbations on small and large scales. The former would result in the production of an excessive number of black holes and the latter excessively large temperature perturbations in the CMB radiation. A power spectrum of the form

$$P(k) \propto k \tag{8.113}$$

does not diverge on large physical scales and so is consistent with the observed large-scale isotropy of the Universe.

In fact we do not observe the initial power spectrum except on the largest physical scales. The *transfer function* $T(k)$ describes how the shape of the initial power spectrum $\Delta_k(z)$ in the dark matter is modified by various physical processes through the relation

$$\Delta_k(z = 0) = T(k) \, f(z) \, \Delta_k(z). \tag{8.114}$$

Here, $\Delta_k(z = 0)$ is the power spectrum at the present epoch and $f(z) \propto a \propto t^{\frac{2}{3}}$ is the linear growth factor between the scale factor at redshift $z$ and the present epoch in the matter-dominated era. In the case of the cold dark matter model, the form of the transfer function is largely determined by the fact that there is a delay in the growth of the perturbations between the time when they came through the horizon and began to grow. In the standard cold dark matter picture, growth only begins after the epoch of equality of matter and radiation, because the dark matter perturbations are dynamically dominant after that epoch; before then, the oscillations in the photon–baryon plasma were dynamically more important. Transfer functions for the adiabatic cold and hot dark matter pictures and the isocurvature model are shown in Fig. 8.41a; the corresponding processed initial power spectrum is shown in Fig. 8.41b. Only on very large scales (small wavenumbers) is the initial mass function unprocessed; on the scale of galaxies and clusters, the spectrum has been strongly modified.

Finally, we need to relate the processed power spectrum in the dark matter to the power spectrum in the baryonic component, in other words, the power spectrum of the distribution of galaxies. Four examples of the transfer functions for models of structure formation with baryons only (top pair of diagrams) and with mixed cold and baryonic models (bottom pair of diagrams) by Eisenstein and Hu (1998) are shown in Fig. 8.42. Their numerical results are shown as solid lines and their fitting functions by dashed lines. The lower small boxes in each diagram show the percentage residuals to their fitting functions, which are always less than 10%. The lower pair of diagrams should be compared with the *baryon acoustic oscillations* observed in the 2dF and Sloan Digital Sky Surveys shown in Figs. 8.28 and 8.29, although the best-fitting parameters are different from those of Fig. 8.42.

We have now assembled most of the ingredients necessary as input parameters for models of structure formation. In typical simulations, the procedure would be as follows:

- Select a cosmological model with values of $\Omega_0$, $\Omega_\Lambda$ and $H_0$.
- Select a value for the density parameter in baryons $\Omega_b$, which is only about 5–10% of the dark matter.
- Adopt a power spectrum of the initial perturbations, which is usually assumed to be of Harrison–Zeldovich form $P(k) = Ak^n$ with random phases. The value of $n$ can be varied to find the best fit to the observations.

Large-scale computer simulations such as the Millennium Simulation, carried out by the Virgo consortium, begin the computations at a redshift $z = 127$, well after the epoch of recombination, and so adopted appropriate transfer functions as an input to the simulations. These also included the gas dynamics of matter responding to the growing dark matter perturbations. A good summary of what is involved and what has been achieved is described by Springel *et al.* (2005).

## 8.14 The non-linear collapse of density perturbations

The large computer simulations of structure formation follow the evolution of structures well into the non-linear regime. It turns out that a number of the important non-linear phenomena can be understood by analytic means and these provide additional insight into the processes of structure formation. A number of useful examples are developed in this section.

The collapse of a uniform spherical density perturbation in an otherwise uniform Universe can be worked out exactly, a model sometimes referred to as *spherical top-hat collapse*, which is similar to that developed in Section 8.9.4. The dynamics are the same as those of a closed Universe with $\Omega_0 > 1$. The variation of the scale factor of the perturbation $a_p$ is given by the parametric solution

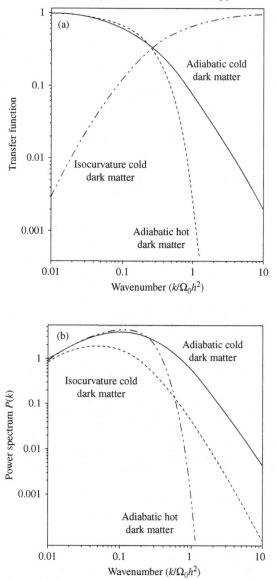

FIGURE 8.41. (a) Examples of transfer functions $T(k)$ for different models of structure formation. The functions are those quoted by Peacock (2000), which were taken from the paper by Bardeen *et al.* (1986). (b) The predicted power spectra $P(k)$ for the models shown in (a). In the cases of the cold dark matter models, a Harrison–Zeldovich power spectrum, $n = 1$ has been assumed. In the case of the isocurvature model, the value $n = -3$ has been adopted. The scaling has been chosen so that the power spectra are the same at small wavenumbers, that is, on very large physical scales. In both cases, the wavenumbers are in $\mathrm{Mpc}^{-1}$. Taken from fig. 14.1a-b in Longair, 2008, with kind permission from Springer Science and Business Media.

$$a_{\mathrm{p}} = A(1 - \cos\theta) \qquad t = B(\theta - \sin\theta),$$

$$A = \frac{\Omega_0}{2(\Omega_0 - 1)} \quad \text{and} \quad B = \frac{\Omega_0}{2H_0(\Omega_0 - 1)^{\frac{3}{2}}}.$$

The perturbation reached maximum radius at $\theta = \pi$ and then collapsed to infinite density at $\theta = 2\pi$. The perturbation stopped expanding, $\dot{a}_{\mathrm{p}} = 0$, and separated out of the

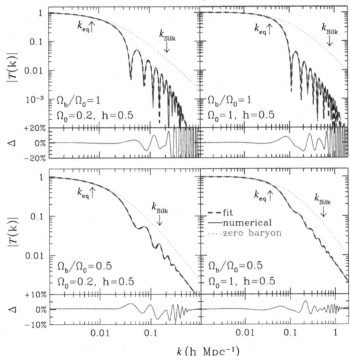

FIGURE 8.42. Four examples of the transfer functions for models of structure formation with baryons only (top pair of diagrams) and with mixed cold and baryonic models (bottom pair of diagrams). Eisenstein and Hu's primary objective was to present fitting functions to the transfer functions derived from numerical solutions to the Boltzmann equation for the development of mixed baryonic and cold dark matter perturbations (Eisenstein and Hu, 1998). The numerical results are shown as solid lines and their fitting functions by dashed lines. The lower small boxes in each diagram show the percentage residuals to their fitting functions, which are always less than 10%. Taken from fig. 3 in Eisenstein and Hu (1998). Reproduced by permission from the AAS.

expanding background at $\theta = \pi$. This occurred when the scale factor of the perturbation was $a = a_{max}$, where

$$a_{max} = 2a = \frac{\Omega_0}{\Omega_0 - 1} \quad \text{at time} \quad t_{max} = \pi b = \frac{\pi \Omega_0}{2H_0(\Omega_0 - 1)^{\frac{3}{2}}}. \quad (8.115)$$

The density of the perturbation at maximum scale factor $\rho_{max}$ can now be related to that of the background $\rho_0$, which, for illustrative purposes, we take to be the critical model, $\Omega_0 = 1$. Recalling that the density within the perturbation was $\Omega_0$ times that of the background model to begin with,

$$\frac{\rho_{max}}{\rho_0} = \Omega_0 \left( \frac{a}{a_{max}} \right)^3 = 9\pi^2/16 = 5.55, \quad (8.116)$$

where the scale factor of the background model has been evaluated at cosmic time $t_{max}$. Thus, by the time the perturbed sphere had stopped expanding, its density was already 5.55 times greater than that of the background density. This epoch is often called the *turnround epoch*.

Interpreted literally, the spherical perturbed region collapsed to a black hole but it is much more likely to form some sort of bound object. This is because, during collapse, the dark matter sphere would begin to fragment into subunits and then, through the process of *violent relaxation*, these regions would come into dynamical equilibrium under the influence of large-scale gravitational potential gradients.

The end result is a system which satisfies the virial theorem. At $a_{max}$, the sphere is stationary and all the energy of the system is in the form of gravitational potential energy. For a uniform sphere of radius $r_{max}$, the gravitational potential energy is $-3GM^2/5r_{max}$. If the system does not lose mass and collapses to half this radius, its gravitational potential energy becomes $-3GM^2/(5r_{max}/2)$ and, by conservation of energy, the kinetic energy, or internal thermal energy, acquired is

$$\text{kinetic energy} = \frac{3GM^2}{5(r_{max}/2)} - \frac{3GM^2}{5r_{max}} = \frac{3GM^2}{5r_{max}}. \tag{8.117}$$

Thus, by collapsing by a factor of two in radius from its maximum radius of expansion, the kinetic energy, or internal thermal energy, becomes half the negative gravitational potential energy, the condition for dynamical equilibrium according to the virial theorem. Therefore, the density of the perturbation increased by a further factor of 8, while the background density continued to decrease.

The scale factor of the perturbation reached the value $a_{max}/2$ at time $t = (1.5 + \pi^{-1})t_{max} = 1.81t_{max}$, when the background density was a further factor of $(t/t_{max})^2 = 3.3$ less than at maximum. The net result of these simple calculations is that, when the collapsing cloud became a bound virialised object, its density was $5.55 \times 8 \times 3.3 \approx 150$ times the background density at that time.

These simple calculations illustrate how structures form in the large-scale simulations. According to Coles and Lucchin (1995), the systems become virialised at a time $t \approx 3t_{max}$ when the density contrast was about 400. Using these arguments, galaxies of mass $M \approx 10^{12} M_\odot$ could not have been virialised at redshifts greater than 10 and clusters of galaxies cannot have formed at redshifts much greater than 1.

The next approximation is to assume that the perturbations were ellipsoidal with three unequal principal axes. In the *Zeldovich approximation*, the development of perturbations into the non-linear regime is followed in Lagrangian coordinates (Zeldovich, 1970). If $x$ and $r$ are the proper and comoving position vectors of the particles of the fluid, the Zeldovich approximation can be written

$$x = a(t)r + b(t)p(r). \tag{8.118}$$

The first term on the right-hand side describes the uniform expansion of the background model and the second term the perturbations of the particles' positions about the Lagrangian (or comoving) coordinate $r$. Zeldovich showed that, in the coordinate system of the principal axes of the ellipsoid, the motion of the particles in comoving coordinates is described by a "deformation tensor" $D$

$$D = \begin{bmatrix} a(t) - \alpha b(t) & 0 & 0 \\ 0 & a(t) - \beta b(t) & 0 \\ 0 & 0 & a(t) - \gamma b(t) \end{bmatrix}. \tag{8.119}$$

Because of conservation of mass, the density $\rho$ in the vicinity of any particle is

$$\rho[a(t) - \alpha b(t)][a(t) - \beta b(t)][a(t) - \gamma b(t)] = \bar{\rho}a^3(t), \tag{8.120}$$

where $\bar{\rho}$ is the mean density of matter in the Universe.

The clever aspect of the Zeldovich solution is that, although the constants $\alpha$, $\beta$ and $\gamma$ vary from point to point in space depending upon the spectrum of the perturbations, the functions $a(t)$ and $b(t)$ are the same for all particles. In the case of the critical model, $\Omega_0 = 1$,

$$a(t) = \frac{1}{1+z} = \left(\frac{t}{t_0}\right)^{\frac{2}{3}} \quad \text{and} \quad b(t) = \frac{2}{5}\frac{1}{(1+z)^2} = \frac{2}{5}\left(\frac{t}{t_0}\right)^{\frac{4}{3}}, \qquad (8.121)$$

where $t_0 = \frac{2}{3}H_0$. The function $b(t)$ has exactly the same dependence upon scale factor (or cosmic time) as was derived from perturbing the Friedman solutions.

If we consider the case in which $\alpha > \beta > \gamma$, collapse occurs most rapidly along the $x$-axis and the density becomes infinite when $a(t) - \alpha b(t) = 0$. At this point, the ellipsoid has collapsed to a "pancake" and the solution breaks down for later times. Although the density becomes formally infinite in the pancake, the surface density remains finite, and so the solution still gives the correct result for the gravitational potential at points away from the caustic surface. The Zeldovich approximation cannot deal with the more realistic situation in which collapse of the gas cloud into the pancake gives rise to strong shock waves, which heat the matter falling into the pancake.

The results of numerical N-body simulations have shown that the Zeldovich approximation is quite remarkably effective in describing the evolution of the non-linear stages of the collapse of large-scale structures up to the point at which caustics are formed (Fig. 8.43).

It is evident from the power-law form of the two-point correlation function for galaxies (8.2) that on scales much larger than the characteristic length scale $r_0 \approx 7$ Mpc, the perturbations are still in the linear stage of development and so provide directly information about the form of the processed initial power spectrum. On scales $r \leq r_0$, the perturbations are non-linear and it might seem more difficult to recover information about the processed initial power spectrum on these scales. An important insight was provided by Hamilton *et al.* (1991) who showed how it is possible to relate the observed spectrum of

(a)                                              (b)

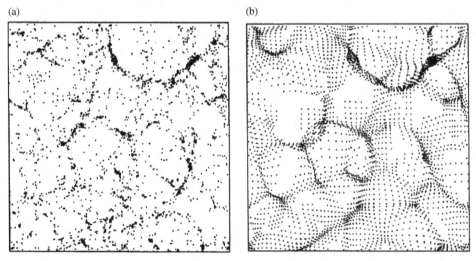

FIGURE 8.43. A comparison between the formation of large-scale structure according to (a) N-body simulations, and (b) the Zeldovich approximation, which began with the same initial conditions, which were assumed to be Gaussian with power spectrum $P(k) \propto k^{-1}$ (from Coles *et al.*, 1993; Coles and Lucchin, 1995). The agreement between the two approaches is very good, particularly when highly non-linear Fourier modes with $k \geq k_{nl} = 8$ are truncated, as has been adopted in (b). Figure taken from fig. 1 in Coles *et al.* 1993, with kind permission from Blackwell Publishing.

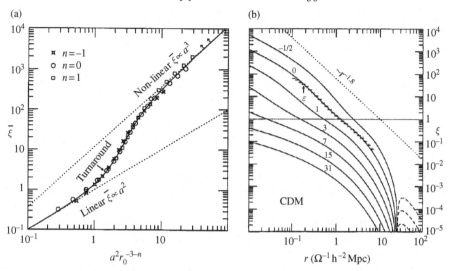

FIGURE 8.44. (a) The variation of the spatial two-point correlation function with the square of the scale factor as the perturbations evolve from linear to non-linear amplitudes. (b) The evolution of the spatial two-point correlation function as a function of redshift. The function has been normalized to result in a correlation function which resembles the observed two-point correlation function for galaxies, which has slope $-1.8$. Non-linear clustering effects, as represented by the function shown in (a), are responsible for steepening the processed initial power spectrum. Taken from Hamilton *et al.* (1991). Reproduced with permission from the AAS.

perturbations in the non-linear regime, $\xi(r) \geq 1$, to the processed initial spectrum in the linear regime.

Figure 8.44a shows the variation of the spatial two-point correlation function with the square of the scale factor as the perturbations evolve from linear to non-linear amplitudes and then to bound systems. It can be seen that the evolution is remarkably independent of the value of the index of the power spectrum. The line $\bar{\xi} \propto a^3$ corresponds to the virialised structure. Fig. 8.44b shows the corresponding evolution of the spatial two-point correlation function as a function of redshift, normalized to result in a two-point correlation function for galaxies, which has slope $-1.8$.

## 8.15 The Press–Schechter formalism

According to the cold dark matter scenario for galaxy formation, galaxies and larger-scale structures are built up by the process of hierarchical clustering. Press and Schechter (1974) provided an analytic formalism for the process of structure formation once the density perturbations had reached such an amplitude that they could be considered to have formed bound objects.

It is assumed that the primordial density perturbations are *Gaussian fluctuations*, that is, the phases of the waves which make up the density distribution are random and the distribution of the amplitudes of the perturbations of a given mass $M$ can be described by a Gaussian function

$$p(\delta) = \frac{1}{\sqrt{2\pi}\sigma(M)} \exp\left[-\frac{\delta^2}{2\sigma^2(M)}\right], \tag{8.122}$$

where $\delta = \delta\rho/\rho$ is the density contrast associated with perturbations of mass $M$. Being a Gaussian distribution, the mean value of the distribution is zero and its variance is $\sigma^2(M)$ is

$$\langle \delta^2 \rangle = \left\langle \left(\frac{\delta\rho}{\rho}\right)^2 \right\rangle = \sigma^2(M). \tag{8.123}$$

It is assumed that, when the perturbations grow in amplitude to a value greater than some critical value $\delta_c$, they develop rapidly into bound objects with mass $M$. The perturbations are assumed to have a power-law power spectrum $P(k) = k^n$ and we know the rules for the growth of the perturbations with cosmic epoch. Let us assume that the background world model is the critical Einstein–de Sitter model, $\Omega_0 = 1, \Omega_\Lambda = 0$, so that the perturbations develop as $\delta \propto a \propto t^{\frac{2}{3}}$.

For fluctuations of a given mass $M$, the fraction $F(M)$ of those which become bound at a particular epoch are those with amplitudes greater than $\delta_c$

$$F(M) = \frac{1}{\sqrt{2\pi}\sigma(M)} \int_{\delta_c}^{\infty} \exp\left[-\frac{\delta^2}{2\sigma^2(M)}\right] d\delta = \tfrac{1}{2}\left[1 - \Phi(t_c)\right], \tag{8.124}$$

where $t_c = \delta_c/\sqrt{2}\sigma$ and $\Phi(x)$ is the probability integral defined by

$$\Phi(x) = \frac{2}{\sqrt{\pi}} \int_0^x e^{-t^2}\, dt. \tag{8.125}$$

We can relate the mean square density perturbation on a particular scale to the power spectrum of the perturbations.

$$\sigma^2(M) = \left\langle \left(\frac{\delta\rho}{\rho}\right)^2 \right\rangle = \langle \delta^2 \rangle = AM^{-(3+n)/3}, \tag{8.126}$$

where $A$ is a constant. We can now express $t_c$ in terms of the mass distribution

$$t_c = \frac{\delta_c}{\sqrt{2}\sigma(M)} = \frac{\delta_c}{\sqrt{2}A^{1/2}}M^{(3+n)/6} = \left(\frac{M}{M^*}\right)^{(3+n)/6}, \tag{8.127}$$

where we have introduced a reference mass $M^* = (2A/\delta_c^2)^{3/(3+n)}$.

Since the amplitude of the perturbation $\delta(M)$ grows as $\delta(M) \propto a \propto t^{\frac{2}{3}}$, it follows that $\sigma^2(M) = \delta^2(M) \propto t^{\frac{4}{3}}$, that is, $A \propto t^{\frac{4}{3}}$. Therefore,

$$M^* = M_0^* \left(\frac{t}{t_0}\right)^{4/(3+n)}, \tag{8.128}$$

where $M_0^*$ is the value of $M^*$ at the present epoch $t_0$.

The fraction of perturbations with masses in the range $M$ to $M + dM$ is $dF = (\partial F/\partial M)\, dM$. In the linear regime, the mass of the perturbation is $M = \bar{\rho}V$ where $\bar{\rho}$ is the mean density of the background model. Once the perturbation becomes non-linear, collapse ensues and ultimately a bound object of mass $M$ is formed. The space density $N(M)\, dM$ of these masses is $V^{-1}$, that is,

$$N(M)\, dM = \frac{1}{V} = -\frac{\bar{\rho}}{M}\frac{\partial F}{\partial M}\, dM, \tag{8.129}$$

the minus sign appearing because $F$ is a decreasing function of increasing $M$.

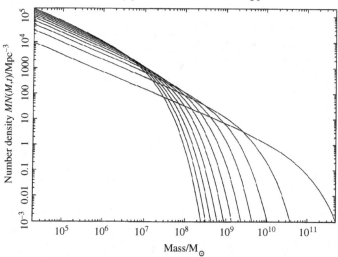

FIGURE 8.45. Illustrating the variation of the form of the Press–Schechter mass function as a function of cosmic time for the processed Harrison–Zeldovich spectrum in the limit of small masses, that is, $n = -3$ (Courtesy of Dr. Andrew Blain). Note that the ordinate is plotted in units of $M N(M, t)$ and that the critical Einstein–de Sitter model is assumed.

We now have everything we need to determine the mass distribution and how it evolves with time. We find

$$N(M) = \frac{1}{2\sqrt{\pi}} \left(1 + \frac{n}{3}\right) \frac{\bar{\rho}}{M^2} \left(\frac{M}{M^*}\right)^{(3+n)/6} \exp\left[-\left(\frac{M}{M^*}\right)^{(3+n)/3}\right],$$

(8.130)

in which all the time dependence of $N(M)$ has been absorbed into the variation of $M^*$ with cosmic epoch. The evolution of the Press–Schechter mass function with cosmic epoch for the $\Omega_0 = 1$ model is illustrated in Fig. 8.45. The usefulness of this function is that it gives a good description of the results of numerical analyses.

A pleasant assessment of the above procedure was the statement by Monaco (1999):

"There is a simple, effective and wrong way to describe the cosmological mass function. Wrong, of course, does not refer to the results, but to the whole procedure."

Comparison of the Press–Schechter formalism with the results of the Millennium supercomputer simulations by Springel *et al.* (2005) is shown in Fig. 8.46. The dotted line shows the Press–Schechter formula and the symbols with error bars the result of the simulations. Despite the many rough approximations involved, the Press–Schechter formalism provides a convenient way of describing the process of hierarchical clustering of galaxies as they form the structures we observe today. For example, the function can be used to predict the development of the mass spectrum of objects with cosmic epoch (Fig. 8.47), an important test of the picture of structure formation.

## 8.16 Dissipation processes

So far we have mostly considered the development of perturbations under the influence of gravity alone. In addition, we need to consider the rôle of dissipation, by which we mean energy loss by radiation, resulting in the loss of thermal energy from the system.

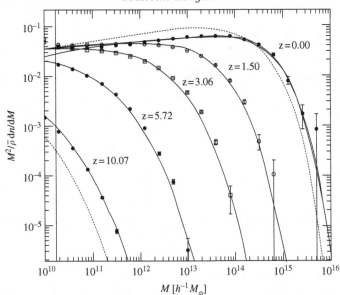

FIGURE 8.46. Differential halo number density as a function of mass and redshift from the Millennium simulation (Springel *et al.*, 2005). The function $n(M, z)$ is the comoving number density of halos with less than masses $M$. What is plotted is the differential halo multiplicity function in the form $(M^2/\bar{\rho})\, \mathrm{d}n/\mathrm{d}M$, where $\bar{\rho}$ is the mean density of the Universe. Springel and his colleagues describe the procedures used to identify bound systems at each epoch. The solid lines represent an analytic fitting function, while the dotted lines show the Press–Schechter function at $z = 10.07$ and $z = 0$. Reprinted with permission from Macmillan Publishers Ltd: *Nature* (Springel *et al.*, 2005), copyright 2005.

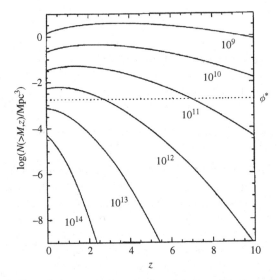

FIGURE 8.47. The evolution of the comoving number density of dark matter haloes with masses greater than $M$ as a function of redshift for a standard cold dark matter model with $\Omega_0 = 1$. The curves have been derived using the Press–Schechter form of evolution of the mass spectrum which is a close fit to the results of N-body simulations. The dotted line labelled $\phi^*$ shows the present number density of $L^*$ galaxies. Taken from Efstathiou (1995).

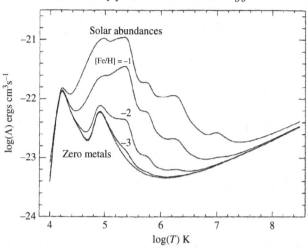

FIGURE 8.48. The cooling rate per unit volume $\Lambda(T)$ of an astrophysical plasma of number density 1 nucleus per cm$^3$ by radiation for different cosmic abundances of the heavy elements, ranging from zero-metals to the present abundance of heavy elements as a function of temperature $T$. In the zero-metal case, the two maxima in the cooling curve are associated with the recombination of hydrogen ions and doubly ionised helium. Taken from Silk and Wyse (1993).

In a number of circumstances, once the gas within the system is stabilised by thermal pressure, loss of energy by radiation can be an effective way of decreasing the internal pressure, allowing the region to contract in order to preserve pressure equilibrium. If the radiation process is effective in removing pressure support from the system, this can result in a runaway situation, known as a *thermal instability*. This is the process that may be responsible for the cooling flows which are present in the hot gas in the central regions of rich clusters of galaxies,

The cooling rate per unit volume $\Lambda(T)$ of an astrophysical plasma with number density 1 nucleus per cm$^3$ by radiation for different cosmic abundances of the heavy elements as a function of temperature is shown in Fig. 8.48 (Silk and Wyse, 1993). This cooling curve can be converted into a number density–temperature diagram on which various loci can be plotted, as well as the baryonic masses of the systems (Fig. 8.49). If the system falls within the cooling locus, cooling is more important than gravitational collapse. These cooling processes are important in understanding the formation of galaxies as we know them within dark matter haloes.

## 8.17 Successes and problems with the standard picture

The application of all the tools we have developed in Sections 8.9 to 8.16 have been subsumed in the large numerical simulations of galaxy and structure formation, the state of the art being the Millennium simulations carried out by the Virgo consortium. The article by Springel, White and their collaborators is strongly recommended as a guide to what can now be incorporated into these simulations, their successes and problems (Springel *et al.*, 2005). There is no doubt that these studies represent a very considerable achievement, although there are many aspects of the modeling which have to be incorporated empirically. The review by Baugh (2006) can be recommended as a warning about the areas of uncertainty in the adoption of the numerous semi-analytic approximations which need to be made. Rather than list the successes, the following is a

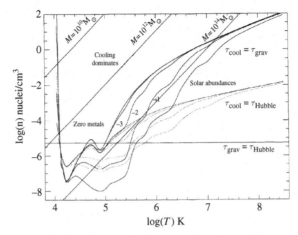

FIGURE 8.49. A number density–temperature diagram showing the locus defined by the condition that the collapse time of a region $t_{\mathrm{dyn}}$ should be equal to the cooling time of the plasma by radiation $t_{\mathrm{cool}}$ for different abundances of the heavy elements. Also shown are lines of constant mass, a cooling time of $10^{10}$ years (dotted lines), and the density at which the perturbations are of such low density that they do not collapse in the age of the Universe. Taken from Silk and Wyse (1993).

brief summary of the areas where there are discrepancies between the modeling and the observations.

- The models predict an excess of dwarf galaxies. It is quite possible that these are not observed because the potential wells of dwarf galaxies are too shallow to prevent the sweeping out of diffuse baryonic gas.
- The simulations of the structures of galaxies and clusters predict that the distribution of matter tends to be more "cuspy" than observed.
- There seems to be little change in the baryonic masses of massive elliptical galaxies out to redshift $z = 2$. The same result is found for the 3CR radio galaxies. There must be some process which prevents the continued growth of massive elliptical galaxies after $z \sim 2$.
- The most massive galaxies have greater metallicities than less massive galaxies. Yet, in the hierarchical clustering picture, the massive galaxies are built out of lower mass galaxies. There would have to be enhancement of the metal abundances during the coalescence process.

It remains to be seen how fundamental these problems are for the overall standard model of galaxy and structure formation. My suspicion is that the problems lie in obtaining a better understanding of the astrophysics of galaxy formation.

## 8.18 Fundamental problems with the standard concordance model

It is remarkable how successful the standard concordance model described by values of the cosmological parameters in Table 8.6 has been in accounting for the wealth of cosmological data now available. Nonetheless, this set of parameters gives rise to a number of the fundamental problems, which need to be confronted within the standard concordance model.

- **The horizon problem** At earlier cosmological epochs, the particle horizon $r \sim ct$ encompassed less and less mass and so the scale over which particles could be causally connected became smaller and smaller. In matter-dominated models, this distance is $r = 3ct$ and so corresponds to an angle $\theta_H \approx 2°$ as observed on the last scattering surface at a redshift of about 1000 on the sky. Thus, regions of the sky separated by greater angular distances could not have been in causal communication. Why then is the CMB radiation so isotropic?

- **The flatness problem** Why is the Universe geometrically flat? This was a real problem when the cosmological constant was assumed to be zero. If the Universe was not set up precisely with the critical density, it would rapidly diverge from that model. The problem is considerably alleviated with the evidence that $\Omega_\Lambda \approx 0.7$ since the dark energy is now dominant and the impending exponential expansion drives the models towards a flat geometry.

- **The baryon-asymmetry problem** The baryon-asymmetry problem arises from the fact that the photon-to-baryon ratio today is $N_\gamma/N_b \approx 1.6 \times 10^9$. If photons are neither created nor destroyed, this ratio is conserved as the Universe expands. At temperature $T \approx 10^{10}$ K, electron–positron pair production takes place from the photon field. At a correspondingly higher temperature, baryon–antibaryon pair production takes place with the result that there must have been a very small asymmetry in the baryon–antibaryon ratio in the very early Universe if we are to end up with the observed photon-to-baryon ratio at the present day. If the Universe had been symmetric with respect to matter and antimatter, the photon-to-baryon ratio would now be about $10^{18}$, in gross contradiction with the observed value.

- **The primordial fluctuation problem** What was the origin of the density fluctuations from which galaxies and large-scale structures formed? The amplitudes of the density perturbations when they came through the horizon had to be of finite amplitude, $\delta\rho/\rho \sim 10^{-5}$, on a very wide range of mass scales. These cannot have originated as statistical fluctuations in the numbers of particles on, say, the scales of superclusters of galaxies. There must have been some physical mechanism which generated finite amplitude perturbations with power spectrum close to $P(k) \propto k$ in the early Universe.

- **The values of the cosmological parameters** The Universe is very close to being geometrically flat and so the sum of the density parameters in the matter and the dark energy must sum to unity, $\Omega_\Lambda + \Omega_0 = 0.72 + 0.28 = 1$. It is a surprise that these are of the same order of magnitude at the present epoch. The matter density evolves with redshift as $(1 + z)^3$, while the dark energy density is unchanging with cosmic epoch. Why then do we live at an epoch when they have more or less the same values?

- **The value of the cosmological constant** A key insight resulted from the introduction of Higgs fields into the theory of weak interactions. The Higgs fields are *scalar* fields, which have negative pressure equations of state, $p = -\rho c^2$. Inserting this relation into Einstein's field equations results in a simple relation between the density parameter in the vacuum fields and the cosmological constant, $\Omega_\Lambda = 8\pi G\rho_v/3H_0^2$ and so $\Lambda = 3H_0^2\Omega_\Lambda$. The theoretical value of $\rho_\Lambda$ can be estimated from quantum field theory with the result $\rho_v = 10^{95}$ kg m$^{-3}$, about $10^{120}$ times greater than the value of $\rho_\Lambda$ at the present epoch. This is a non-trivial discrepancy.

These problems are compounded by the fact that the nature of the dark matter and the dark energy are unknown. Thus, one of the consequences of precision cosmology is the remarkable result that we do not understand the nature of about 95% of the material which drives the large-scale dynamics of the Universe. The concordance values for the cosmological parameters are really very strange. Rather than being causes for despair, however, these problems should be seen as the great challenges for the astrophysicists and cosmologists of the twenty-first century. It is not too far-fetched to see an analogy with Bohr's theory of the hydrogen atom, which was an uncomfortable mix of classical and primitive quantum ideas, but which was ultimately to lead to completely new and deep insights with the development of quantum mechanics.

### 8.18.1 Possible solutions

I have suggested five possible approaches to solving these problems.

- That is just how the Universe is – the initial conditions were set up that way.
- There are only certain classes of Universe in which intelligent life could have evolved. The Universe has to have the appropriate initial conditions and the fundamental constants of nature should not be too different from their measured values or else there would be no chance of life forming as we know it. This approach involves the *anthropic cosmological principle* according to which the Universe is as it is because we are here to observe it.
- We invoke the inflationary scenario for the early Universe.
- Seek clues from particle physics and extrapolate that understanding beyond what has been confirmed by experiment to the earliest phases of the Universe. My guess is that some new form of symmetry breaking is needed to account for the observed photon-to-baryon ratio.
- There is something else we have not yet thought of. This returns us to Donald Rumsfeld cosmology.

### 8.18.2 The limits of observation

Even the first, somewhat defeatist, approach might be the only way forward if it turned out to be just too difficult to disentangle convincingly the physics responsible for setting up the initial conditions from which our Universe evolved. In 1970, William McCrea considered the fundamental limitations involved in asking questions about the very early Universe, his conclusion being that we can obtain less and less information the further back in time one asks questions about the early Universe. A modern version of this argument would be framed in terms of the limitations imposed by the existence of a last scattering surface for electromagnetic radiation at $z \approx 1000$ and those imposed on the accuracy of observations of the CMB radiation and the large-scale structure of the Universe because of their cosmic variances.

In the case of the CMB radiation, the observations made by the WMAP experiment are already cosmic variance limited for multipoles $\ell \leq 354$ – we will never be able to learn much more than we know already about the amplitude of the power spectrum on these scales. Observations by the Planck satellite will provide independent validation of the results of WMAP for these multipoles and extend the cosmic variance limit to $\ell \approx 2000$. In these studies, the search for new physics will depend upon discovering the discrepancies between the standard concordance model and future observations. The optimists, of whom the present author is one, would argue that the advances will come through extending our technological capabilities so that new classes of observation become cosmic variance limited. For example, the detection of primordial gravitational waves through their polarisation signature at small multipoles in the CMB radiation, the nature of dark matter particles and even the nature of the vacuum energy are the cutting edge of fundamental issues for astrophysical cosmology in the twenty-first century. These approaches will be accompanied by discoveries in particle physics with the coming generations of ultra-high-energy particle experiments. There will undoubtedly be surprises which open up new ways of tackling problems that seem to be intractable today – for example, what will be discovered in ultra-high-energy cosmic ray experiments, such as those now being carried out with the Auger array? What will neutrino astrophysics tell us of cosmological importance?

### 8.18.3 The anthropic cosmological principle

There is certainly some truth in the fact that our ability to ask questions about the origin of the Universe says something about the sort of Universe we live in. In this line of reasoning, there are only certain types of Universe in which life as we know it could have formed. For example, the stars must live long enough for there to be time for biological life to form and evolve into sentient beings. Part of the problem stems from the fact that we have only one Universe to study.

It is a matter of taste how seriously one wishes to take this line of reasoning. To many cosmologists, it is not particularly appealing because it suggests that it will never be possible to find physical reasons for the initial conditions from which the Universe evolved, or for the values of the fundamental constants of nature. But some of these problems are really hard. Steven Weinberg (1997), for example, found it such a puzzle that the vacuum energy density $\Omega_\Lambda$ is so very much smaller than the values expected according to current theories of elementary particles, that he invoked anthropic reasoning to account for its smallness. Another manifestation of this type of reasoning is to invoke the range of possible initial conditions that might come out of the picture of chaotic inflation and argue that, if there were at least $10^{120}$ of them, then we live in one of the few which has the right conditions for life to develop as we know it. Again, I leave it to the reader how seriously these ideas should be taken. I prefer to regard the anthropic cosmological principle as the very last resort if all other physical approaches fail.

## 8.19 The inflationary scenario

At first sight, the inflationary model of the very early Universe does not seem to offer a real solution to the basic problems – it simply pushes the set of problems onto the definition of the inflaton potential which is specifically chosen to overcome the causality and flatness problems. On its own, I do not find this argument particularly compelling. But the scenario has the remarkable feature of accounting rather naturally for the origin of the Harrison–Zeldovich spectrum of initial perturbations.

The simplest inflationary solution to some of these problems involves the exponential expansion of the Universe during its very early stages. If the exponential expansion took place over at least 60 e-folding times in the early Universe, the Universe would be inflated far beyond its present size and the horizon and flatness problems can be solved *without any physics* (Fig. 8.50). These solutions can be appreciated from the extended conformal diagrams shown in Figs. 8.9 and 8.10. The time coordinate is set to zero at the end of the inflationary era at, say, $10^{-32}$ s and evolution of the Hubble sphere and the past light cone at recombination extrapolated back to the inflationary era. The inflationary expansion also drives the geometry to a flat geometry since its radius of curvature $\Re \to \infty$, solving the flatness problem.

### 8.19.1 Scalar fields

In many ways, the story of inflation up to this point has been remarkably physics-free. All that has been stated is that an early period of rapid exponential expansion can overcome a number of the fundamental problems of cosmology. The next step involves real physics, but it is not the type of physics familiar to most astrophysical cosmologists.

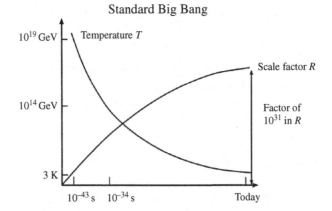

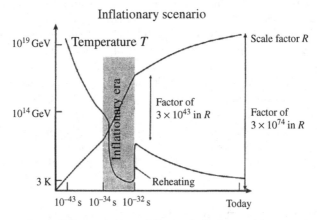

FIGURE 8.50. Comparison of the evolution of the scale factor and temperature in the standard Big Bang and inflationary cosmologies. In the latter, it is assumed that inflation took place over 100 e-folding time. Taken from fig. 20.1 in Longair, 2008, with kind permission from Springer Science and Business Media.

The key role is played by *scalar fields*, which have quite different properties from the vector and tensor fields familiar in electrodynamics and general relativity.

To quote the words of Daniel Baumann (2007),

'Although no fundamental scalar field has yet been detected in experiments, there are fortunately plenty of such fields in theories beyond the standard model of particle physics. In fact, in string theory for example there are numerous scalar fields (moduli), but it proves very challenging to find just one with the right characteristics to serve as an inflaton candidate.'

Here are the results of calculations of the properties of the scalar field $\phi(t)$ which is assumed to be homogeneous at a given epoch. There is a kinetic energy term $\dot{\phi}^2/2$ and a potential energy term $V(\phi)$ associated with the field. Putting these through the machinery of field theory results in expressions for the density and pressure of the scalar field:

$$\rho_\phi = \tfrac{1}{2}\dot{\phi}^2 + V(\phi), \tag{8.131}$$

$$p_\phi = \tfrac{1}{2}\dot{\phi}^2 - V(\phi). \tag{8.132}$$

Thus, the scalar field can result in a negative pressure equation of state, provided the potential energy of the field is very much greater than its kinetic energy. In the limit in which the kinetic energy is neglected, we obtain the equation of state $p = -\rho c^2$, where I have restored the $c^2$ which is set equal to one by professional field theorists. This is exactly what we need to mimic the cosmological constant.

To find the time evolution of the scalar field, we need to combine the above expressions with Einstein's field equations. The results are:

$$H^2 = \frac{1}{3}\left(\frac{1}{2}\dot{\phi}^2 + V(\phi)\right),$$ (8.133)

$$\ddot{\phi} + 3H\dot{\phi} + V(\phi)_{,\phi} = 0,$$ (8.134)

where $V(\phi)_{,\phi}$ means the derivative of $V(\phi)$ with respect to $\phi$. To obtain the inflationary expansion over many e-folding times, the kinetic energy term must be very small compared with the potential energy and the potential energy term must be very slowly varying with time. This is formalised by requiring the two *slow-roll parameters* $\epsilon(\phi)$ and $\eta(\phi)$ to be very small during the inflationary expansion.

These parameters set constraints upon the dependence of the potential energy function upon the field $\phi$ and are formally written:

$$\epsilon(\phi) \equiv \frac{1}{2}\left(\frac{V_{,\phi}}{V}\right)^2 \quad ; \quad \eta(\phi) \equiv \frac{V_{,\phi\phi}}{V} \quad \text{with} \quad \epsilon(\phi), |\eta(\phi)| \ll 1,$$ (8.135)

where $V(\phi)_{,\phi\phi}$ means the second derivative of $V(\phi)$ with respect to $\phi$. Under these conditions, we obtain what we need for inflation, namely,

$$H^2 = \frac{1}{3}V(\phi) = \text{constant} \quad \text{and} \quad a(t) \propto e^{Ht}.$$ (8.136)

At this stage, it may appear that we have not really made much progress since we have adjusted the theory of the scalar field to produce what we know we need. The bonus comes when we consider fluctuations in the scalar field and their role in the formation of the spectrum of primordial perturbations.

### 8.19.2 The quantised harmonic oscillator

To obtain some insight into this remarkable argument, we revise a result which can be derived from the elementary quantum mechanics of a harmonic oscillator.[3] The solutions of Schrödinger's equation for a harmonic potential have quantised energy levels

$$E = \left(n + \tfrac{1}{2}\right)\hbar\omega$$ (8.137)

and the wavefunctions of these stationary states are

$$\psi_n = H_n(\xi)\exp\left(-\tfrac{1}{2}\xi^2\right),$$ (8.138)

where $H_n(\xi)$ is the Hermite polynomial of order $n$ and $\xi = \sqrt{\beta}x$. For the simple harmonic oscillator, $\beta^2 = am/\hbar^2$, where $a$ is the constant in the expression for the harmonic potential $V = \frac{1}{2}ax^2$ and $m$ is the reduced mass of the oscillator. Then, the angular frequency $\omega = \sqrt{a/m}$ is exactly the same as is found for the classical harmonic oscillator.

We are interested in fluctuations about the zero-point energy, that is, the stationary state with $n = 0$. The zero-point energy and Hermite polynomial of order $n = 0$ are

$$E = \tfrac{1}{2}\hbar\omega \quad \text{and} \quad H_0(\xi) = \text{constant}.$$ (8.139)

---

[3] I have simplified even further the excellent article by Baumann (2007), which can be thoroughly recommended.

The first expression is the well-known result that the oscillator has to have finite kinetic energy in the ground state. The underlying cause of this is the need to satisfy Heisenberg's uncertainty principle.

Part of the package of quantum mechanics is that there must be quantum fluctuations in the stationary states. It is straightforward to work out the variance of the position coordinate $x$ of the oscillator. First, we need to normalise the wavefunction so that

$$\int_{-\infty}^{+\infty} \psi\psi^* \, \mathrm{d}x = 1. \tag{8.140}$$

It is straightforward to show that

$$\psi = \left(\frac{am}{\hbar^2\pi^2}\right)^{1/8} \exp\left(-\tfrac{1}{2}\xi^2\right). \tag{8.141}$$

To find the variance of the position coordinate of the oscillator, we form the quantity

$$\langle x^2 \rangle = \int_{-\infty}^{+\infty} \psi\psi^* \, x^2 \, \mathrm{d}x. \tag{8.142}$$

Carrying out this integral, we find the result

$$\langle x^2 \rangle = \frac{\hbar}{2\sqrt{am}} = \frac{\hbar}{2\omega m}. \tag{8.143}$$

These are the fluctuations that must necessarily accompany the zero-point energy of the vacuum fields.

### 8.19.3 The spectrum of fluctuations in the scalar field

We need only one more equation – the expression for the evolution of the vacuum fluctuations in the inflationary expansion. If $k$ is the comoving wavenumber and $\lambda_0$ the wavelength at the present epoch, the proper wavelength of the perturbation is $\lambda = a\lambda_0 \sim a/k$ and the proper wavenumber at scale factor $a$ is $k_{\mathrm{prop}} = k/a$. Then, the evolution of $\delta\phi_k$ is given by the differential equation

$$\ddot{\delta\phi}_k + 3H\dot{\delta\phi}_k + \frac{k^2}{a^2}\delta\phi_k = 0, \tag{8.144}$$

where $H = \dot{a}/a$. Baumann discusses the many technical complexities that need to be dealt with in a rigorous derivation of this equation, which is valid on superhorizon as well as subhorizon scales.

The equation (8.144) bears a strong resemblance to those we derived for the evolution of the density contrast $\Delta$ of non-relativistic and relativistic material in the standard world models. For example, (8.61) reads

$$\frac{\mathrm{d}^2\Delta}{\mathrm{d}t^2} + 2\left(\frac{\dot{a}}{a}\right)\frac{\mathrm{d}\Delta}{\mathrm{d}t} = \Delta(4\pi G\rho_0 - k^2 c_{\mathrm{s}}^2), \tag{8.145}$$

where $k$ is the proper wavenumber and $c_{\mathrm{s}}$ is the speed of sound.

It is a reasonably straightforward calculation to convert this equation into one describing the evolution of the Fourier components of the potential perturbation associated with $\Delta$, in which case one finds

$$\ddot{\delta\phi}_k + 3\left(\frac{\dot{a}}{a}\right)\dot{\delta\phi}_k + (k_c^2 c_{\mathrm{s}}^2 - 2\kappa)\delta\phi_k = 0, \tag{8.146}$$

where $\kappa$ is the curvature of space at the present epoch (Bertschinger, 1996). This equation is remarkably similar to (8.144), which comes out of the full general relativistic treatment of the problem. The analogy becomes very close indeed when we realize that the dynamics are dominated by the vacuum fields and so we can neglect the term in $G$ in the right-hand side. The speed of sound $c_s$ is the speed of light, which is set equal to unity.

We recognise that (8.144) is the equation of motion for a damped harmonic oscillator. If the "damping term" $3H\delta\dot{\phi}_k$ is set equal to zero, we find harmonic oscillations, just as in the case of the Jeans' analysis. For scales much greater than the radius of the Hubble sphere, $\lambda \gg c/H$, an order of magnitude calculation shows that the damping term dominates and the velocity $\delta\dot{\phi}_k$ tends exponentially to zero, corresponding to the "freezing" of the fluctuations on superhorizon scales.

Both $x$ and $\delta\phi_k$ have zero-point fluctuations in the ground state. In the case of the harmonic oscillator, we found $\langle x^2 \rangle \propto (\omega)^{-1}$. In exactly the same way, we expect the fluctuations in $\delta\phi_k$ to be inversely proportional to the "angular frequency" in (8.144), that is,

$$\langle (\delta\phi_k)^2 \rangle \propto \frac{1}{k/a} \propto \lambda, \tag{8.147}$$

recalling that $\lambda$ is the proper wavelength. Since $\lambda \propto a$, the "noise-power" $\langle (\delta\phi_k)^2 \rangle$ increases linearly, proportional to the scale factor, until the wavelength is equal to the dimensions of the Hubble sphere when the noise-power stops growing. Therefore, the power spectrum is given by the power within the horizon when $\lambda = c/H$, that is, when $k = a_* H_*$ where $a_*$ and $H_*$ are the values of the scale factor and Hubble's constant when the wavelength is equal to the radius of the Hubble sphere.

Therefore, per unit volume, the primordial power spectrum on superhorizon scales is expected to have the form

$$\langle (\delta\phi_k)^2 \rangle \propto \frac{1}{a_*^3 (k/a_*)} \propto \frac{H_*^2}{k^3}. \tag{8.148}$$

In the simplest approximation, $H_* = H = $ constant throughout the inflationary era. Now, (8.148) is the power spectrum in Fourier space and to convert it into a real-space power spectrum we perform a Fourier inversion which amounts to multiplying by $k^3$.

$$\langle (\delta\phi)^2 \rangle \propto \int \langle (\delta\phi_k)^2 \rangle \, e^{i\mathbf{k}\cdot\mathbf{r}} \, k^2 \, dk = \int \langle (\delta\phi_k)^2 \rangle \left( \frac{\sin kr}{kr} \right) k^2 \, dk \approx \langle (\delta\phi_k)^2 \rangle k^3. \tag{8.149}$$

Thus, we obtain the important result

$$\boxed{\langle (\delta\phi)^2 \rangle \propto H^2.} \tag{8.150}$$

At the end of the inflationary era, the scalar field is assumed to decay into the types of particles which dominate our Universe at the present epoch, releasing a vast amount of energy, which reheats the contents of the Universe to a very high temperature.

The final step in the calculation is to relate the fluctuations $\delta\phi$ to the density perturbations in the highly relativistic plasma in the post-inflation era. We can think of this transition as occurring abruptly between the era when $p = -\rho c^2$ to that in which the standard relativistic equation of state $p = \frac{1}{3}\rho c^2$ applies. Guth and Pi (1982) introduced what is known as the *time-delay formalism* which enables the density perturbation to be related to the inflation parameters. The perturbation in the scalar field $\delta\phi$ results in a time delay

$$\delta t = \frac{\delta\phi}{\dot{\phi}}. \tag{8.151}$$

This should be evaluated at the time the fluctuation in $\phi$ is frozen-in at horizon crossing.

At the end of the inflationary era, this time delay translates into a perturbation in the density in the radiation-dominated era. Since $\rho \propto t^{-2}$ and $H \propto t^{-1}$,

$$\frac{\delta\rho}{\rho} \propto H \,\delta t. \tag{8.152}$$

Hubble's constant must be continuous across the discontinuity at the end of the inflationary era and must have roughly the same value at horizon crossing. It follows that

$$\boxed{\frac{\delta\rho}{\rho} \propto \frac{H_*^2}{\dot{\phi}_*}.} \tag{8.153}$$

Thus, quantum fluctuations in the scalar field $\phi$ can result in density fluctuations in the matter, which all have the same amplitude when they passed through the horizon in the very early Universe, in other words, the Harrison–Zeldovich spectrum. It should be emphasised that, although the correct spectral shape is obtained, the amplitude of the power spectrum is model-dependent.

Nonetheless, this is a remarkable result. According to Liddle and Lyth (2000):

> Although introduced to resolve problems associated with the initial conditions needed for the Big Bang cosmology, inflation's lasting prominence is owed to a property discovered soon after its introduction: it provides a possible explanation for the initial inhomogeneities in the Universe that are believed to have led to all the structures we see, from the earliest objects formed to the clustering of galaxies to the observed irregularities in the microwave background.

## REFERENCES

BAHCALL, N. A. (2000). *Phys. Rep.*, **333**, 233–244.

BARDEEN, J. M., BOND, J. R., KAISER, N. and SZALAY, A. S. (1986). *Astrophys. J.* **304**, 15.

BAUGH, C. M. (2006). *Rep. Prog. Phys.*, **69**, 3101.

BAUMANN, D. (2007). arXiv:0710.3187 [hep-th].

BENNETT, C. L., BANDAY, A. J., GORSKI, K. M. *et al.* (1996). *Astrophys. J.*, **464**, L1.

BENNETT, C., HALPERN, M., HINSHAW, G. *et al.* (2003). *Astrophy. J. Suppl. Ser.*, **148**, 1.

BERTSCHINGER, E. (1996). In *Cosmology and Large Scale Structure: Proceedings of the "Les Houches Ecole d'Ete de Physique Theorique"*, eds R. Schaeffer, J. Silk, M. Spiro, and J. Zinn-Justin, pp. 273–346. Elsevier Scientific Publishing Company.

BEST, P. N., LONGAIR, M. S. and RÖTTGERING, H. J. A. (1996). *Mon. Not. R. Astron. Soc.*, **280**, L9.

BOLTE, M. (1997). In *Critical Dialogues in Cosmology*, ed. N. Turok, pp. 156–168. World Scientific.

BONAMENTE, M., Joy, M. K., LaRoque, S. J. *et al.* (2006). *Astrophys. J.*, **647**, 25.

CARROLL, S. M., Press, W. H. and TURNER, E. L. (1992). *Annu. Rev. Astron. Astrophys.*, **30**, 499.

CHABOYER, B. (1998). *Phys. Rep.*, **307**, 23–30.

CHAND, H., SRIANAND, R., PETITJEAN, P. and ARACIL, B. (2004). *Astron. Astrophys.*, **417**, 853.

COLE, S., PERCIVAL, W. J., PEACOCK, J. A. *et al.*, (2005). *Mon. Not. R. Astron. Soc.*, **362**, 505.

COLES, P. and LUCCHIN, F. (1995). *Cosmology – the Origin and Evolution of Cosmic Structure.* John Wiley & Sons.

COLES, P., MELOTT, A. L. and SHANDARIN, S. F. (1993). *Mon. Not. R. Astron. Soc.*, **260**, 765.

COLLESS, M., DALTON, G., MADDOX, S. *et al.* (2001). *Mon. Not. R. Astron. Soc.*, **328**, 1039.

DASHEVSKY, V. M. AND ZELDOVICH, Y. B. (1964). *Astron. Zh.* **41**, 1071. Translation: (1965), *Sov. Astron.* **8**, 854.

DAVIS, T. M. and LINEWEAVER, C. H. (2004). *Pub. Astron. Soc. Australia* **21**, 97.

DYER, C. C. and ROEDER, R. C. (1972). *Astrophys. J. Lett.* **174**, L115.

DYER, C. C. and ROEDER, R. C. (1973). *Astrophys. J. Lett.* **180**, L31.

EFSTATHIOU, G. (1995). In *Galaxies in the Young Universe*, eds. H. Hippelein, K. Meissenheimer, and H. J. Röser, pp. 299–314. Springer-Verlag, Lecture Notes in Physics, vol. 463.

EISENSTEIN, D. J. and HU, W. (1998). *Astrophys. J.* **496**, 605.

EISENSTEIN, D. J., ZEHAVI, I., HOGG, D. W. *et al.* (2005). *Astrophys. J.* **633**, 560.

FIXSEN, D., CHENG, E., GALES, J. *et al.* (1996). *Astrophys. J.*, **473**, 576.

FREEDMAN, W. L., MADORE, B. F., GIBSON, B. K. *et al.* (2001). *Astrophys. J.*, **533**, 47.

GE, J., BECHTOLD, J. and BLACK, J. (1997). *Astrophys. J.* **474**, 67.

GELLER, M. J. and HUCHRA, J. P. (1989). *Science* **246**, 897.

GOLDHABER, G., GROOM, D. E., KIM, A. *et al.* (2001). *Astrophys. J.* **558**, 359.

GUNN, J. E. (1978). In *Observational Cosmology: 8th Advanced Course, Swiss Society of Astronomy and Astrophysics, Saas-Fee 1978*, eds. A. Maeder, L. Martinet, and G. Tammann, pp. 1–121. Geneva Observatory Publications.

GUTH, A. H. and PI, S.-Y. (1982). *Phys. Rev. Lett.*, **49**, 1110.

HAMILTON, A. J. S., KUMAR, P., LU, E. and MATTHEWS, A. (1991). *Astrophys. J.* **374**, L1.

HARRISON, E. (1970). *Phys. Rev.* **D1**, 2726.

HEATH, D. J. (1977). *Mon. Not. R. Astron. Soc.* **179**, 351.

HESSER, J. E., HARRIS, W. E., VANDENBERG, D. A. *et al.* (1987). *Publ. Astron. Soc. Pacific* **99**, 739.

HUDSON, M. J., DEKEL, A., COURTEAU, S., FABER, S. M., and WILLICK, J. A. (1995). *Mon. Not. R. Astron. Soc.* **274**, 305.

INSKIP, K. J., BEST, P. N., LONGAIR, M. S. and MACKAY, D. J. C. (2002). *Mon. Not. R. Astron. Soc.* **329**, 277.

KOMATSU, E., DUNKLEY, J., NOLTA, M. R. *et al.* (2008). ArXiv e-prints, **803**.

KUNDIC, T., TURNER, E. L., COLLEY, W. N. *et al.* (1997). *Astrophys. J.* **482**, 75.

LEDOUX, C., PETITJEAN, P. and SRIANAND, R. (2006). *Astrophys. J.* **640**, L25.

LIDDLE, A. R. and LYTH, D. (2000). *Cosmological Inflation and Large-Scale Structure*. Cambridge University Press.

LIGHTMAN, A. P. and SCHECHTER, P. L. (1990). *Astrophys. J. Suppl. Ser.* **74**, 831.

LILLY, S. J. and LONGAIR, M. S. (1984). *Mon. Not. R. Astron. Soc.* **211**, 833.

LONGAIR, M. S. (2006). *The Cosmic Century: A History of Astrophysics and Cosmology*. Cambridge University Press.

LONGAIR, M. S. (2008). *Galaxy Formation*. Springer-Verlag.

LYNE, A. G., BURGAY, M., KRAMER, M. *et al.* (2004). *Science* **303**, 1153.

MA, C.-P. and BERTSCHINGER, E. (1995). *Astrophys. J.*, **455**, 7.

MADDOX, S. J., EFSTATHIOU, G., SUTHERLAND, W. G. and LOVEDAY, J. (1990). *Mon. Not. R. Astron. Soc.* **242**, 43P.

MCCREA, W. (1970). *Nature*, **228**, 21–24.

MITCHELL, J. L., KEETON, C. R., FRIEMAN, J. A. and SHETH, R. K. (2005). *Astrophys. J.* **622**, 81.

MONACO, P. (1999). In *Observational Cosmology: The Development of Galaxy Systems*, eds. G. Giuricin, M. Mezzetti, and P. Salucci, pp. 186–197. Astronomical Society of the Pacific Conference Series No. 176.

OGURI, M. (2007). *Astrophys. J.* **660**, 1.

PAGE, L., HINSHAW, G., KOMATSU, E. *et al.* (2007). *Astrophys. J. Suppl.*, **170**, 335.

PEACOCK, J. (2000). *Cosmological Physics.* Cambridge University Press.

PEACOCK, J. A., COLE, S., NORBERG, P. *et al.* (2001). *Nature* **410**, 169.

PRESS, W. and SCHECHTER, P. (1974). *Astrophys. J.* **187**, 425.

SCHNEIDER, P., KOCHANEK, C. S. and WAMBSGANSS, J. (2006). *Gravitational Lensing: Strong, Weak and Micro*, eds. G. Meylan, and P. Jetzer, and P. North. Springer Verlag. Saas-Fee Advanced Course 33.

SCHRAMM, D. N. (1997). In *Critical Dialogues in Cosmology*, ed. Turok, N., pp. 81–91. World Scientific.

SHAPIRO, I. I. (1964). *Phys. Rev. Lett.* **13**, 789.

SILK, J. and WYSE, R. F. G. (1993). *Phys. Rep.* **231**, 293.

SNEDEN, C., MCWILLIAM, A., PRESTON, G. W. *et al.* (1992). *Astrophys. J.* **467**, 819.

SONGAILA, A., COWIE, L. L., VOGT, S. *et al.* (1994). *Nature* **371**, 43.

SPERGEL, D. N., BEAN, R., DORÉ, O. *et al.* (2007). *Astrophys. J. Suppl.* **170**, 377.

SPRINGEL, V., WHITE, S. D. M., JENKINS, A. *et al.* (2005). *Nature* **435**, 629.

STEIGMAN, G. (2004). In *Measuring and Modeling the Universe*, ed. Freedman, W. L., pp. 169–195. Cambridge University Press.

STEIGMAN, G. (2007). *Annu. Rev. Nuclear Particle Systems* **57**, 463.

SUNYAEV, R. A. and ZELDOVICH, Y. B. (1970a). *Astrophys. Space Sci.* **7**, 3.

SUNYAEV, R. A. and ZELDOVICH, Y. B. (1970b). *Astrophys. Space Sci.* **9**, 368.

TAYLOR, J. H. (1992). *Phil. Trans. R. Soc.* **341**, 117.

WEBB, J. K., MURPHY, M. T., FLAMBAUM, V. V. *et al.* (2001). *Phys. Rev. Lett.* **87**.

WEINBERG, S. (1997). In *Critical Dialogues in Cosmology*, ed. Turok, N., pp. 195–203. World Scientific.

WILL, C. M. (2006). *Living Reviews in Relativity*, **9**. Online article: cited on 21 June 2006; see http://www.livingreviews.org/lrr-2006-3.

WOOD-VASEY, W. M., MIKNAITIS, G., STUBBS, C. W. *et al.* (2007). *Astrophys. J.* **666**, 694.

ZELDOVICH, Y. B. (1970). *Astron. Astrophys.* **5**, 84.

ZELDOVICH, Y. B. (1964). *Astron. Zh.* **41**, 19. [Translation: (1964), *Sov. Astron.* **8**, 13].

Printed in the United States
by Baker & Taylor Publisher Services